CELLULOSE - FUNDAMENTAL ASPECTS AND CURRENT TRENDS

Edited by **Matheus Poletto** and **Heitor Luiz Ornaghi Júnior**

Cellulose - Fundamental Aspects and Current Trends
http://dx.doi.org/10.5772/59889
Edited by Matheus Poletto and Heitor Luiz Ornaghi Júnior

Contributors

Gunnar Westman, Mikaela Borjesson, J. Vincent Edwards, John Rojas, Tamilselvan Mohan, Karin Stana Kleinschek, Silvo Hribernik, Rupert Kargl, Guillermo Riva, Jose A. Silva, Fernando Lopez-Dellamary, Maria E. Hernandez, Joaquin García-Estrada, Brenda Vega, Bruno Medronho, Poonam Singh, Hugo Duarte, Filipe Antunes, Nicolas Le Moigne, Jan Dormanns, Benoît Duchemin, Mark Staiger, Luis Alves, Nevin Çankaya, Roosevelt Delano, Paulo Ronaldo, Ana Siqueira, Carla Eiras, Josy Anteveli, Edson Cavalcanti, Josy Osajima, Fabricia Silva, Luciano Lima, Hernane Da Silva Barud, Wilton R Lustri, Helida G Oliveira Barud, Maristela Fs Peres, Junkal Gutierrez, Agnieszka Tercjak, Osmir B Oliveira Junior, Sidney Jl Ribeiro

Published by InTech

Janeza Trdine 9, 51000 Rijeka, Croatia

Publishing Process Manager Iva Lipovic

Technical Editor InTech DTP team

Cover InTech Design team

First published December, 2015

Printed in Croatia

Additional hard copies can be obtained from orders@intechopen.com

Cellulose - Fundamental Aspects and Current Trends, Edited by Matheus Poletto and Heitor Luiz Ornaghi Júnior

p. cm.

ISBN 978-953-51-2229-6

For EU product safety concerns:
IN TECH d.o.o., Prolaz Marije Krucifikse Kozulić 3, 51000 Rijeka, Croatia,
info@intechopen.com or visit our website at intechopen.com.

Meet the editor

Matheus Poletto is professor and researcher at Universidade de Caxias do Sul. He is an expert in composite science, working with thermoplastic composites and cellulosic materials. He has published over 80 scientific articles and conference papers and several book chapters. He obtained his bachelor's degree in Chemical Engineering and master's degree in Materials Science and Engineering from Universidade de Caxias do Sul, Brazil, and his PhD in Materials Engineering from Universidade Federal do Rio Grande do Sul, Brazil. Professor Poletto currently works with cellulosic and lignocellulosic materials, studying the effect of fiber composition on the thermal and mechanical properties of polymer composite materials. He is also member of the Polymer Laboratory Research Group at Universidade de Caxias do Sul.

Heitor Luiz Ornaghi Júnior is professor and researcher in PGMAT (Programa de Pós-Graduação em Engenharia e Ciências dos Materiais) at Universidade de Caxias do Sul (UCS). He obtained his bachelor's degree in Polymer Technology from Universidade de Caxias do Sul, Brazil, and master's degree and PhD in Materials Engineering from Universidade Federal do Rio Grande do Sul (PPGEM), Brazil. He is an expert in polymeric composite materials, working mainly with thermosetting resins reinforced with cellulosic vegetal and/or synthetic fibers. He has published over 23 scientific papers and several book chapters.

He currently works with mechanical, thermal dynamic mechanical, thermal and rheological properties of composites and nanocomposites, degradation kinetics of polymeric and cellulosic materials, polymer processing, and condensed matter physics of composite materials.

Contents

Preface

Environmental concerns and global warming are the two main reasons for developing new materials mainly from renewable sources and studying the chemical and energy trends involved. Due to this fact, after more than 170 years of the discovery of the "sugar of the plant cell wall", there is an increasing demand of cellulose products from consumers, industries, and governments. The importance of cellulose as a potential material is widely recognized, and research in this field is expanding around the world. At this point, one question can be proposed: What makes cellulose such an important material? Cellulose is the most abundant biopolymer on earth and presents unique properties associated with its specific structure. Therefore, a solid knowledge is necessary to transform cellulose into useful materials and chemicals. Owing to this, this book intends to develop a deeper understanding about the fundamental aspects and current applications of cellulose and its derivatives in the interested reader. *Cellulose - Fundamental Aspects and Current Trends* consists of ten chapters related to the cellulose and the nanocellulose structure, modification, production, dissolution, and application. This is a useful book for readers from diverse areas, such as physics, chemistry, biology, materials science, and engineering. It is hoped that this book will expand the reader's knowledge about this fascinating polymer.

Matheus Poletto
Center of Exact Sciences,
Nature and Technology (CENT),
Caxias do Sul University (UCS)
Brasil

Heitor Luiz Ornaghi Júnior
Materials Science and Engineering Graduate Program (PGMAT),
Caxias do Sul University (UCS)
Brasil

Chapter 1

From Cellulose Dissolution and Regeneration to Added Value Applications — Synergism Between Molecular Understanding and Material Development

Poonam Singh, Hugo Duarte, Luís Alves, Filipe Antunes, Nicolas Le Moigne, Jan Dormanns, Benoît Duchemin, Mark P. Staiger and Bruno Medronho

Additional information is available at the end of the chapter

http://dx.doi.org/10.5772/61402

Abstract

Modern society is now demanding "greener" materials due to depleting fossil fuels and increasing environmental awareness. In the near future, industries will need to become more resource-conscious by making greater use of available renewable and sustainable raw materials. In this context, agro-forestry and related industries can indeed contribute to solve many resource challenges for society and suppliers in the near future. Thus, cellulose can be predicted to become an important resource for materials due to its abundance and versatility as a biopolymer. Cellulose is found in many different forms and applications. However, the dissolution and regeneration of cellulose are key (and challenging) aspects in many potential applications. This chapter is divided into two parts: (i) achievements in the field of dissolution and regeneration of cellulose including solvents and underlying mechanisms of dissolution; and (ii) state-of-the-art production of value-added materials and their applications including manmade textile fibers, hydrogels, aerogels, and all-cellulose composites, where the latter is given special attention.

Keywords: Cellulose, dissolution and regeneration, fiber, hydrogels, all-cellulose composites

1. Introduction

Cellulose was isolated for the first time by the French chemist Anselme Payen in 1838 [1], who extracted it from green plants and reported its elemental composition four years later [2]. Cellulose is the main component of the cell wall in higher plants, typically combined with lignin, hemicelluloses, pectins, proteins, and water. Apart from higher plants, cellulose can be synthesized by bacteria or be found in algae and tunicates. This readily available and renewable biopolymer is widely used in many applications such as paper, textiles, membranes, or packaging [3]. Cellulose is the most abundant and studied biorenewable material with an estimated annual production of 7.5 x 10^{10}t [4]. After more than 170 years of research into the "sugar of the plant cell wall", consumers, industries, and governments are increasingly demanding products from renewable and sustainable resources that are biodegradable, non-petroleum based, carbon neutral, and, at the same time, generating low environmental, animal/human health and safety risks [5].

Regarding its basic structure, cellulose is a linear syndiotactic homopolymer composed of D-anhydroglucopyranose units (AGU), that are connected by β(1–4)-glycosidic bonds (Figure 1) [5].

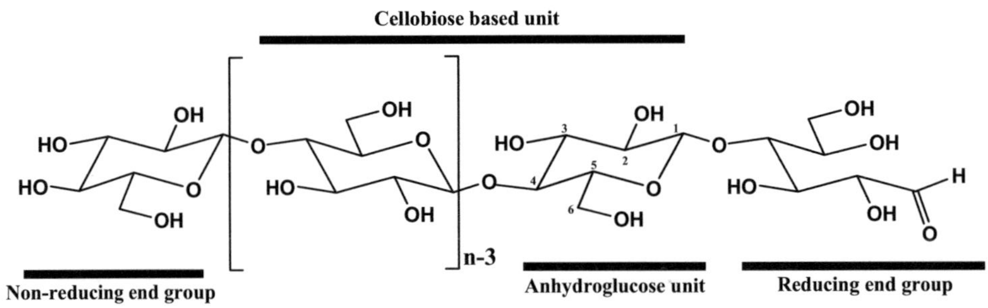

Figure 1. Molecular structure of cellulose (n = value of DP).

The size of the cellulose molecules can be defined by the average degree of polymerization (DP). The average molecular weight is estimated from the product of the DP and the molecular mass of a single AGU. Each AGU bears three hydroxyl groups (one primary and two secondary moieties that represent more than 30 % by weight), with the exception of the terminal ones. These structural features make cellulose surface chemistry quite intriguing and opens a broad spectrum of potential reactions, which typically occur in the primary and secondary hydroxyl groups [3].

From the single AGU up to the cellulose micro- and macrofibrils that constitute the cell walls, cellulose organizes in a rather dense and highly hierarchical fashion where an extended intra- and intermolecular network of hydrogen bonds is believed to constitute the basis of cohesion between cellulose molecules [6]. At the beginning of the biosynthesis, cells in higher plants are surrounded by a very thin outer wall, the primary (P) wall (0.1–0.5 μm thick), which envelops the cytoplasm. After completion of the elongation growth stage of the cells, the thickness of the cell wall increases significantly by the successive deposition of concentric inner layers, constituting the secondary (S1) and (S2) walls (0.1–0.3 μm and 1–10 μm thick, respectively). The last layer deposited, called the tertiary layer (T) in the case of wood fibers and (S3) layer in the case of cotton, is not always present and is very thin (< 100 nm). At the end of the biosynthesis, the cytoplasm dies, and the resulting central channel within the cells, so-called the lumen, is more or less narrow depending on the maturity of the fibers. In this layered structure, the different chemical components are distributed and organized, forming a complex and tri-dimensional composite microstructure [4, 6]. Cellulose microfibrils are reasonably oriented and hold together due to the cooperative function of the hemicelluloses, lignin, and pectins that act as matrix and adhesive components. The resulting fibrillar cells, also called elementary fibers, are usually gathered in fiber bundles as for wood or flax and hemp stems, or can be eventually found individualized in the elementary fibers as is in the case of cotton fibers.

A complementary perspective highlights the amphiphilic nature of cellulose [7–12]; the equatorial direction of a glucopyranose ring has a hydrophilic character because all three hydroxyl groups are located on the equatorial positions of the ring. On the other hand, the axial direction of the ring is hydrophobic since the hydrogen atoms of C–H bonds are located on the axial positions of the ring. Thus, cellulose molecules have an intrinsic structural anisotropy and due to intra- and intermolecular hydrogen bonding, there is a formation of rather flat ribbons, with sides that differ markedly in their polarity [10, 13, 14]; this is expected to considerably influence both the microscopic (e.g. interactions) and macroscopic (e.g. solubility) properties of cellulose.

As a semi-crystalline polymer, cellulose is able to adopt different forms in the cell wall of a plant such that amorphous regions (lower order) coexist with crystalline domains (higher order) [4]. The degree of crystallinity of cellulose, usually in the range of 40–60 %, depends on the origin and pre-treatment of the sample [6]. Interestingly, the parallel arrangement found in nature (so called cellulose I), is not the most stable structure for a cellulose crystal. Thus, when cellulose I is dissolved and recrystallized, cellulose chains may adopt an anti-parallel arrangement known as the cellulose II type crystal [15, 16]. This intriguing process is still not understood in detail. However, it has been postulated that the transition from cellulose I to cellulose II crystals does not require full chain swelling and coiling in solution; this transition could be reached by a simple translational movement of molecules, in particular during mercerization [4].

2. Cellulose dissolution

2.1. Solvents

The first attempts to dissolve cellulose or cellulose-containing materials were described about 150 years ago [3]. In many important cellulose applications, it is necessary to dissolve the biopolymer and normally this is a challenging step. Due to the complexity of such a biopoly meric network, the partial crystalline structure and the extended non-covalent interactions among molecules, chemical processing of cellulose is rather difficult. Cellulose is neither meltable nor soluble in water or a large range of organic compounds [17]. However, this biopolymer is soluble in more exotic media with no apparent common properties [18].

Cellulose solvents are usually divided in two main groups: non-derivatizing and derivatizing solvents [19]. Historically, derivatization came first. As the name suggests, the derivatizing solvent group comprises all the systems where cellulose dissolution occurs via covalent modification giving an "unstable" ether, ester, or acetal intermediate, such as cellulose nitrate, cellulose xanthate or cellulose carbamate. The viscose process ($NaOH + CS_2$) is by far the most common derivatizing cellulose solvent system used [20].

The "non-derivatizing solvent" class, comprise systems capable of dissolving cellulose only via physical intermolecular interactions. This class of solvents is quite relevant for the organic chemistry of cellulose under homogeneous conditions. Historically significant, and of practical relevance for analytical purposes, is the system introduced by Schweizer in 1857 [21]. It was found that copper salts and concentrated ammonia effectively dissolve cotton. Among the most popular is cuprammonium hydroxide or simply "cuam". Fifty years later, ethylenediamine was found to be a good alternative to ammonia and also new complexing solvents were designed, such as the cadmium hydroxide in aqueous ethylenediamine, "cadoxen", or nickel oxide in the same aqueous ethylenediamine, "nioxen" [18, 22, 23]. Similar alternative systems have been reported using mainly other transition metals (such as palladium and zinc) and an amine or ammonium compound. However, none of these systems have been able to achieve full commercial success [18].

Since the work of Sobue et al., it is known that cellulose is soluble in aqueous NaOH below -5 °C within a specific concentration range of NaOH (7–10 %) [24, 25]. This system is potentially cheap, non-polluting, easy to handle and uses common chemicals. The aqueous alkali systems often, however, do not completely disrupt the semicrystalline regions of cellulose and the solubility is limited to cellulose of relatively low DP. The apparent solubility also depends on the degree of crystallinity and crystal type. Pre-treatments, such as steam explosion of the dissolving pulp, have been successfully used [26]. Other aqueous bases, such as LiOH [27], or quaternary ammonium hydroxides are also capable of dissolving cellulose [4, 18, 28]. More recently, the combination of aqueous NaOH solutions with other additives such as polyethylene glycol, PEG [29], urea [30], and thiourea [31] has been explored.

In 1932, Letters et al. investigated the swelling and dissolution of cellulose in aqueous zinc chloride; dissolution was only observed at concentrations as high as 63 % (w/w) [32]. However, in later publications only three water/salt systems were described as effective cellulose solvents in more detail: $Ca(SCN)_2/H_2O$, $LiSCN/H_2O$, and $ZnCl_2/H_2O$. Among such classes of salts, $Ca(SCN)_2\,3H_2O$ showed a strong swelling and solvent action on cellulose [33]. Only the mixture of NaSCN/KSCN with $Ca(SCN)_2\,3H_2O$ or dimethyl sulfoxide (DMSO) was found to be able to dissolve cellulose. In addition, molten LiSCN $2H_2O$ is described as cellulose solvent [19]. Recently, Leipner et al. reported that LiClO4 $3H_2O$ is a very effective solvent where transparent cellulose solutions can be obtained within a few minutes without pre-treatment or activation [34]. Moreover, mixtures of $LiClO_4\,3H_2O$ with $Mg(ClO_4)_2\,H_2O$ or $MgCl_2.6H_2O$ are also regarded as promising solvents.

Non-aqueous systems are also suitable to dissolve cellulose. Again, the solvent spectrum is broad and the number of possible combinations is high. From a historical point of view, combinations of some simple inorganic compounds, such as SO_2 and NH_3 with a suitable ammonium salt, can indeed be considered as the origin of two large groups of non-derivatizing non-aqueous solvent systems [35]. The first group worthy of mention generally comprises mixtures of a polar organic liquid, SO_2, and primary, secondary, or tertiary aliphatic or secondary alicyclic amines [36]. Alternatively, the sulfur component may be modified to $SOCl_2$. Suitable polar liquids employed are for instance N,N-dimethylformamide (DMF), DMSO, N,N-dimethylacetamide (DMAc) or formamide. From the wide variety of possible mixtures, DMSO/SO_2/diethylamine is one of the most versatile [35].

Relevant mixtures of two-component solvents containing DMSO [19] such as DMSO/methylamine, DMSO/KSCN, DMSO/$CaCl_2$, DMSO/formaldehyde, and DMSO/substituted ammonium fluorides, such as tetrabutylammonium fluoride, TBAF, a recently discovered powerful solvent capable of dissolving cellulose of reasonably high DP (650) within a few minutes without any pre-treatment at room temperature [37]. Non-aqueous solutions based on lithium salts are also very relevant for the analysis of cellulose and for the preparation of a wide variety of derivatives. In this regard, the DMAc/LiCl mixture, developed by McCormick in the late 1970s is noteworthy [38]. N-methylmorpholine-N-oxide (NMMO) emerged as the best of the amine-oxides in the late 1970's [3, 39]. Solutions up to 23 wt. % of cellulose were obtained by dissolving the polymer in NMMO/water mixtures, subsequently removing water under vacuum. This constitutes the basis of the Lyocell process whose commercial potential has been demonstrated and is now applied in large scale. Two main problems are still connected with the NMMO process; the instability of the solvent (demands costly investments for safety reasons) and the tendency of the regenerated fiber towards fibrillation [40].

Ionic liquids (ILs) are a very promising group of compounds in cellulose dissolution. These systems were first employed by Graenacher, in the form of N-alkylpyridinium salts, for the dissolution of cellulose and as media for homogeneous chemical reactions [41]. Almost seventy years passed before the work of Swatloski et al. where several low melting ionic liquids (below 100 °C) were reported as cellulose solvents [42]. This work triggered a new and exciting field in cellulose research and, since then, a huge variety of ILs has been developed. Note that the number of potential ion combinations available is estimated to provide around 10^{12} ILs [43]. Presently, the most efficient ILs for cellulose dissolution are mainly composed of a salt with halide [44], phosphonate [45], formate [46] or acetate [47] as anion, and imidazolium [48], pyridinium [6], choline [49], or phosphonium [50] as cation.

2.2. Swelling and dissolution mechanisms: From the cell wall structure down to the molecular level

2.2.1. *Evidence of a heterogeneous swelling and role of the solvent quality*

When placed in contact with a solvent, synthetic polymers dissolve in three main steps, *(i)* the solvent swells the solid phase that goes above its glass transition T_g, *(ii)* the swelling increases up to the disentanglement of the polymer chains, and finally *(iii)* the chains move out of the swollen phase to the solvent phase and the dissolution is completed [51, 52]. The dissolution occurs step by step from the outside to the inside of the bulk of the polymer, being the kinetics considerably different regarding the amorphous and crystalline zones.

Due to their complex hierarchical structures, cellulose fibers show a different picture characterized by a heterogeneous swelling and dissolution. The most peculiar effect of this heterogeneous swelling is the ballooning phenomenon, in which swelling occurs in specific zones along the fibers (Figure 2). The ballooning phenomenon has been observed and described long ago, first in 1864 by Nägeli [53], then by Pennetier [54], Flemming and Thaysen [55, 56], Rollins and Tripp [57, 58], Hock [59], Warwicker et al. [60]. According to these authors, this phenomenon is assumed to be caused by the swelling of the cellulose contained in the secondary wall of the fibers that leads to the bursting of the primary wall. As the cellulose swells, the primary wall rolls up in such a way as to form collars, rings, or spirals that restricts the uniform expansion of the fiber forming balloons.

Further studies of Chanzy et al. [61] and Cuissinat and Navard [62] showed that the swelling and dissolution mechanisms are strongly influenced by the solvent quality. Based on optical microscopy observations of the dissolution of wood and cotton fibers in a wide range of solvent qualities (i.e. in NMMO at 90 °C with increasing amount of water), the authors identified four main dissolution modes: *(i)* fast dissolution by fragmentation in good solvent (NMMO - < 17 % w/w water), *(ii)* swelling by ballooning and full dissolution in moderate solvent (NMMO - 19 to 24 % w/w water), *(iii)* swelling by ballooning and no complete dissolution in bad solvent (NMMO - 25 to 35 % w/w water), *(iiii)* low homogeneous swelling and no dissolution in non-solvent (NMMO - > 35 % w/w water).

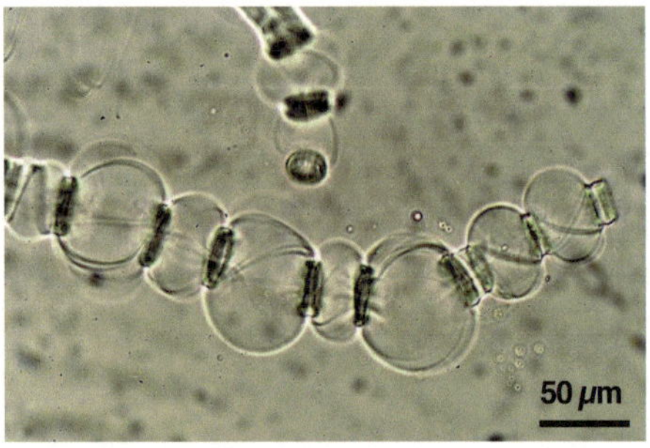

Figure 2. Wood fiber swollen by "ballooning" in NMMO – 20 % w/w water at 90 °C (magnification of 200 times). (Pictures were taken by NLM at CEMEF - MINES ParisTech)

These mechanisms have also been observed in NaOH-water [63], ionic liquids [64], and for a wide range of other plant fibers [65] and some cellulose derivatives prepared by heterogeneous derivatization [66]. All these studies pointed out the leading roles of the solvent quality and the fiber microstructure on the dissolution mechanisms. As long as the native cell wall structure of cellulosic fibres is preserved, the swelling and dissolution mechanisms are found to be similar.

2.2.2. Gradient of dissolution within the cell wall structure and role of the chemical environment of cellulose chains

The role of the cell wall structure was investigated in detail by Le Moigne et al. [67, 68] by studying the dissolution mechanisms of cotton fibers extracted from cotton bolls at successive growth stages. When observing fiber dissolution in various solvent qualities, the authors showed that the primary P wall and the secondary S1 wall and S2 wall, which are successively deposited in concentric layers during biosynthesis, behave very differently upon dissolution. In good solvents (NMMO - 16 % w/w water), the whole fiber breaks into rod-like fragments and dissolves fast. In contrast, cellulose fibres in moderate solvent (NMMO – 20 % w/w water) dissolve by ballooning following the steps (1, 2, and 3) illustrated in Figure 3. Three main areas are implied in the ballooning phenomenon: *(i)* the fragments that are made of the secondary S2 wall and dissolve easily and quickly within the balloons, *(ii)* the membrane of the balloons that is made of the swollen secondary S1 wall and dissolves slowly, and *(iii)* the helices and collars surrounding the balloons that are made of the primary P wall do not dissolve easily and break under the expansion of the secondary wall. These observations revealed a centripetal radial gradient of the dissolution capacity from the inside to the outside of the fiber, being the inside layers from the secondary S2 wall easier to dissolve.

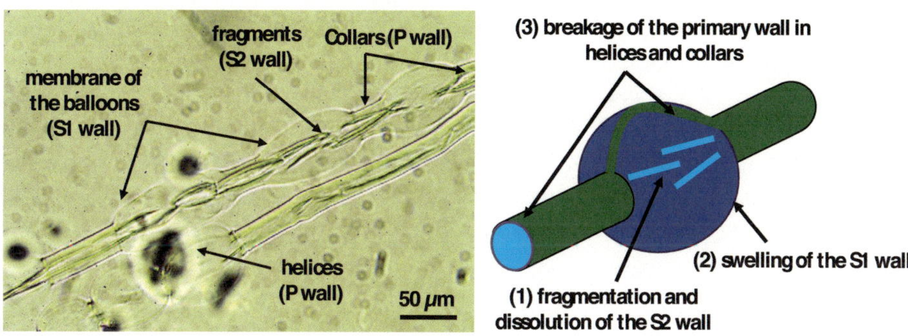

Figure 3. Optical microscopy observation of a cotton fiber in NMMO – 20 % w/w water at 90 °C: (left) description of the zones implied in the ballooning phenomenon; (right) successive dissolution steps (1, 2, and 3). (Pictures were taken by NLM at CEMEF - MINES ParisTech)

Considering that the degree of polymerization and the crystallinity of cellulose increase during the biosynthesis of the secondary S2 wall [69, 70], the dissolution capacity of cellulose fibers does not depend, in a first instance, on thermodynamics (molar mass), or kinetic (crystalline or amorphous) parameters. The gradient in dissolution capacity of the successively deposited cell wall layers has to be related with the specific hierarchical and multi-component structure of cellulose fibres. The primary wall is indeed composed of cellulose and hemicellulose as well as pectins and proteins, while the secondary S2 wall contains almost only cellulose and hemicelluloses [71]. Dynamic FTIR spectroscopy experiments showed that all these components are strongly linked together [72]. As described by Klemm et al. [4], the cellulose microfibrils are also differently arranged and packed within the cell wall structure. The dissolution capacity of cellulose fibers must thus be considered at two structural levels, *(i)* the cell wall structure that needs to be dismantled for improving dissolution, and *(ii)* the chemical environment of cellulose chains. In this sense, it has been shown that the dismantlement of the cell wall layers induced by steam explosion, acidic hydrolysis, or a simple shearing of the solution eases the dissolution of wood pulps in bad quality solvents such as NaOH-water mixtures [73–76]. Based on a selective separation by centrifugation of insoluble and soluble cellulose fractions in cellulose-NaOH-water mixtures and further analyses by size exclusion chromatography, Le Moigne et al. [76] also showed that cellulose chains of similar molar masses can be either dissolved or remain as insoluble fractions. Some of the cellulose chains are thus less accessible than others and embedded in structural regions difficult to dissolve. The carbohydrate composition analysis of the various insoluble fractions revealed that it may contain higher amount of non-cellulosic components than the soluble ones [76, 77]. Beyond thermodynamic considerations, the dissolution capacity of cellulose chains is thus highly influenced by their localization in the cell wall structure. Additionally, their chemical environment has to be regarded as an important parameter in the dissolution efficiency.

2.2.3. Molecular mobility and cellulose/solvent interactions

The dissolution of cellulose substrates implies that the solvating molecules have sufficient accessibility within the fiber structure to approach cellulose chains, and that the latter are able to gain conformational entropy to be further dissolved. However, the spinning process used for regenerated cellulose fibers as well as the naturally occurring bio-deposition mechanism in native cellulose fibers imply that cellulose chains are well aligned and strongly oriented in an extended conformational state, out of equilibrium. Considering the large number of hydroxyl groups available per AGU, this results in a strong intra- and intermolecular hydrogen bond network [7] that links all the cellulose chains together achieving a long range order. As postulated by O'Sullivan [1], even the chains in non-crystalline regions most likely still possess a certain degree of order. All these features strongly restrict the molecular mobility of cellulose chains.

Evidence of this long range order is the large contraction usually observed during swelling and dissolution of cellulose fibres in moderate quality solvent [68], which can be interpreted as the release of the stresses stored within the fiber structure during fiber processing or biosynthesis, and the return to the equilibrium conformational of individual chains that force the whole fiber structure to contract. If this contraction mechanism is restricted by holding the fiber under tension [78], it has been shown that full dissolution can be prevented and a large decrease of the chemical derivatization efficiency was also observed. The acetylation of Lyocell fibers without tension indeed led to degree of substitution (DS) values up to 1.3, while under tension, lower DS values of ca. 0.2–0.5 were estimated. Thus, the local molecular motions needed for the solvatation and the chemical reactions of cellulose chains were inhibited under tension. These observations highlight the key role of the molecular mobility and the need to find efficient ways to break the abovementioned intra- and intermolecular hydrogen bond network and prevent its reformation [79, 78] for improving cellulose dissolution and reactivity.

Apart from the hydrogen bonding role, there are also several indications regarding cellulose amphiphilicity, and thus a careful examination of the interactions involved suggests that hydrophobic interactions may also play a significant role in governing cellulose solubility. In fact, hydrophobic interactions have been shown to markedly contribute to the crystal-like structure of cellulose and its stability over a hypothetical solution state; from free energy simulations in oligomers it has been estimated that there is a 2.0 kcal/mol/residue contribution for hydrophobic stacking, while the estimated hydrogen bonding contribution is about eight times less [7].

From a thermodynamic point of view, the dissolution of a polymer, such as cellulose, in a solvent is, of course, governed by the free energy of mixing [18]. A negative value of the free energy change on mixing means that the mixing process will occur spontaneously. Charging up a polymer is always expected to help solubility in many solvents; thus dissociated counterions contribute strongly to the translational entropy of mixing. This seems to be the main reason why cellulose tends to be more soluble or be more penetrated by water at either high or low pH. However, the pK values are such that rather extreme conditions are needed for either deprotonation or protonation of the hydroxyls; a pKa of ca. 13.3 has been found assuming that only one hydroxyl group per AGU dissociates [80].

A reasonable solvent for cellulose dissolution must be able to overcome the low entropy gain by favorable solvent/polymer interactions, and better dissolution results are obtained using amphiphilic solvents that are not only able to eliminate hydrogen bonding but also eliminate hydrophobic interactions. Both the amorphous and the crystalline regions can be affected by the solvent. However, the amorphous domains, due to a higher free energy, are preferentially and more easily accessed [8]. Therefore, it is not surprising that quite often, prior to dissolution, different activation processes are applied to mainly transform the more ordered and less accessible (crystalline) domains of cellulose into disordered and more accessible regions. It is argued that these alterations of the cellulose structure facilitate the solvent molecules to get access to the cellulose chains [81]. The crystallinity effect in dissolution is still controversial despite some supportive evidences of its effect. For instance, it is argued that sisal pulp (e.g. fibers extracted from the leaves of the *Agave sisalana* plant) dissolves more readily than cotton linters in a particular solvent as its crystallinity index and crystallite size are smaller than that of the latter. After pre-treatment of cotton linter (e.g. mercerization), dissolution was considerably improved and this observation was related to the decrease in both the crystallinity index and average crystal size [82].

In the majority of the cases, cellulose is not dissolved down to a molecular level but rather forms stable colloidal dispersions where ordered cellulose aggregates of, at least, several hundred chains, are present. The structure in solution has been proposed to consist of aggregates of "fringed micelle" type characterized by a highly ordered cylindrical core of aligned chains, which is insoluble in the solvent, and two spherical coronas surrounding the core ends [83]. Reaching molecularly dispersed systems has been challenging for nearly all known solvent systems. Recently, Cohen et al. showed that ionic liquids are able to dissolve cellulose down to a molecular level [84]. Typically, the cations of IL are bulky species with amphiphilic properties. Proof of this is that most literature agrees on the formation of aggregates or micelles of ionic liquids in water, similar to a surfactant behavior [85]. Such amphiphilicity is normally not considered when discussing the mechanism of dissolution of cellulose. This is particularly relevant since crystalline cellulose has an amphipathic-like structure; hydrophobic surfaces consisting of pyranose ring hydrogens and hydrophilic regions arising from the hydroxyl groups directed towards the sides of the ring. In fact, this also follows the earlier discussion on the effect of additives such as PEG, urea, and thiourea on NaOH solutions. Recent molecular dynamics simulation reveals that urea is preferentially adsorbed on the hydrophobic faces of the anhydroglucose rings but has the same affinity as water to the hydroxyl groups. Thus, the simulations suggest that urea acts primarily by mitigating the effect of the hydrophobic portions of the cellulose molecule [86, 87]. In the same direction, a recent molecular dynamics simulations study carried out on cellulose oligomers and 1-ethyl-3-methylimidazolium acetate (C2mimOAc) indeed suggest that the cations are in close contact with the cellulose through hydrophobic interactions [88].

2.2.4. Improving cellulose dissolution: Towards targeted activation treatments

The challenge faced by generations of scientists and numerous companies is to find a simple, cheap, and non-polluting cellulose dissolution process. As pointed out in the previous sections, one of the crucial aspects in cellulose dissolution is the capacity of the fibers to be accessible for reagents in order to perform the subsequent transformation stages. Consequently, several chemical and physical methods, so-called "activation", have been developed to ease cellulose dissolution, e.g. acid hydrolysis [24, 89], ball or vibration milling [4, 90], steam explosion [73–75], and electron beam irradiation [91–93]. The main goal is to disrupt the cell wall structure and the hydrogen bond network in order to increase the accessibility of the solvent within the fiber structure, while preserving the original macromolecular structure of cellulose chains. Although such methods improve cellulose dissolution, typically these activation treatments are accompanied by a strong degradation of the cellulose chains leading to low DPs, ca. 600–1200 for dissolving wood pulps and ca. 100–200 for cellulose powders. Such reduction in molecular weight may adversely compromise the properties of the final cellulose-based products [4]. This is mainly due to the lack of selectivity of the activation treatments that strongly affect the integrity of cellulose substrates at the different scales of their structure depending on the procedure employed. Thus, there is a need for more targeted activation methods.

In this respect, the recent advances in biotechnology with the use of enzymes are expected to bring new insights for the development of targeted activation treatments. As it was previously discussed, the chemical environment of the cellulose chains, as well as their localization within the cell wall structure, plays a key role in their dissolution capacity. Due to their specificity, enzymes should thus be an alternative strategy to improve the selectivity of activation treatments towards "non-dissolving" areas within the fiber structure. Several studies have showed that enzymatic treatments can increase cellulose solubility basically due to changes in DP and hydrogen bond density [94, 95]. However, in these studies, the potential specific action of enzymes within the fiber structure for the improvement of their dissolution capacity was not exploited.

Based on the observations that the external cell wall layers, i.e. the primary and S1 walls, prevent the efficient dissolution of cellulose fibers [67], Le Moigne et al. have recently investigated the possibility to perform a targeted enzymatic peeling treatment in order to remove the external walls and to improve the dissolution capacity of cellulose fibers without degrading their whole structure [96]. The alkaline solubility of the different enzymatically treated samples was investigated in NaOH–water solution. The results showed that the enzymatic peeling has two effects at short treatment times, *(i)* a digestion of the primary wall that was assessed by the near absence of ballooning and *(ii)* a degradation of the fiber structure that was supported by the decrease in the DP. At longer treatment times, the external walls were totally digested and the fiber structure totally dismantled. As expected, a direct correlation was found between the cellulose solubility and its DP and degree of crystallinity. Nevertheless, the removal of the external walls and the macrostructural disruption of the fibers must be regarded as important factors for the improvement of cellulose dissolution. The alkaline solubility of wood pulps having similar intrinsic viscosity values was indeed two times higher when the external walls were removed by the enzymatic peeling.

Interesting results were also obtained by the use of endopectinase and endoglucanase mixtures to improve dissolution of wood pulps in NaOH-water solution [97]. Although the mechanisms are not fully elucidated, it was suggested that endopectinases have a targeted hydrolyzing action on the pectic network present in the primary and S1 walls. This resulted in an enhanced accessibility and diffusion of the solvent through the external walls of the fibers and a significant increase in their swelling and dissolution capacity, while having a limited effect on the cellulose DP.

The use of enzymes for the improvement of cellulose dissolution is thus of great interest due to their high specificity and potential targeted action on the fiber structure that should allow to preserve the solid state of cellulose while improving its processing and the properties of final cellulose based products. As underlined by Koivula et al. [98], the future role of enzymes in lignocellulosic polysaccharide processing will depend on their added value and economical feasibility, as well as the wide industrial acceptance of biotechnologies.

3. Understanding the cellulose regeneration process

Typically, the regeneration of cellulose occurs when contacting the cellulose solution with a coagulation medium. The polymer profile at the point of precipitation exhibits a very high interfacial concentration, thus favoring the formation of a dense polymer "skin". The bulk of the sample is at near the initial concentration and is in a fluid state. Thus, a rapid inflow of the coagulant can take place through the weak points at the skin interface. Rapid growth of finger-like voids in the fluid region is expected to occur due to the moving interface created by the coagulant (less viscous) and solution (more viscous). The kinetics of regeneration is mainly controlled by the relative velocities of the counter-diffusion process; diffusion of the solvent from the solution into the coagulation bath and the coagulant from the bath into the solution [99].

The exchange of solvent with non-solvent leads to a desolvation of the cellulose molecules and supposed reformation of the intra- and intermolecular hydrogen bonds [100]. The regenerated mechanical and surface chemical properties are known to depend strongly on the type of cellulose solvent and coagulant. The reader is directed to the work of Isobe et al. and references therein [101].

Regenerated cellulose is highly wettable and has been identified as one of the most hydrophilic polymers. The contact angles of water droplets on regenerated cellulose such as cellophane and cuprophane are 11.6° and 12.2°, respectively. In contrast, the wetting angle of commodity polymers such as poly(styrene) are 83 and 108.5°, respectively. The high hydrophilicity of cellulose is due to 3 hydroxyls per glucose unit, although this alone is not sufficient to explain the high wettability. A further reason for high hydrophilicity could be due to the uniplanar orientation of the (1 $\bar{1}$ 0) crystal plane when parallel to the material surface as many hydroxyl groups are on this plane [102]. Despite its hydrophobic nature, the structural anisotropy of cellulose may provide regenerated cellulose with hydrophilic properties [10]. As described by Yamane et al. [103] through wide angle X-ray diffraction (WAXD) analysis, glucopyranoses in the (1 $\bar{1}$ 0) crystal plane are stacked together by hydrophilic interactions and van der Waals forces. As a result, the density of hydroxyls on this plane surface is very high, resulting in the highest surface energy of crystal planes among cellulose I and cellulose II polymorphs. Molecular dynamics (MD) simulations are often used to elucidate on the mechanistic aspects of cellulose regeneration. Cousins and Brown have [104] suggested that a molecular sheet is the starting structure for the formation of native cellulose. Additionally, it is stated that the potential energy of the molecular sheets formed by hydrophobic interaction is far lower than that of hydrogen-bonded molecular sheets in water and therefore the former case was suggested to constitute the initial structure of crystallization.

Cellulose chains exist in a folded form and once chains are packed into particles their length can be more than ten-fold larger than their diameter [105]. Therefore, it is possible that the regenerated cellulose chains are also folded and, in the case of the cellulose II polymorph, chains align in an anti-parallel fashion. Studying the cellulose chain folded structure and ring conformers through MD, Yamane et al, proposed a mechanism for the structural formation of regenerated cellulose from a polar solution based on three stages, as schematically illustrated in Figure 4 [106].

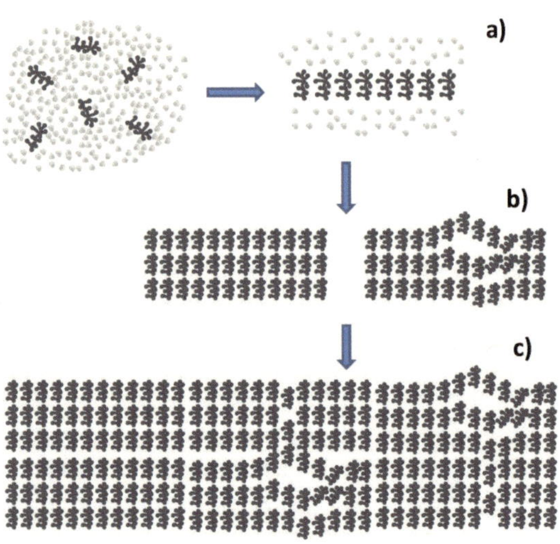

Figure 4. Schematic model of the structural formation of regenerated cellulose in an aqueous environment. The dark grey symbols schematically represent cellulose molecules, while the light grey represent water molecules. a) Glucopyranose rings stacking by hydrophobic interactions, forming a sheet-like structure; b) ordered and less ordered domains; c) regenerated material with a mixture of crystalline and amorphous regions (adapted from reference [106]).

Firstly, when the aqueous cellulose solution is set under unstable conditions (i.e. addition of a coagulant or via marked changes in temperature), cellulose molecules appear to aggregate side by side, with glucopyranose rings stacking by hydrophobic interactions, forming a sheet-like structure (Figure 4.a). Secondly, as the coagulation process proceeds, many sheet-like structures are progressively stacked by hydrogen bonds to form thin planar crystals incorporating amorphous regions. These structural domains are simultaneously laid down parallel to the surface of the shrinking cellulose gel due to their sheet-like shapes. Some aggregates that are tightly stacked with each other and free of defects transform into the crystalline regions (Figure 4.b on the left), while those that incorporate defects in their structure become less ordered domains or amorphous regions (Figure 4.b on the right). This mechanism suggests that the amorphous regions in regenerated cellulose could be defined as being composed of molecular sheets with distances that have substantial distributions, as originally described by Hermans in the late 1940s [107]. In the final stage, these randomly dispersed structural units make contact with other units and, by a diffusion cluster-cluster aggregation mechanism, form a regenerated material with a mixture of crystalline and amorphous regions (Figure 4.c) [108, 109].

As previously mentioned, in the course of coagulation the solvent is squeezed out from the precipitating cellulose gel, resulting in shrinkage, which causes the uniplanar orientation of the $(1\bar{1}0)$ crystal plane. There is also recent experimental evidence for the molecular sheet formation process in cellulose regeneration [101]. The regeneration of cellulose triggered either by a coagulant or upon heating was followed in an aqueous alkali-urea solvent and monitored by time-resolved synchrotron X-ray scattering. It is suggested that when the medium surrounding the cellulose molecules becomes energetically unfavorable for molecular dispersion, the regeneration starts, and the initial process would consist of stacking the hydrophobic glucopyranoside rings (driven by hydrophobic interactions) to form monomolecular sheets, which then would line up by hydrogen bonding to form Na-cellulose type IV crystallites, a hydrate form of cellulose II (Figure 4) [110]. This may constitute one of the first experimental findings that is evidence of the development of hydrophobically stacked monomolecular sheets as hypothesized first by Hermans [107], later by Hayashi [111], and also discussed by Yamane et al. The work of Isobe et al. not only elucidates the regeneration mechanism of cellulose but also provides an alternative vision to the typical regeneration mechanism found in literature that essentially focuses on the reformation of the broken inter- and intramolecular hydrogen bonds among cellulose molecules [101]. Independently of the solvent system used, different studies concluded that both the coagulation medium and the post-treatment influence the regenerated cellulose films regarding the pore size distribution and crystallinity [112, 14, 113, 114]. Furthermore, the crystallinity, porosity and density of regenerated cellulose depend on the temperature and solvent concentration of the coagulation bath, which affect the diffusion rate and slower regeneration leads to higher crystallinity and density [115, 116]. The precipitation method affects cellulose formation as well, with remarkable differences between immersion and vapor precipitation. Song et al. show that crystalline cellulose spherulites can grow from concentrated cellulose solution in IL by very slow precipitation in vapor [117].

Typically, water is the most suitable coagulation agent but often the presence of certain additives (such as salts and amphiphiles) is required in order to tune the mechanical and morphological properties of the regenerated material.

4. Brief overview of some cellulose applications

4.1. Fibers for textile and non-woven applications

The forest industry is facing a paradigm shift, where fast growing short fibers from the southern hemisphere are claiming more room in a matured paper pulp market. Margins are under pressure and profitability will inevitably decrease. Simultaneously, paper consumption is ever decreasing. For the world leaders within the area of forestry and refining of forest raw materials, identifying new products and processes has become paramount. One very interesting alternative is the use of cellulose for the production of man-made fibers. Even though regenerated cellulose fiber has confirmed its position as one of the most dynamic segments of the century, achieving an average annual growth rate of 5.4 % since the year 2000, the demand for cellulose fibers is predicted to exceed the available supply by 3.3 million tons in the year 2020! This is due to population growth and increased prosperity megatrends, along with stagnating cotton production. Currently the global fiber consumption within the textile industry is about 80 million tons where only 5 % of the fibers are originated from dissolving pulp. Since petrochemically-derived fibrs are not sustainable in the long run and the cotton production is being increasingly questioned, sourcing sustainable raw materials has become increasingly difficult for global players in the textile market. In this respect, regenerated cellulose is regarded to potentially play a leading role as a replacement for cotton and synthetic fibres for textile and non-woven applications. Cellulose fibres are known to have good strength properties and a relatively high stiffness, while being only of moderate density, due to the hollow nature of the fiber composition. While the properties of natural fibers have an inherent variability in properties due to differences in species, as well as growing and harvesting conditions, the properties of regenerated cellulose fibers are homogeneous and can be adjusted by processing parameters. However, the successful use of this rapidly renewable, abundant, and biodegradable material will depend on the development of economically and environmentally advantageous processing technologies. Historically, one of the first solvent systems developed for cellulose was based on aqueous complexing agents (such as cupper salts in concentrated ammonia solutions, also referred to as cuprammonium solution). This discovery by Schweizer essentially triggered the cellulose fiber industry [21]. Three years later, H. Despassie developed another process to produce fibers from the same solvent system (aqueous cuprammonium solution-Cupro). A few years later, with the "stretch-spinning process" invention by Thiele, the process became commercialized. A competitor to this process, was the viscose route developed by Cross, Bevan, and Beadke in 1891 that is based on a reaction between cellulose and carbon disulfide (CS_2) in alkali media. A labile and soluble intermediate in aqueous sodium hydroxide solution is formed (e.g. cellulose xanthogenate) and during precipitation of the shaped product, the substituent is cleaved off, and high purity non-modified cellulose is regenerated [102, 118]. The viscose method is still the most important procedure for cellulose shaping and dominates the supply of regenerated cellulose fiber accounting for more than 90 % of the annual production volume.

This route is technologically complex, requiring reasonably high-quality dissolving pulp and leading to problematic environmental loads from the use of CS_2, heavy metal compounds in the precipitation process, and resultant by-products [119]. Due to the emergence of the synthetic fiber industry in the 1940s and to the environmental pollution during the production of regenerated cellulose fibers, caused by the generation of harmful gases, the production of fibers based on Viscose and Cupro processes decreased in the 1960s. In order to overcome these disadvantages, many new solvents have been discovered and studied [120, 121]. A reasonably successful example in the commercial scale production is the Lyocel, a type of rayon fiber [39,122]. The underlying process has been in industrial use for several years, but the production is relatively small. One of the reasons is the thermal instability of the solvent (N-methylmorpholine-N-oxide), which demands major investments in safety technology [40]. A very interesting alternative based on the viscose spinning technology is the CarbaCell process, which does not require the use of sulfur-containing compounds and has been developed to commercial maturity [119]. In this process, a new reaction variant employs the urea to convert cellulose into cellulose carbamate, which can be subsequently processed on existing viscose spinning systems.

As discussed before, the use of ionic liquids to dissolve and produce cellulose based fibers seems very attractive. An example of a recently synthesized ionic liquid with great potential for fiber spinning is the 1,5-diaza-bicyclo[4.3.0]non-5-enium acetate ([DBNH][OAc]), which is described as a powerful direct cellulose solvent [123]. The development of a process designated as Ioncell-F produces regenerated cellulose fiber with properties comparable (or even superior) to Lyocell [124].

4.2. Bioethanol

Cellulose dissolution is typically required for one of the following reasons: analytical characterization of cellulose; shaping in a different format (such as fibers); homogenous chemical modification or degradation to obtain the elementary sugars. In the latter case, it is therefore possible to obtain ethanol from cellulose. In theory, this can be simply achieved by hydrolysis of cellulose-based materials using, for instance, enzymes to break cellulose into simple sugars, followed by fermentation and distillation. In reality, the process is not that straightforward. A pretreatment is typically necessary as cellulose usability is limited due to its structural rigidity, partial crystallinity, and complex network of interactions. This pretreatment also liberates cellulose from the lignin so it becomes more accessible for hydrolysis [125]. During enzymatic hydrolysis, the cellulose chains are broken into small oligomers or even into the elementary sugar units through cellulase enzymes in a similar process as occurring in the stomach of some ruminants. Some species of bacteria, such as *Clostridium thermocellum*, can promote an immediate conversion of cellulose into ethanol. These species break down cellulose and synthesize ethanol. Research in this area has also focused on genetically engineering bacteria that are capable of optimizing ethanol production and inhibit the formation of side products [126]. The environmental benefits of cellulosic ethanol are considerable as it can reduce carbon dioxide emissions to nearly zero and it can impressively reduce green house gas emissions compared with fossil fuels [127]. For instance, corn ethanol reduces green house gas emissions by 13 %, while this value exceeds 85 % if ethanol is of cellulosic origin [128]. The long-predicted arrival of cellulose-based fuels might be very close as several companies are initializing the production of ethanol from biomass while others are producing efficient enzymes.

4.3. Aerogels

Aerogels are low-density solids (0.004–0.15 g/cm3) with high porosity (90–99.8 % air) and a large internal surface area [129–131]. Being one of the most attractive materials, aerogels are also referred to as frozen smoke, solid smoke, or solid air. Some of the most interesting applications include NASA missions, where silica aerogels are used to collect comet particles, insulators in semi-transparent roofs, thickening agents in paints and cosmetics, thermal protections for clothes and materials, drug delivery systems and water purification systems. The low density and high porosity (air takes up the majority of the space in an aerogel) permit the aerogel to be almost weightless. Most aerogels exhibit a slight color due to Rayleigh scattering and they can show a thermal conductivity smaller than the gas they contain, due to the Knudsen effect [132]. Aerogel shows superior insulation and can be used as alternative to the current home insulation. One windowpane (2.5 cm thick) composed of silica aerogel is equivalent to the insulation provided by 20 windowpanes of glass [133]. This will have a strong impact on energy saving to keep houses warm. Additionally, it has obvious impact in the worldwide production of carbon dioxide and other greenhouse gases. Aerogels can be extremely tough, able to hold up to 4,000 times their weight. These systems are typically prepared by a sol-gel process, involving crosslink formation between the components. Firstly, a colloidal suspension or solution of particles with diameter in the range of 1–1,000 nm are subjected to polymerization reactions and/or crosslinking that leads to the formation of a gel with a three dimensional network. The pores of the network contain the solvent of the suspension, usually water. The solvent in the gel can then be removed by exchanging with a CO_2 miscible liquid such as ethanol, followed by liquid CO_2 and then tuning CO_2 above its critical point. An alternative method is to direct inject the supercritical CO_2 into the gel. The pore solvent is replaced with air without altering the gel network structure or the volume of the gel. An alternative way to remove the solvent from the pores is by freeze drying the suspension. Such materials are typically called "cryogels" [134]. Normal evaporation of the pore solvent is inadequate to obtain common aerogels with high porosity: such methods result in xerogels that exhibit large amounts of shrinkage after drying, since the magnitude of the surface tensions of the liquid-solid interfaces can destroy the network.

Aerocellulose is a biodegradable aerogel composed of cellulose. The first aerocellulose known was prepared in 1971 from cellulose pulp [135]. The water from the wet cellulose pulp was exchanged with ethanol and then with CO_2, which was removed by critical point drying, as described above. A cellulose aerogel with a specific surface area of 206 m^2/g was obtained. Since then many aerocellulose derivatives were reported. Tan et al. have developed cellulose acetate aerogels [136]. Jin et al. have developed highly porous cellulose aerogels consisting of cellulose nanofibrils with an approximate pore size of 500 nm [132]. Transparent aerocellulose that combine mechanical toughness and good heat insulation was reported by Kobayashi et al. [137]. Novel aerogels (or aerocellulose) based on all-cellulose composites were prepared by partially dissolving microcrystalline cellulose (MCC) in LiCl/DMAc solution [138]. The inclusion of nanocellulose as a building block for aerogels has also motivated several studies in particular because nanocellulose is able to provide mechanical reinforcement to the material among other advantages [139–145]. Also aerogels of regenerated cellulose prepared through molecular dissolution processes have attracted attention [146, 147]. It is clear that the interest in these materials is considerable and this is probably motivated from the vast range applica-

http://dx.doi.org/10.5772/61402

tions and unique features of aerogels. Thus, it is not surprising that these materials have already started being produced on an industrial scale.

4.4. Hydrogels

Renewable raw materials have immense potential for the development and research of new products. Development of new techniques together with the knowledge of traditional methods can facilitate the production of, for example, super-sorbent gels or hydrogels [148]. Cellulose-based hydrogels are biocompatible and biodegradable materials which are promising for a large number of industrial uses, in particular the cases where environmental issues are concerned, as well as in biomedical applications. Several water-soluble cellulose derivatives can be used, alone or combined, to form robust hydrogel networks possessing highly specific properties in terms of swelling capability and sensitivity to external stimuli. The current trend in the design of cellulose hydrogels mostly related to the use of non-toxic crosslinking chemicals or treatments, to further improve the safeness of both the end product and the productions processes. Structurally, due to the extended crosslinked network of polymers (intimately related to previously discussed aerogels), hydrogels are similar to the extracellular matrix of the human body [149]. These structures may be modified or designed in accordance to the external physical and chemical factors that may influence their activity during a certain application. Hydrogels are generally categorized into functional and synthetic on the basis of their formation [150]. Self-assembled natural polymers are termed as functional hydrogels whereas synthetic hydrogels are programmed to have desired functionality with multiple interactions.

The acrylamide-based hydrogels are currently all over the market but such systems raise serious environmental and societal concerns [148]. As stated above, cellulose is naturally available and abundant resource in nature. Even though cellulose is a biocompatible, biodegradable, and sustainable polymer, cellulose itself cannot be reabsorbed by human cells due to the lack of cellulases.

The stability of hydrogels is altered by changing the type of cross-linker (in the case of chemical hydrogels) and number of cross-linking sites per unit volume. The free hydroxyl groups present on the cellulose backbone enable a rich surface chemistry and cellulose derivatives that are typically water-soluble [151]. Common examples are carboxymethyl cellulose (CMC), hydroxypropyl methylcellulose (HPMC), and hydroxyethyl cellulose (HEC) and these systems have the capacity to produce ultra fine gel networks, particularly when mixed with other polymers, resulting in high water retaining capacities and remarkable mechanical properties [152]. The water holding capacity is determined by the chemical composition and the manufacturing procedures of the starting material. The "smart" behavior of some cellulose-based systems regarding response to physiologically relevant variables (e.g. pH, ionic strength, temperature) makes the resulting hydrogels particularly appealing for in vivo applications.

4.4.1. Hydrogels in agriculture

A common concern nowadays in agriculture is related to the adequate supply and use of pesticides, herbicides, and fungicides, and the use of non-effective agrochemical products since these are devastating the agricultural economy. Additionally, the crop uptake of micronutrient should be balanced. Encapsulation in agriculture is a delivering method that seems to hold a promising targeted approach. This will enable, for instance, the decrease of losses of agrochemicals by erosion, being advantageous from an economic and sustainability point of view. Hydrogels can be specifically synthesized to give a sustainable delivery of water, nutrients, or agrochemicals in adequate amounts in an undisturbed fashion [153]. In this respect, cellulose-based systems (e.g. CMC- and HEC-based hydrogels) can be, for example, effective water reservoirs in agriculture. For instance, tomatoes have been successfully cultivated using these hydrogels in soil without any other water supply. The water absorption and conserving capacity seems to be enhanced. The development of super absorbent polymer (SAP) gels aim at reducing the release of water from the network. It has been reported that CMC and starch, when cross-linked with aluminium sulphate octadecahydrate, produce super absorbent gels [154] that can even be used in extreme drought conditions [155]. Other successful applications in agriculture include the so-called root-targeted delivery vehicle systems. These systems can be formed by dissolving CMC, iron, and calcium salts and are used, for instance, to grow wheat in nutrient depleted soil media. Additionally, when mixed with fertilizers, an efficient and increased growth rate is often observed. Another promising cellulose-based hydrogel is composed of methyl cellulose, polyacrylamide, and calcium montmorillonite that has been shown to be an efficient nutrient carrier. The addition of urea can make the hydrogel more porous, thus enhancing its capability for absorption [156].

4.4.2. Hydrogels in biomedicine

Biomedical, healthcare, and pharmaceutical industries extensively use cellulose derivatives as release matrices, tablets, granules, delivery systems, stabilizers, semi solid gelling agents, artificial wound dressing, for cell encapsulation, and in many other applications [148]. A cellulose hybrid hydrogel having doped phosphor (PP) and epichlorohydrin as a cross-linker was reinforced in a lightly alkaline solution. This PP-based cellulose hydrogel was found to have great potential for bioimaging applications [157]. This system is particularly promising due to its high intensity in green florescence and extended glowing periods. This is relevant because it not only avoids the use of harmful radiation but can also be detected under the skin and/or in the stomach. A hydrophobic hydrogel network formed by cellulose nanowhiskers (extracted crystalline fractions of cellulose), acrylamide, and stearyl methacrylate possess unique properties of self-healing and remolding. This system has a wide extensibility and good mechanical strength making it highly suitable for different applications within the biomedical field [158]. Zhou et al. found that cellulose extracted from bamboo (*Phyllostachys heterocycla)* shows high swelling rates at human body temperature and intestinal pH. Thus, hydrogels based on this cellulose exhibit an immense potential as an oral drug model [159]. According to Sannino et al., edema problems can also be treated using cellulose-based hydrogels as these systems have improved water absorbing and holding capacity [160]. In several examples, no arachidonic acid release is identified confirming that inflammation does not occur and the cellulose-based hydrogel is biocompatible. Different studies have also given the prospect of using cellulose-based hydrogels as encapsulating agents for probacteria and its sustainable delivery after incorporation in different food matrix systems [161–163].

4.5. Cellulose-based composite materials

Composite materials have gained tremendous interest over the last decades and are found in numerous lightweight, structural applications ranging from aerospace to automotive and sports equipment. A large body of research has focused on replacing the traditional reinforcement of composites, such as glass fibers, with lignocellulosic fibers [164–167]. The most commonly used natural fibers are flax, hemp, sisal, jute, and wood [168, 167]. The high specific strength and stiffness of bast fibers, such as hemp, flax, and ramie, give natural fiber reinforced composite properties that compare favorably with glass fiber reinforced composites [169]. The advantages of natural fibers over synthetic fibers, include lower energy usage during production, reduced hazards during handling, less abrasion on machinery, lower density, and a neutral CO_2-balance [169, 164]. However, a drawback of natural fiber is the large inherent scatter in properties due to differences in growing and harvesting conditions, necessitating rigorous quality management to ensure consistent composite properties [167].

In addition to bio-based reinforcements, there has been much effort on the development of naturally derived biopolymers as matrix systems [166, 170] in order to achieve a completely bio-based, biodegradable, and CO_2-neutral material. Composites that are a mix of bio-based and synthetic materials are referred to as *biocomposites* or *eco-composites*, while those completely composed of bio-based materials are elevated to *green composites*.

4.5.1. All-cellulose composite materials

The replacement of synthetic fiber by natural fiber is a promising advancement. However, chemical incompatibility may exist between a hydrophilic fiber reinforcement and hydrophobic polymer matrix that leads to weak fiber-matrix bonding. Poor fiber-matrix bonding leads to inefficient load transfer between the two phases such that the high mechanical properties of natural fibers are not fully exploited [168]. The relatively poor mechanical performance of biocomposites is significantly improved by chemically treating the fiber and/or matrix. Common procedures include chemical grafting, corona discharge and silane or alkaline treatments. However, such procedures add process complexity and costs [171–173].

Single polymer composites (SPCs) overcome the issue of chemical incompatibility by using the same polymer for both fiber and matrix. The main advantage of using the same polymer for both phases is the high level of bonding, resulting in high interfacial strength and efficient load transfer. A further advantage of SPCs is the ease of recycling as the fiber and matrix components do not require extraction and separation. An overview of the field of SPCs is presented in several reviews [174–176].

All cellulose composites (ACCs) are green composites that have been developed using the SPC concept. In ACCs, both the reinforcing and matrix phases are based on cellulose, leading to high chemical compatibility at the fiber-matrix interface. Although high performance ACCs have recently appeared in the literature [177–179], there are historically a number of materials that could be categorized as ACCs. For example, currently used materials such as vulcanized fiber and paper were patented in 1859 [180], cellophane was patented in 1918 [181], and the processing of vegetable parchment is a technology dating back to the 19th century [182].

4.5.2. Processing of all-cellulose composites

Generally, ACCs are processed *via* one of two different pathways, as illustrated in Figure 5. The first pathway involves two distinct processing steps in which (i) cellulose is first dissolved in a solvent and then (ii) regenerated in the presence of an undissolved cellulose reinforcement. The first example of this two-step method was reported by Nishino et al. [177], in which unidirectional ramie fiber was combined with fully dissolved wood pulp. The regeneration of the wood pulp created a matrix phase that provided adhesion between the undissolved ramie fibers. The second pathway is essentially a single processing step that involves wetting a cellulose reinforcement with a solvent and partially dissolving the cellulose. Regeneration of the dissolved portion of cellulose leads to the *in situ* formation of a matrix phase that binds together the undissolved fraction of cellulose. Gindl and Keckes [183] were the first to report the use of the one-step method for creating an ACC, and it has been variously referred to as partial dissolution [184], surface-selective dissolution [185] and natural fiber welding [186]. The one-step method results in ACCs with a relatively high volume fraction of fibers (e.g. ≤ 88 vol. %) [179]. Thus, the volume fraction of fibers in an ACC approaches the theoretical maximum of 90.7 % that exists for a hexagonal packing arrangement [187].

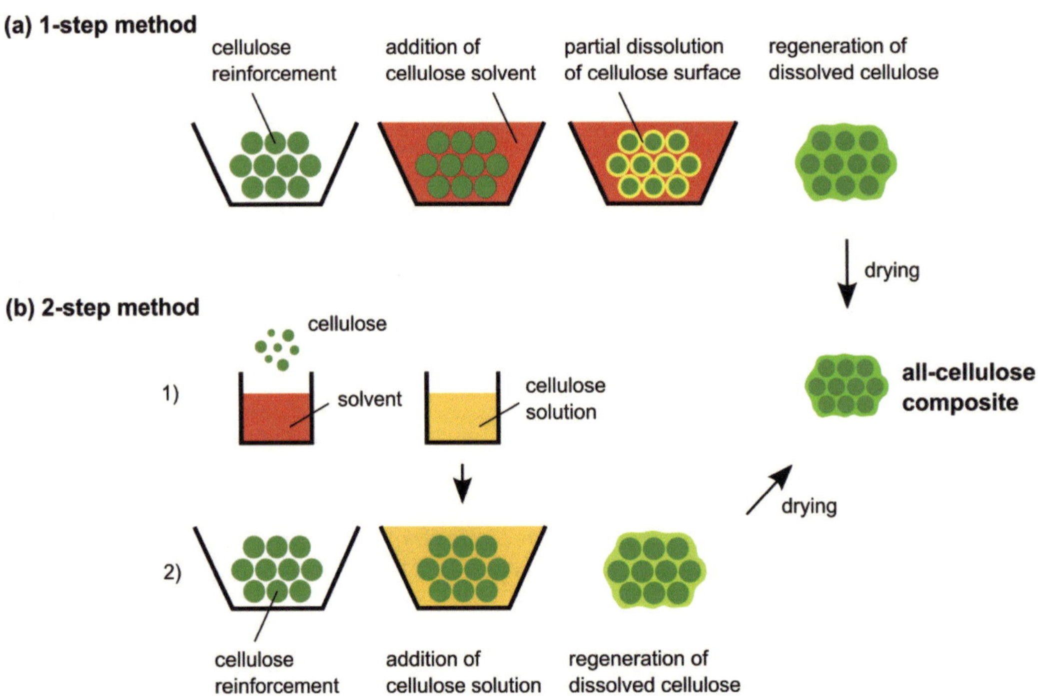

Figure 5. Schematic of processing pathways of ACCs: (a) One-step method and (b) Two-step method. Adapted from Nishino et al. [177].

A number of different cellulose solvents and cellulose types have been used for processing ACCs. The most commonly used solvents are NMMO, LiCl/DMAc, IL, and NaOH aqueous solution [188], of which NMMO, IL, and NaOH based solvents are advantageous because they do not require an activation step before dissolution.

The choice of cellulose source mainly determines the type of composite that will be obtained. Isotropic ACCs have been created from micro- or nanofibrillated cellulose [183, 189–192], wood pulp [193], bacterial cellulose [194, 195] and filter paper [189]. Uni- or multidirectional ACCs have been prepared from natural fibre including ramie [177, 178, 185] and flax [196], and regenerated cellulose fibers including Lyocell [196], Cordenka [184, 197], and Bocell [179].

The use of man-made regenerated fiber in ACCs is beneficial for several reasons: man-made regenerated cellulose fiber can be synthesised to be highly consistent in terms of purity, microstructure, morphology, and mechanical properties [198, 199]. Moreover, the homogeneous microstructure of man-made regenerated cellulose fiber results in a uniform dissolution from the skin to the core [200]. In contrast, the dissolution of natural fiber can be inhomogeneous, with ballooning, contraction, and rotation of cotton and wood fibers as an example [63, 68]. Furthermore, the structural integrity of the cell wall components of natural fiber is severely affected by partial dissolution, which is detrimental to the mechanical properties of flax fiber-based ACCs [196]. ACCs based on regenerated fibers (Lyocell) have a higher mechanical performance, and their mechanical properties compare favorably with epoxy-based composites [196].

The structural form (or polymorph) of the precursor cellulose may undergo a phase transformation during the processing of ACCs. An obvious example is the transformation of cellulose I to cellulose II following dissolution and regeneration. Thus, it is possible for a mixture of different polymorphs to be present in the final ACC. However, the transformation of cellulose I following dissolution and regeneration may follow different pathways depending on the starting materials and processing conditions. For example, Duchemin et al. reported that the partial dissolution and regeneration of MCC leads to the formation of a paracrystalline matrix phase, rather than the expected cellulose II [201].

The nature of the interface between different polymorphs within an ACC has not been studied in detail. Interestingly, Qin et al. observed improved mechanical properties of ramie fiber-based ACCs after a mercerisation treatment of the composites [178]. During mercerisation, cellulose I is transformed to cellulose II, which is concomitant with a decrease in the mechanical properties of cellulose [202]. Therefore, it is supposed that the resulting increase in mechanical properties is due to improved chemical compatibility between the cellulose II in the fiber and that of the matrix formed after partial dissolution. However, several other factors may play a role including the removal of hemicellulose, fiber strengthening, or elimination of defects [178].

Pullawan et al. investigated the micromechanics of the fiber-matrix interface of all-cellulose nanocomposites [203, 204]. Interestingly, the overall crystallinity of the cellulose increased from 26 % in the pure matrix film to 48 % in the ACC with the addition of 1 vol. % of highly crystalline cellulose nanowhiskers. However, a simple calculation based on the rule of mixtures shows that the upper limit of crystallinity of the nanocomposite is 27 %. Thus, the presence of nanowhiskers appears to assist in the nucleation of crystallites during the regeneration of the dissolved cellulose [203]. Chemical compatibility between cellulose I and II is also suggested by the successful stress transfer from fibre to matrix [203, 204].

4.5.3. Developing industrial scale manufacturing pathways for all-cellulose composites

The majority of literature studies of ACCs have produced and characterized films of ACCs with thicknesses that are ca. 0.5 mm. However, the expansion of ACCs into different applications will likely require greater thicknesses of material. The manufacture of ACCs relies on wet processing and necessitates a washing and drying step. The removal of solvent and subsequent drying result in a volumetric shrinkage approximately equal to the ratio of cellulose to solvent. ACC films are typically cast from 5 to 25 vol. % cellulose and hence a shrinkage of > 80 % is to be expected, that is significantly higher than the shrinkage of 1 to 3 % observed during injection moulding of thermoplastics [205]. Furthermore, the removal of solvent from the dissolved portion of cellulose also results in *differential* shrinkage due to large differences in the shrinkage of the reinforcing and regenerated matrix phases. Differential shrinkage is problematic for two reasons: (i) internal residual stresses are generated that compromise the mechanical performance of the composite; and (ii) dimensional stability of the material is decreased following the final drying step (i.e. warpage occurs). Hence, it is clear that extensive experimental studies and development of predictive models of shrinkage in ACCs are still required in order to meet the requirements of the composite industry. Finally, the disposal and/or recycling of the solvent and identification of cost-effective sources of cellulose are important aspects in the context of industrial manufacturing that require further research and development.

Processes based on the one-step method seem more promising for larger scale manufacture of ACCs, due to the higher cellulose/solvent ratio and minimal transformation of cellulose precursor to matrix, which limits the overall and differential shrinkage. Industrial-scale composite manufacturing techniques (e.g., compression moulding, resin infusion) have been explored and adapted for larger scale one-step manufacturing of ACCs, although the technology is still in its infancy. A common characteristic of these processing routes is the application of pressure to consolidate the material during all stages of processing so as to manage the shrinkage and ensure dimensional stability. Conventional compression moulding of composites involves the use of a rigid double-sided mould through which pressure and heat are applied to consolidate the reinforcement and matrix materials [206]. Compression moulding of ACC laminates was carried out by Huber et al. [197]. Initially, several layers of a woven regenerated cellulose fiber textile were impregnated with an ionic liquid as the solvent, followed by stacking and compression of the layers. The application of heat and pressure (110 °C, < 2.5 MPa, 80 min) leads to the partial dissolution of the fibers within the textile layers, resulting in the in situ formation of the matrix phase. The compression-moulded ACC laminates were formed into dimensionally-stable, flat sheets with a final thickness of 2 to 8 mm, tensile strength of 70 MPa, and Young's modulus of 2.5 GPa.

Vacuum-assisted resin transfer moulding (VARTM) is a liquid moulding process that is used to fabricate complex-shaped, high quality composite laminate parts [207, 208]. Typically, a woven textile preform is placed on a one-sided rigid mould that is then covered with a vacuum bag. A low pressure vacuum forces resin to flow through the textile preform, while also acting to remove voids and compact the laminate stack. As an adaptation of the VARTM process, a method called solvent infusion processing (SIP) was developed for the fabrication of dimensionally-stable ACC laminates. During SIP, a cellulosic textile is homogeneously wetted by a solvent (rather than resin). Further compaction and partial dissolution of the cellulose reinforcement are then achieved using external pressure and heat. SIP results in ACC laminates with a high fiber volume fraction (70 to 90 vol. %) with minimal void content [184]. The thickness of the ACC laminates is easily adjusted by varying the number of textile layers. The mechanical properties of the final laminate can be tailored through the choice of precursor (i.e. fiber preform). ACCs based on rayon fiber processed *via* SIP have been shown to exhibit high tensile and flexural strength of 95 MPa and 135 MPa, respectively, and outstanding impact resistance with a puncture impact strength of 2 kN mm^{-2} and an unnotched Charpy impact strength of 42 kJ m^{-2} [184, 209].

4.5.4. Mechanical properties of all-cellulose composites

The mechanical properties of ACCs strongly depend on the type of reinforcement and processing conditions. One of the main factors is the dissolution time and its effect is best explained with the example of a unidirectional ACC. With increasing dissolution times, a higher fraction of the fibers is transformed into the matrix, which decreases the longitudinal tensile strength due to the reduction of the fiber cross-sectional area, and simultaneously leads to an increase in transverse strength, due to a higher matrix-volume fraction and better interfacial adhesion [177, 185]. The dissolution time has a pronounced effect on the microstructure and crystallinity of ACCs, as well as the DP of the processed cellulose [185, 190]. A similarly important parameter is the regeneration rate, with ACCs formed by slow precipitation displaying higher crystallinity, Young's modulus, and tensile strength [190].

The mechanical properties of ACCs can also be influenced by wet drawing. Stretching a regenerated, but still wet ACC leads to a preferred orientation of the cellulose crystallites in the direction of stretching [210, 192]. The crystalline orientation is maintained after drying and a linearly increasing relationship of tensile strength and Young's modulus with the applied draw ratio was found. Applying a draw ratio of 1.5 leads to an increase in tensile strength from 202 to 428 MPa and Young's modulus from 9.9 to 33.5 MPa [210]. Similarly, the orientation of nanowhiskers within an ACC can be influenced by a magnetic field to achieve an increase in mechanical properties in a preferred direction [211, 212].

The comparison of mechanical properties of ACCs in the literature is problematic due to variations in cellulose type and processing parameters. A broad overview of the mechanical properties of ACCs was presented by Huber et al. [188]. A comparison of ACCs with other biocomposites shows that ACCs are highly competitive in terms of their properties (Figure 6a). The Young's modulus of isotropic ACCs is relatively high when compared to conventional biocomposites; this can be attributed to the high properties of the regenerated cellulose matrix [188]. The situation is different for unidirectional composites, where the Young's modulus of ACCs falls in the same range as that of biocomposites (Figure 6b). This may be due to either a decrease in the fiber modulus caused by solvent interaction in the processing of ACCs or the modulus of biocomposites is mainly determined by the fiber modulus and the interfacial strength is of little influence [188]. When it comes to ultimate tensile strength, ACCs perform significantly better than isotropic and unidirectional biocomposites.

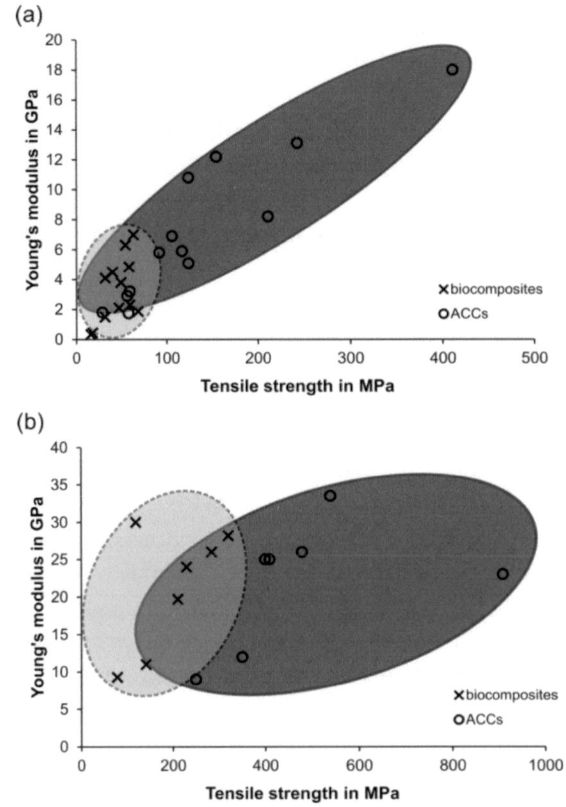

Figure 6. Comparison of mechanical properties of (a) isotropic and (b) unidirectional biocomposites and ACCs. Underlying data is presented in the review by Huber et al. [188].

In terms of viscoelastic behavior, ACCs exhibit a storage modulus that decreases with increasing temperatures. Nevertheless, the storage modulus remains considerably high, up to a temperature of around 250 °C (Figure 7), indicating that ACCs may have relatively high thermal performance as polymer-based composites [190, 213]. The thermal stability strongly depends on the cellulose source. For example, ramie-based ACCs have a higher thermal resistance [177] than flax and Lyocell-based ACCs [196]. Dynamic mechanical analysis of cellulose reveals three separate glass transitions, labeled as α_1, α_2, and α_3 (Figure 7). The α_3 transition occurs at ca. 30 °C and has been assigned to cooperative motions of cellulose chains and water molecules in the non-crystalline regions. The transitions α_2 and α_1 are observed around 140 °C and 300 °C, respectively, and have been assigned to micro-Brownian motions of polymer segments in non-crystalline regions [214]. Duchemin et al. observed that the α_1 transition is composed of two contributions, $\alpha_{1,1}$, and $\alpha_{1,2}$ [213]. Interestingly, the amplitude of the $\alpha_{1,2}$ viscoelastic relaxation at 303 ± 2 °C decreases with increasing crystallinity in ACCs made by partial dissolution of MCC. This relaxation was assigned to the molecular motion of non-crystalline domains during thermal decomposition. Composites with a higher crystallinity thus showed a better retention of the dynamic storage modulus when tested at 300 °C, which translates to a better thermo-mechanical stability in this temperature range.

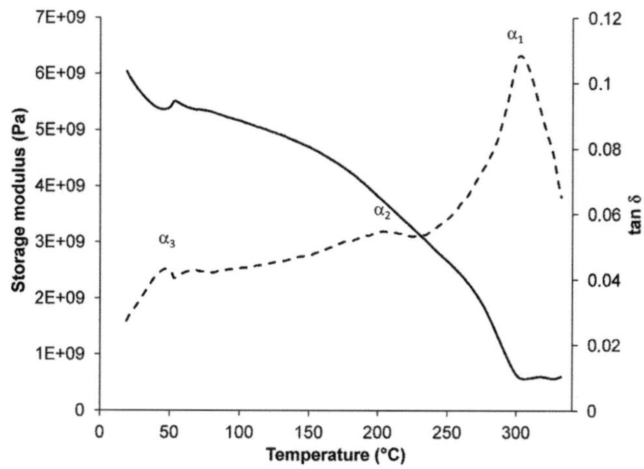

Figure 7. Storage modulus (solid line) and damping (dashed line) of dry all-cellulose composites prepared by partial dissolution of 15 wt. % MCC in DMAc/LiCl for 8 h and followed by precipitation in water. Adapted from Duchemin et al. [213].

4.5.5. Biodegradability

A key advantage of green composites that are based on biopolymers is their inherent biodegradability. However, many synthetic biopolymers (e.g. polylactides, PLA) require energy-intensive composting temperatures (ca. 60 °C) for biodegradation to occur rapidly [215, 216]. In contrast, native cellulose and regenerated cellulose can degrade rapidly at ambient temperatures (i.e. 20–30 °C) due to the hydrophilicity of cellulose [217, 218].

Kalka et al. were the first to demonstrate the rapid biodegradation of ACCs using soil burial trials. Initially, the ACC laminates increased in mass due to moisture uptake, although this was followed by significant mass loss as the material degraded. The percentage mass loss of the ACCs following 70 days of soil burial varied between 39 and 73 % at temperatures of 20.8 and 33.5 °C, respectively. Comparatively, the mass loss of a PLA-matrix composite using the same rayon fiber as reinforcement was minimal (< 2.4 %) under the same conditions [219].

Kalka et al. observed delamination and discoloration of the surface of the ACCs as evidence of microbial activity during the soil burial trials. Furthermore, the surface of ACCs was found to be colonized by filamentous bodies of cellulose decaying fungi and spherical fruit bodies. Microscopic analysis also revealed that clusters of fibers had separated from the composite. The mechanism of fiber-matrix debonding may be the preferential degradation of a more disordered matrix phase that permits greater accessibility to microorganisms compared with that of the original rayon fiber [219].

The rapid biodegradation of ACCs is a distinct advantage over other bio-based composites. However, the high potential for degradation also limits the use of ACCs in some applications. The development of bulk modifications or surface treatments are likely to be required to protect the high mechanical properties of ACCs from the influence of water. For example, Yousefi et al. found that immersion treatment of ACCs with a silane coupling agent (3 wt. % dodecyltriethoxysilane solution) leads to an increase in water contact angle from 59 to 93° and reduces the percentage water uptake from > 4 % to < 1 %. These results demonstrate the efficacy of a silane treatment for the protection of ACCs from moisture ingress, while additionally resulting in higher mechanical properties due to a lower equilibrium moisture content of treated samples and the closure of voids by the silane [220].

5. Conclusions

The dependence of mankind on depleting fossil resources is a problematic issue, and based upon its criticism, a high demand for more sustainable engineering materials has arisen. Cellulose is a decidedly interesting alternative raw material to petrochemical-derived materials due to its high strength and stiffness, wide availability, and biodegradability. Due to the complexity of the biopolymeric network, as well as the partially crystalline structure and extended noncovalent interactions among molecules, chemical processing of cellulose is rather difficult. A wide variety of suitable solvents for cellulose is already available. Nevertheless, most solvent systems have important limitations and there is an intense activity in both industrial and academic research aiming to optimize existing solvents and develop new ones. The problem of obtaining a picture of molecular processes is not trivial since cellulose solvents are of remarkably different nature and thus the understanding of the subtle balance between the different interactions involved becomes non-trivial. The dissolution and regenerative precipitation of cellulose appear to both depend upon a subtle interplay of forces, among which hydrogen bonds are only one of the important factors, and do not seem to be the most significant driving force. Key factors instead include hydrophobic interactions, due to the amphiphilic character of cellulose, as well as the entropy of the counterions for the case that cellulose molecules have a net charge; this net charge can arise from protonation or deprotonation as well as association with charged species. Cellulose can find applications in many and different areas as the ones briefly discussed in this treatise. However, it becomes clear that the understanding of the interplay between all the interactions is very important not only from a fundamental point but particularly for the creation and development of enhanced materials.

Acknowledgements

The authors acknowledge support from the Portuguese Foundation for Science and Technology (project PTDC/AGR-TEC/4049/2012, PTDC/AGR-TEC/4814/2014, PhD and post-doc grants assigned to Luis Alves: SFRH/BD/80556/2011 and Bruno Medronho: SFRH/BPD/

74540/2010, respectively). Jan Dormanns thanks the University of Canterbury for funding a Doctoral Scholarship. Poonam Singh acknowledges the received funding from the People Programme (Marie Curie Actions) of the European Union's Seventh Framework Programme FP7/2007-2013/ under REA grant agreement n° 606713.

Author details

Poonam Singh[1], Hugo Duarte[2], Luís Alves[1], Filipe Antunes[1], Nicolas Le Moigne[3], Jan Dormanns[4,5], Benoît Duchemin[6], Mark P. Staiger[4,5] and Bruno Medronho[2*]

*Address all correspondence to: bfmedronho@ualg.pt

1 University of Coimbra, Department of Chemistry, Coimbra, Portugal

2 Faculty of Sciences and Technology, Centre for Mediterranean Bioresources and Food, University of Algarve, Campus de Gambelas, Faro, Portugal

3 Centre des Matériaux des Mines d'Alès (C2MA), Ecole des Mines d'Alès, France

4 Department of Mechanical Engineering, University of Canterbury, Christchurch, New Zealand

5 MacDiarmid Institute for Advanced Materials and Nanotechnology, Wellington, New Zealand

6 Laboratoire Ondes et Milieux Complexes, Normandie Université, Le Havre, France

References

[1] O'Sullivan AC (1997) Cellulose: The structure slowly unravels. Cellulose 4:173-208

[2] Krässig H, Schurz J, Steadman RG, Schliefer K, Albrecht W, Mohring M, Schlosser H (2000) Cellulose. In: Ullmann's Encyclopedia of Industrial Chemistry. Wiley-VCH Verlag GmbH & Co. KGaA. doi:10.1002/14356007.a05_375.pub2.

[3] Liebert T (2009) Cellulose Solvents: For Analysis, Shaping and Chemical Modification. In: Cellulose Solvents: For Analysis, Shaping and Chemical Modification, vol 1033. ACS Symposium Series. pp 3-54.

[4] Klemm D, Philipp B, Heinze T, Heinze U, Wagenknecht W (1998) In: Fundamentals and analytical methods. Comprehensive Cellulose Chemistry (vol 1). Wiley - VCH, Weinheim.

[5] Festucci-Buselli RA, Otoni WC, Joshi CP (2007) Structure, organization, and functions of cellulose synthase complexes in higher plants. Brazilian Journal of Plant Physiology 19:1-13.

[6] Klemm D, Heublein B, Fink HP, Bohn A (2005) Cellulose: Fascinating biopolymer and sustainable raw material. Angew Chem Int Edit 44 (22):3358-3393. doi:DOI 10.1002/anie.200460587.

[7] Bergenstråhle M, Wohlert J, Himmel ME, Brady JW (2010) Simulation studies of the insolubility of cellulose. Carbohyd Res 345 (14):2060-2066. doi:DOI 10.1016/j.carres.2010.06.017.

[8] Lindman B, Karlström G, Stigsson L (2010) On the mechanism of dissolution of cellulose. J Mol Liq 156 (1):76-81. doi:DOI 10.1016/j.molliq.2010.04.016.

[9] Medronho B, Romano A, Miguel MG, Stigsson L, Lindman B (2012) Rationalizing cellulose (in)solubility: Reviewing basic physicochemical aspects and role of hydrophobic interactions. Cellulose 19 (3):581-587. doi:DOI 10.1007/s10570-011-9644-6.

[10] Yamane C, Aoyagi T, Ago M, Sato K, Okajima K, Takahashi T (2006) Two different surface properties of regenerated cellulose due to structural anisotropy. Polym J 38 (8):819-826. doi:DOI 10.1295/polymj.PJ2005187.

[11] Yamane C, Miyamoto H, Hayakawa D, Ueda K (2013) Folded-chain structure of cellulose II suggested by molecular dynamics simulation. Carbohyd Res 379:30-37. doi:DOI 10.1016/j.carres.2013.06.012.

[12] Kalashnikova I, Bizot H, Cathala B, Capron I (2012) Modulation of Cellulose Nanocrystals Amphiphilic Properties to Stabilize Oil/Water Interface. Biomacromolecules 13 (1):267-275. doi:Doi 10.1021/Bm201599j.

[13] Biermann O, Hadicke E, Koltzenburg S, Muller-Plathe F (2001) Hydrophilicity and lipophilicity of cellulose crystal surfaces. Angew Chem Int Edit 40 (20):3822-3825. doi:Doi 10.1002/1521-3773(20011015)40:20<3822::Aid-Anie3822>3.0.Co;2-V.

[14] Miyamoto H, Umemura M, Aoyagi T, Yamane C, Ueda K, Takahashi K (2009) Structural reorganization of molecular sheets derived from cellulose II by molecular dynamics simulations. Carbohyd Res 344 (9):1085-1094. doi:DOI 10.1016/j.carres.2009.03.014.

[15] Kolpak FJ, Blackwell J (1975) Structure of Regenerated Cellulose. Macromolecules 8 (4):563-564. doi:Doi 10.1021/Ma60046a037.

[16] Raymond S, Kvick A, Chanzy H (1995) The Structure of Cellulose-Ii - a Revisit. Macromolecules 28 (24):8422-8425. doi:Doi 10.1021/Ma00128a063.

[17] Marsh JT (1942) An introduction to the chemistry of cellulose. 2nd edn Chapman & Hall, London.

[18] Hudson SM, Cuculo JA (1980) The Solubility of Unmodified Cellulose - a Critique of the Literature. J Macromol Sci R M C C18 (1):1-82.

[19] Heinze T, Koschella A (2005) Solvents Applied in the Field of Cellulose Chemistry - A Mini Review. Polímeros: Ciência e Tecnologia 15:84-90.

[20] Olsson C, Westman G (2013) Direct Dissolution of Cellulose: Background, Means and Applications. Cellulose - Fundamental Aspects. doi:45614.

[21] Schweitzer EJ (1857) Das Kupferoxyd-Ammoniak, ein Auflösungsmittel für die Pflanzenfaser. Journal für Praktische Chemie 72:109-111.

[22] Saalwachter K, Burchard W, Klufers P, Kettenbach G, Mayer P, Klemm D, Dugarmaa S (2000) Cellulose solutions in water containing metal complexes. Macromolecules 33 (11):4094-4107. doi:Doi 10.1021/Ma991893m.

[23] Achwal WB, Sai BS (1970) Studies in a modified cadoxen solvent. Angew Macromol Chem 14:165.

[24] Kamide K, Okajima K, Kowsaka K (1992) Dissolution of Natural Cellulose into Aqueous Alkali Solution—Role of Super-Molecular Structure of Cellulose. Polym J 24 (1): 71-86. doi:Doi 10.1295/Polymj.24.71.

[25] Sobue H, Kiessig H, Hess K (1939) The System: Cellulose-Sodium Hydroxide-Water in Relation to the Temperature. Zeitschrift für Physikalishe Chemie B43:309.

[26] Yamashiki T, Matsui T, Saitoh M, Matsuda Y, Okajima K, Kamide K, Sawada T (1990) Characterization of Cellulose Treated by the Steam Explosion Method.3. Effect of Crystal Forms Cellulose-1, Cellulose-2, Cellulose-3 of Original Cellulose on Changes in Morphology, Degree of Polymerization, Solubility and Supermolecular Structure by Steam Explosion. Brit Polym J 22 (3):201-212. doi:DOI 10.1002/pi.4980220305.

[27] Yamashiki T, Kamide K, Okajima K, Kowsaka K, Matsui T, Fukase H (1988) Some Characteristic Features of Dilute Aqueous Alkali Solutions of Specific Alkali Concentration (2.5moll-1) Which Possess Maximum Solubility Power against Cellulose. Polym J 20 (6):447-457. doi:DOI 10.1295/polymj.20.447.

[28] Bock LH (1937) Water-Soluble Cellulose Ethers. Ind Eng Chem Proc Dd 29:985-987.

[29] Yan LF, Gao ZJ (2008) Dissolving of cellulose in PEG/NaOH aqueous solution. Cellulose 15 (6):789-796. doi:DOI 10.1007/s10570-008-9233-5.

[30] Cai J, Zhang L (2005) Rapid dissolution of cellulose in LiOH/Urea and NaOH/Urea aqueous solutions. Macromol Biosci 5 (6):539-548. doi:DOI 10.1002/mabi.200400222.

[31] Zhang LN, Ruan D, Gao SJ (2002) Dissolution and regeneration of cellulose in NaOH/thiourea aqueous solution. J Polym Sci Pol Phys 40 (14):1521-1529. doi:Doi 10.1002/Polb.10215.

[32] Fischer S, Leipner H, Thummler K, Brendler E, Peters J (2003) Inorganic molten salts as solvents for cellulose. Cellulose 10 (3):227-236. doi:Doi 10.1023/A:1025128028462.

[33] Kuga S (1980) The Porous Structure of Cellulose Gel Regenerated from Calcium Thiocyanate Solution. J Colloid Interf Sci 77 (2):413-417. doi:Doi 10.1016/0021-9797(80)90311-2.

[34] Leipner H, Fischer S, Brendler E, Voigt W (2000) Structural changes of cellulose dissolved in molten salt hydrates. Macromol Chem Physic 201 (15):2041-2049. doi:Doi 10.1002/1521-3935(20001001)201:15<2041::Aid-Macp2041>3.3.Co;2-5.

[35] Isogai A, Ishizu A, Nakano J (1987) Dissolution mechanism of cellulose in SO2–amine–dimethylsulfoxide. J Appl Polym Sci 33:1283-1290.

[36] Heinze T, Koschella A (2005) Solvents applied in the field of cellulose chemistry: A mini review. Polímeros 15:84-90.

[37] Heinze T, Dicke R, Koschella A, Kull AH, Klohr E-A, Koch W (2000) Effective preparation of cellulose derivatives in a new simple cellulose solvent. Macromol Chem Phys 201:627-631.

[38] McCormick CL, Lichatowich DK (1979) Homogeneous solution reactions of cellulose, chitin, and other polysaccharides to produce controlled-activity pesticide systems. J Polym Sci, Part B: Polym Lett Ed 17:479-484.

[39] Fink HP, Weigel P, Purz HJ, Ganster J (2001) Structure formation of regenerated cellulose materials from NMMO-solutions. Prog Polym Sci 26 (9):1473-1524. doi:Doi 10.1016/S0079-6700(01)00025-9.

[40] Perepelkin KE (2007) Lyocell fibres based on direct dissolution of cellulose in N-methylmorpholine N-oxide: Development and prospects. Fibre Chem+ 39 (2): 163-172. doi:DOI 10.1007/s10692-007-0032-9.

[41] Graenacher C (1934) U.S. Patent 1,943,176.

[42] Swatloski RP, Spear SK, Holbrey JD, Rogers RD (2002) Dissolution of cellose with ionic liquids. J Am Chem Soc 124 (18):4974-4975. doi:Doi 10.1021/Ja025790m.

[43] Pinkert A, Marsh KN, Pang SS, Staiger MP (2009) Ionic Liquids and Their Interaction with Cellulose. Chem Rev 109 (12):6712-6728. doi:Doi 10.1021/Cr9001947.

[44] Zhang H, Wu J, Zhang J, He JS (2005) 1-Allyl-3-methylimidazolium chloride room temperature ionic liquid: A new and powerful nonderivatizing solvent for cellulose. Macromolecules 38 (20):8272-8277. doi:Doi 10.1021/Ma0505676.

[45] Fukaya Y, Hayashi K, Wada M, Ohno H (2008) Cellulose dissolution with polar ionic liquids under mild conditions: Required factors for anions. Green Chem 10 (1):44-46. doi:Doi 10.1039/B713289a.

[46] Fukaya Y, Sugimoto A, Ohno H (2006) Superior Solubility of Polysaccharides in Low Viscosity, Polar, and Halogen-Free 1,3-Dialkylimidazolium Formates. Biomacromolecules 7 (12):3295-3297. doi:10.1021/bm060327d.

[47] Sun N, Rahman M, Qin Y, Maxim ML, Rodriguez H, Rogers RD (2009) Complete dissolution and partial delignification of wood in the ionic liquid 1-ethyl-3-methylimidazolium acetate. Green Chem 11 (5):646-655. doi:Doi 10.1039/B822702k.

[48] Phillips DM, Drummy LF, Conrady DG, Fox DM, Naik RR, Stone MO, Trulove PC, De Long HC, Mantz RA (2004) Dissolution and regeneration of Bombyx mori Silk fibroin using ionic liquids. J Am Chem Soc 126 (44):14350-14351. doi:Doi 10.1021/Ja046079f.

[49] Garcia H, Ferreira R, Petkovic M, Ferguson JL, Leitao MC, Gunaratne HQN, Seddon KR, Rebelo LPN, Pereira CS (2010) Dissolution of cork biopolymers in biocompatible ionic liquids. Green Chem 12 (3):367-369. doi:Doi 10.1039/B922553f.

[50] Zhao H, Baker GA, Song ZY, Olubajo O, Crittle T, Peters D (2008) Designing enzyme-compatible ionic liquids that can dissolve carbohydrates. Green Chem 10 (6):696-705. doi:Doi 10.1039/B801489b.

[51] Miller-Chou BA, Koenig JL (2003) A review of polymer dissolution. Prog Polym Sci 28 (8):1223-1270. doi:Doi 10.1016/S0079-6700(03)00045-5.

[52] Narasimhan B, Peppas NA (1998). J Polym Sci, Part B: Polym Phys 36:2607-2614.

[53] Nägeli C, (1864) Über den inneren Bau der vegetabilischen Zellmembranen. Sitzber Bay Akad Wiss München 1:282-323.

[54] Pennetier G (1883) Note micrographique sur les altérations du coton. Bull Soc Ind Rouen 11:235–237.

[55] Fleming N, Thaysen AC (1920) On the deterioration of cotton on wet storage. Biochem J 14 (1):25-U23.

[56] Fleming N, Thaysen AC (1921) On the deterioration of cotton on wet storage. Biochem J 15 (3):407-U402.

[57] Tripp VW, Rollins ML (1952) Morphology and Chemical Composition of Certain Components of Cotton Fiber Cell Wall. Anal Chem 24 (11):1721-1728. doi:Doi 10.1021/Ac60071a008.

[58] Rollins ML, Tripp VW (1954) Optical and Electron Microscopic Studies of Cotton Fiber Structure. Text Res J 24 (4):345-357. doi:Doi 10.1177/004051755402400407.

[59] Hock CW (1954) Microscopic structure. In: Ott E, Spurlin HM, Grafflin MW (eds.) Cellulose and cellulose derivatives (part 1), 2nd edn. Interscience, New York, pp 347-392.

[60] Warwicker JO, Jeffries R, Colbran RL, Robinson RN (1966) A review of the literature on the effect of caustic soda and other swelling agents on the fine structure of cotton. Shirley Institute Pamphlet No. 93, St Ann's Press, England.

[61] Chanzy H, Noe P, Paillet M, Smith P (1983) Swelling and dissolution of cellulose in amine oxide/water systems. J Appl Polym Sci 37:239-259.

[62] Cuissinat C, Navard P (2006) Swelling and dissolution of cellulose, Part I: Free floating cotton and wood fibres in N-methylmorpholine-N-oxide-water mixtures. Macromol Symp 244:1-18.

[63] Cuissinat C, Navard P (2006) Swelling and dissolution of cellulose Part II: Free floating cotton and wood fibres in NaOH-water-additives systems. Macromol Symp 244:19-30. doi:DOI 10.1002/masy.200651202.

[64] Cuissinat C, Navard P, Heinze T (2008) Swelling and dissolution of cellulose. Part IV: Free floating cotton and wood fibres in ionic liquids. Carbohyd Polym 72 (4):590-596. doi:DOI 10.1016/j.carbpol.2007.09.029.

[65] Cuissinat C, Navard P (2008) Swelling and dissolution of cellulose, Part III: Plant fibres in aqueous systems. Cellulose 15 (1):67-74. doi:DOI 10.1007/s10570-007-9158-4.

[66] Cuissinat C, Navard P, Heinze T (2008) Swelling and dissolution of cellulose, Part V: Cellulose derivatives fibres in aqueous systems and ionic liquids. Cellulose 15 (1): 75-80. doi:DOI 10.1007/s10570-007-9159-3.

[67] Le Moigne N, Montes E, Pannetier C, Hoefte H, Navard P (2008) Gradient in dissolution capacity of successively deposited cell wall layers in cotton fibres. Macromol Symp 262:65-71. doi:DOI 10.1002/masy.200850207.

[68] Le Moigne N, Bikard J, Navard P (2010) Rotation and contraction of native and regenerated cellulose fibers upon swelling and dissolution: The role of morphological and stress unbalances. Cellulose 17 (3):507-519. doi:DOI 10.1007/s10570-009-9395-9.

[69] Basra AS (1999). In: Cotton Fibers: Developmental Biology, Quality Improvement and Textile Processing, Food Products Press, Haworth Press, Inc., New York.

[70] Marx-Figini M (1966) Comparison of the biosynthesis of cellulose in vitro and in vivo in cotton bolls. Nature 210:754-755.

[71] Klemm D, Schmauder HP, Heinze T (2002) In Biopolymers; De Baets, S.; Vandamme, E. J.; Steinbüchel, A., Eds; Wyley-VCH: Weinheim; Vol. 6, pp 275-319.

[72] Stevanic JS, Salmen L (2008) Characterizing wood polymers in the primary cell wall of Norway spruce (Picea abies (L.) Karst.) using dynamic FT-IR spectroscopy. Cellulose 15 (2):285-295. doi:DOI 10.1007/s10570-007-9169-1.

[73] Yamashiki T, Matsui T, Saitoh M, Okajima K, Kamide K, Sawada T (1990) Characterization of cellulose treated by the steam explosion method. Part 1. Influence of cellu-

lose resources on changes in morphology, degree of polymerization, solubility and solid structure. Br Polym J 22:73–83.

[74] Yamashiki T, Matsui T, Saitoh M, Okajima K, Kamide K, Sawada T (1990) Characterization of cellulose treated by the steam explosion method. Part 2: Effect of treatment conditions on changes in morphology, degree of polymerization, solubility in aqueous sodium hydroxide, and supermolecular structure of soft wood pulp during steam explosion. Br Polym J 22:121–128.

[75] Yamashiki T, Matsui T, Saitoh M, Okajima K, Kamide K, Sawada T (1990) Characterization of cellulose treated by the steam explosion method. Part 3: Effect of crystal forms (cellulose I, II and III) of original cellulose on changes in morphology, degree of polymerization, solubility and supermolecular structure by steam explosion. Br Polym J 22:201–212.

[76] Le Moigne N, Navard P (2010) Dissolution mechanisms of wood cellulose fibres in NaOH-water. Cell Chem Technol 17:31-45.

[77] Isogai A, Atalla RH (1998) Dissolution of cellulose in aqueous NaOH solutions. Cellulose 5 (4):309-319. doi:Doi 10.1023/A:1009272632367.

[78] Le Moigne N, Spinu M, Heinze T, Navard P (2010) Restricted dissolution and derivatization capacities of cellulose fibres under uniaxial elongational stress. Polymer 51 (2):447-453. doi:DOI 10.1016/j.polymer.2009.11.053.

[79] Zhang L, Ruan D, Gao S (2002) Dissolution and regeneration of cellulose in NaOH/thiourea aqueous solution. Journal of Polymer Science Part B: Polymer Physics 40 (14):1521-1529. doi:10.1002/polb.10215.

[80] Saric SP, Schofield RK (1946) The Dissociation Constants of the Carboxyl and Hydroxyl Groups in Some Insoluble and Sol-Forming Polysaccharides. Proc R Soc Lon Ser-A 185 (1003):431-447. doi:DOI 10.1098/rspa.1946.0029.

[81] Klemm D, Philipp B, Heinze T, Heinze U, Wagenknecht W (1998) Fundamentals and Analytical Methods. Comprehensive Cellulose Chemistry, vol Volume 1. Wiley-VCH Verlag GmbH & Co. KGaA. doi:10.1002/3527601929.ch1.

[82] Morgado DL, Frollini E (2011) Thermal decomposition of mercerized linter cellulose and its acetates obtained from a homogeneous reaction. Polímeros 21:111-117.

[83] Morgenstern B, Kammer H-W (1999) On the particulate structure of cellulose solutions. Polymer 40 (5):1299-1304. doi:http://dx.doi.org/10.1016/S0032-3861(98)00267-5.

[84] Rein DM, Khalfin R, Szekely N, Cohen Y (2014) True molecular solutions of natural cellulose in the binary ionic liquid-containing solvent mixtures. Carbohydrate Polymers 112 (0):125-133. doi:http://dx.doi.org/10.1016/j.carbpol.2014.05.059.

[85] Le K, Sescousse R, Budtova T (2012) Influence of water on cellulose-EMIMAc solution properties: A viscometric study. Cellulose 19 (1):45-54. doi:10.1007/s10570-011-9610-3.

[86] Wernersson E, Stenqvist B, Lund M (2015) The mechanism of cellulose solubilization by urea studied by molecular simulation. Cellulose:1-11. doi:10.1007/s10570-015-0548-8.

[87] Bergenstråhle-Wohlert M, Berglund L, Brady J, Larsson PT, Westlund P-O, Wohlert J (2012) Concentration enrichment of urea at cellulose surfaces: Results from molecular dynamics simulations and NMR spectroscopy. Cellulose 19 (1):1-12. doi:10.1007/s10570-011-9616-x.

[88] Liu H, Sale KL, Holmes BM, Simmons BA, Singh S (2010) Understanding the Interactions of Cellulose with Ionic Liquids: A Molecular Dynamics Study. The Journal of Physical Chemistry B 114 (12):4293-4301. doi:10.1021/jp9117437.

[89] Trygg J, Fardim P (2011) Enhancement of cellulose dissolution in water-based solvent via ethanol-hydrochloric acid pretreatment. Cellulose 18 (4):987-994. doi:DOI 10.1007/s10570-011-9550-y.

[90] Wormald P, Wickholm K, Larsson PT, Iversen T (1996) Conversions between ordered and disordered cellulose. Effects of mechanical treatment followed by cyclic wetting and drying. Cellulose 3 (3):141-152. doi:Doi 10.1007/Bf02228797.

[91] Fischer K, Goldberg W, Wilke M (1985) Strahlvorbehandlung von zellstoff für die regeneratfaserherstellung. Lenzinger Berichte 59:32-39.

[92] Kraft G, Schelosky N (2000) Irradiation of Dissolving Pulp by Electron Beams. Lenzinger Berichte 79:65-70.

[93] Driscoll M, Stipanovic A, Winter W, Cheng K, Manning M, Spiese J, Galloway RA, Cleland MR (2009) Electron beam irradiation of cellulose. Radiat Phys Chem 78 (7-8): 539-542. doi:DOI 10.1016/j.radphyschem.2009.03.080.

[94] Rahkamo L, SiikaAho M, Vehvilainen M, Dolk M, Viikari L, Nousiainen P, Buchert J (1996) Modification of hardwood dissolving pulp with purified Trichoderma reesei cellulases. Cellulose 3 (3):153-163. doi:Doi 10.1007/Bf02228798.

[95] Cao Y, Tan HM (2006) Improvement of alkali solubility of cellulose with enzymatic treatment. Appl Microbiol Biot 70 (2):176-182. doi:DOI 10.1007/s00253-005-0069-8.

[96] Le Moigne N, Jardeby K, Navard P (2010) Structural changes and alkaline solubility of wood cellulose fibers after enzymatic peeling treatment. Carbohyd Polym 79 (2): 325-332. doi:DOI 10.1016/j.carbpol.2009.08.009.

[97] Dos Santos NM (2013) Influence of chemical and enzymatic treatments on a variety of wood pulps on their dissolution in NaOH-water. Doctoral dissertation, Ecole Nationale Supérieure des Mines de Paris.

[98] Navard P (2012) (ed.) The European Polysaccharide Network of Excellence (EPNOE), DOI 10.1007/978-3-7091-0421-7, Springer-Verlag, Wien.

[99] Shen TC, Cabasso I (1982) Macromolecular solutions, solvent-property relationship in polymers. In: R. B. Seymour GASE, New York: Pergamon Press. (ed.).

[100] Zhang S, Fu CF, Li FX, Yu JY, Gu LX (2009) Direct Preparation of a Novel Membrane from Unsubstituted Cellulose in NaOH Complex Solution. Iran Polym J 18 (10): 767-776.

[101] Isobe N, Kimura S, Wada M, Kuga S (2012) Mechanism of cellulose gelation from aqueous alkali-urea solution. Carbohyd Polym 89 (4):1298-1300. doi:DOI 10.1016/j.carbpol.2012.03.023.

[102] Yamane C (2015) Structure formation of regenerated cellulose from its solution and resultant features of high wettability: A review. Nordic Pulp & Paper Research Journal 30 (1).

[103] Yamane C, Mori M, Saito M, Okajima K (1996) Structures and mechanical properties of cellulose filament spun from cellulose/aqueous NaOH solution system. Polym J 28 (12):1039-1047. doi:DOI 10.1295/polymj.28.1039.

[104] Cousins SK, Brown RM (1995) Cellulose-I Microfibril Assembly - Computational Molecular Mechanics Energy Analysis Favors Bonding by Vanderwaals Forces as the Initial Step in Crystallization. Polymer 36 (20):3885-3888. doi:Doi 10.1016/0032-3861(95)99782-P.

[105] Shiomi Y, Mimura N Cellulose fine particles, and liquid or solid dispersions thereof; U.S. Patent, 8,629,187 B2, 2014.

[106] Yamane C, Miyamoto H, Hayakawa D, Ueda K (2013) Folded-chain structure of cellulose II suggested by molecular dynamics simulation. Carbohydr Res 379:30-37. doi: 10.1016/j.carres.2013.06.012.

[107] Hermans PH (1949) Degree of Lateral Order in Various Rayons as Deduced from X-Ray Measurements. J Polym Sci 4 (2):145-151. doi:DOI 10.1002/pol.1949.120040204.

[108] Kiyose M, Yamamoto E, Yamane C, Midorikawa T, Takahashi T (2007) Structure and properties of low-substituted hydroxypropylcellulose films and fibers regenerated from aqueous sodium hydroxide solution. Polym J 39 (7):703-711. doi:DOI 10.1295/polymj.PJ2006206.

[109] Kolb M, Botet R, Jullien R (1983) Scaling of Kinetically Growing Clusters. Phys Rev Lett 51 (13):1123-1126. doi:DOI 10.1103/PhysRevLett.51.1123.

[110] Medronho B, Lindman B (2014) Brief overview on cellulose dissolution/regeneration interactions and mechanisms. Advances in colloid and interface science. doi:10.1016/j.cis.2014.05.004.

[111] Hayashi J, Masuda S, Watanabe S (1974) Plane Lattice Structure in Amorphous Region of Cellulose Fibers. Nippon Kagaku Kaishi (5):948-954.

[112] Liu ZH, Sun XF, Hao MY, Huang CY, Xue ZM, Mu TC (2015) Preparation and characterization of regenerated cellulose from ionic liquid using different methods. Carbohyd Polym 117:99-105. doi:DOI 10.1016/j.carbpol.2014.09.053.

[113] Gupta KM, Hu ZQ, Jiang JW (2013) Cellulose regeneration from a cellulose/ionic liquid mixture: The role of anti-solvents. Rsc Adv 3 (31):12794-12801. doi:Doi 10.1039/C3ra40807h.

[114] Östlund A, Idstrom A, Olsson C, Larsson PT, Nordstierna L (2013) Modification of crystallinity and pore size distribution in coagulated cellulose films. Cellulose 20 (4): 1657-1667. doi:DOI 10.1007/s10570-013-9982-7.

[115] Bang YH, Lee S, Park JB, Cho HH (1999) Effect of coagulation conditions on fine structure of generated cellulosic films made from cellulose/N-methylmorpholine-N-oxide/H2O systems. J Appl Polym Sci 73:2681-2690.

[116] Jiang G, Huang W, Zhu T, Zhang C, Kumi AK, Zhang Y, Wang H, Hu l (2011) Diffusion dynamics of 1-Butyl-3-methylimidazolium chloride from cellulose filament during coagulation process. Cell Chem Technol 18:921-928.

[117] Song h, Niu Y, Yu J, Zhang J (2013) Preparation and morphology of different types of cellulose spherulites from concentrated cellulose ionic liquid solutions. Soft Matter 9:3013-3020.

[118] Gotze K, Chemiefasern nach dem Viskoseverfahren, 3rd ed.,Vol. 1, Springer, Heidelberg, 1967, pp. 1 - 778.

[119] Klemm D, Heublein B, Fink HP, Bohn A (2005) Cellulose: Fascinating biopolymer and sustainable raw material. Angew Chem Int Ed Engl 44 (22):3358-3393. doi: 10.1002/anie.200460587.

[120] Yamane C (2015) Structure formation of regenerated cellulose from its solution and resultant features of high wettability: A review. Nordic Pulp and Paper Research Journal 30 (1):78-91.

[121] Medronho B, Lindman B (2014) Competing forces during cellulose dissolution: From solvents to mechanisms. Curr Opin Colloid Interface Sci 19:32-40.

[122] Firgo H, Eibl M, Eichinger D (1995). Lenzinger Ber 75:47-50.

[123] Sixta H, Michud A, Hauru L, Asaadi S, Ma Y, King AWT, Kilpeläinen I, Hummel M (2015) Ioncell-F: A High-strength regenerated cellulose fibre. Nordic Pulp and Paper Research Journal 30 (1):43-57.

[124] MICHUD A, King A, Parviainen A, Sixta H, HAURU L, Hummel M, KILPELÄINEN I (2014) Process for the production of shaped cellulose articles (WO 2014162062 A1).

[125] Mosier N, Wyman C, Dale B, Elander R, Lee YY, Holtzapple M, Ladisch M (2005) Features of promising technologies for pretreatment of lignocellulosic biomass. Bioresource Technol 96 (6):673-686. doi:DOI 10.1016/j.biortech.2004.06.025.

[126] Ohgren K, Bengtsson O, Gorwa-Grauslund MF, Galbe M, Hahn-Hagerdal B, Zacchi G (2006) Simultaneous saccharification and co-fermentation of glucose and xylose in steam-pretreated corn stover at high fiber content with Saccharomyces cerevisiae TMB3400. J Biotechnol 126 (4):488-498. doi:DOI 10.1016/j.jbiotec.2006.05.001.

[127] Farrell AE, Plevin RJ, Turner BT, Jones AD, O'Hare M, Kammen DM (2006) Ethanol can contribute to energy and environmental goals. Science 311 (5760):506-508. doi:DOI 10.1126/science.1121416.

[128] Richard G, Lugar R, Woolsey J (1999) The New Petroleum. Foreign Affairs 78:88.

[129] Kistler S (1931). Nature 127:741.

[130] Husing N, Schubert U (1998) Aerogels airy materials: Chemistry, structure, and properties. Angew Chem Int Edit 37 (1-2):23-45.

[131] Pierre AC, Pajonk GM (2002) Chemistry of aerogels and their applications. Chem Rev 102 (11):4243-4265. doi:Doi 10.1021/Cr0101306.

[132] Jin H, Nishiyama Y, Wada M, Kuga S (2004) Nanofibrillar cellulose aerogels. Colloid Surface A 240 (1-3):63-67. doi:DOI 10.1016/j.colsurfa.2004.03.007.

[133] Mullen L (1999) Aerogel rides again. Nasa Science News (HTTP://SCIENCE.NASA.GOV/SCIENCE-NEWS/SCIENCE-AT-NASA/1999/MSAD05FEB99_1/).

[134] Bhaduri SB (1993) THE PHYSICS AND CHEMISTRY OF SOL-GEL PROCESSING edited by C.J. Brinker and G.W. Scherer Academic Press, Inc., San Diego, CA 908 pages, hard cover, 1990. Materials and Manufacturing Processes 8 (3):391-392. doi: 10.1080/10426919308934843.

[135] Weatherwax RC, Caulfield DF (1971) Cellulose aerogels: An improved method for preparing a highly expanded form of dry cellulose. Tappi J 54:985-986.

[136] Tan CB, Fung BM, Newman JK, Vu C (2001) Organic aerogels with very high impact strength. Adv Mater 13 (9):644-646. doi:Doi 10.1002/1521-4095(200105)13:9<644::Aid-Adma644>3.0.Co;2-#.

[137] Kobayashi Y, Saito T, Isogai A (2014) Aerogels with 3D Ordered Nanofiber Skeletons of Liquid-Crystalline Nanocellulose Derivatives as Tough and Transparent Insulators. Angew Chem Int Edit 53 (39):10394-10397. doi:DOI 10.1002/anie.201405123.

[138] Duchemin BJC, Staiger MP, Tucker N, Newman RH (2010) Aerocellulose Based on All-Cellulose Composites. J Appl Polym Sci 115 (1):216-221. doi:Doi 10.1002/App.31111.

[139] Olsson RT, Samir MASA, Salazar-Alvarez G, Belova L, Strom V, Berglund LA, Ikkala O, Nogues J, Gedde UW (2010) Making flexible magnetic aerogels and stiff magnetic nanopaper using cellulose nanofibrils as templates. Nat Nanotechnol 5 (8):584-588. doi:DOI 10.1038/Nnano.2010.155.

[140] Wang M, Anoshkin IV, Nasibulin AG, Korhonen JT, Seitsonen J, Pere J, Kauppinen EI, Ras RHA, Ikkala O (2013) Modifying Native Nanocellulose Aerogels with Carbon Nanotubes for Mechanoresponsive Conductivity and Pressure Sensing. Adv Mater 25 (17):2428-2432. doi:DOI 10.1002/adma.201300256.

[141] Hamedi M, Karabulut E, Marais A, Herland A, Nystrom G, Wagberg L (2013) Nanocellulose Aerogels Functionalized by Rapid Layer-by-Layer Assembly for High Charge Storage and Beyond. Angew Chem Int Edit 52 (46):12038-12042. doi:DOI 10.1002/anie.201305137.

[142] Paakko M, Vapaavuori J, Silvennoinen R, Kosonen H, Ankerfors M, Lindstrom T, Berglund LA, Ikkala O (2008) Long and entangled native cellulose I nanofibers allow flexible aerogels and hierarchically porous templates for functionalities. Soft Matter 4 (12):2492-2499. doi:Doi 10.1039/B810371b.

[143] Sehaqui H, Zhou Q, Berglund LA (2011) High-porosity aerogels of high specific surface area prepared from nanofibrillated cellulose (NFC). Compos Sci Technol 71 (13): 1593-1599. doi:DOI 10.1016/j.compscitech.2011.07.003.

[144] Chen WS, Li Q, Wang YC, Yi X, Zeng J, Yu HP, Liu YX, Li J (2014) Comparative Study of Aerogels Obtained from Differently Prepared Nanocellulose Fibers. Chemsuschem 7 (1):154-161. doi:DOI 10.1002/cssc.201300950.

[145] Heath L, Thielemans W (2010) Cellulose nanowhisker aerogels. Green Chem 12 (8): 1448-1453. doi:Doi 10.1039/C0gc00035c.

[146] Sehaqui H, Salajkova M, Zhou Q, Berglund LA (2010) Mechanical performance tailoring of tough ultra-high porosity foams prepared from cellulose I nanofiber suspensions. Soft Matter 6 (8):1824-1832. doi:Doi 10.1039/B927505c.

[147] Cai J, Liu SL, Feng J, Kimura S, Wada M, Kuga S, Zhang LN (2012) Cellulose-Silica Nanocomposite Aerogels by In Situ Formation of Silica in Cellulose Gel. Angew Chem Int Edit 51 (9):2076-2079. doi:DOI 10.1002/anie.201105730.

[148] Chang CY, Zhang LN (2011) Cellulose-based hydrogels: Present status and application prospects. Carbohyd Polym 84 (1):40-53. doi:DOI 10.1016/j.carbpol.2010.12.023.

[149] Vashist A, Vashist A, Gupta YK, Ahmad S (2014) Recent advances in hydrogel based drug delivery systems for the human body. J Mater Chem B 2 (2):147-166. doi:Doi 10.1039/C3tb21016b.

[150] Qiu XY, Hu SW (2013) "Smart" Materials Based on Cellulose: A Review of the Preparations, Properties, and Applications. Materials 6 (3):738-781. doi:Doi 10.3390/ Ma6030738.

[151] Sannino A, Demitri C, Madaghiele M (2009) Biodegradable Cellulose-based Hydrogels: Design and Applications. Materials 2 (2):353-373. doi:Doi 10.3390/Ma2020353.

[152] Chang CY, Duan B, Cai J, Zhang LN (2010) Superabsorbent hydrogels based on cellulose for smart swelling and controllable delivery. Eur Polym J 46 (1):92-100. doi:DOI 10.1016/j.eurpolymj.2009.04.033.

[153] Nnadi F, Brave C (2011) Environmentally friendly super absorbent polymers for water conservation in agricultural lands. Soil Science and Environmental Management 2:206-211.

[154] Tongdeesoontorn W, Mauer LJ, Wongruong S, Sriburi P, Rachtanapun P (2011) Effect of carboxymethyl cellulose concentration on physical properties of biodegradable cassava starch-based films. Chem Cent J 5. doi:Artn 6 Doi 10.1186/1752-153x-5-6.

[155] Seki Y, Altinisik A, Demircioglu B, Tetik C (2014) Carboxymethylcellulose (CMC)-hydroxyethylcellulose (HEC) based hydrogels: Synthesis and characterization. Cellulose 21 (3):1689-1698. doi:DOI 10.1007/s10570-014-0204-8.

[156] Davidson D, Verma M, Gu F (2013) Controlled root targeted delivery of fertilizer using an ionically crosslinked carboxymethyl cellulose hydrogel matrix. SpringerPlus 2 (1):318.

[157] Wang ZG, Fan XL, He M, Chen ZQ, Wang YF, Ye QF, Zhang HJ, Zhang LN (2014) Construction of cellulose-phosphor hybrid hydrogels and their application for bioimaging. J Mater Chem B 2 (43):7559-7566. doi:Doi 10.1039/C4tb01240b.

[158] Duan JF, Jiang JX, Li JZ, Liu LJ, Li YQ, Guan C (2015) The Preparation of a Highly Stretchable Cellulose Nanowhisker Nanocomposite Hydrogel. J Nanomater. doi:Artn 963436 Doi 10.1155/2015/963436.

[159] Zhou H, Zhu H, Yang X, Zhang Y, Zhang X, Cui K, Shao L, Yao J (2014) Temperature/pH Sensitive Cellulose-based Hydrogel: Synthesis, Characterization, Loading, and Release of Model Drugs for Potential Oral Drug Delivery. BioResources; Vol 10, No 1 (2015).

[160] Sannino A, Esposito A, Nicolais L, Del Nobile MA, Giovane A, Balestrieri C, Esposito R, Agresti M (2000) Cellulose-based hydrogels as body water retainers. J Mater Sci-Mater M 11 (4):247-253. doi:Doi 10.1023/A:1008980629714.

[161] Buhus G, Peptu C, Popa M, Desbrieres J (2009) Controlled Release of Water Soluble Antibiotics by Carboxymethylcellulose- and Gelatin-Based Hydrogels Crosslinked with Epichlorohydrin. Cell Chem Technol 43 (4-6):141-151.

[162] Prakash S, Soe-Lin H (2004) Strategy for Cell Therapy: Polymers for Live Cell Encapsulation and Delivery. Trends Biomater Artif Organs 18:24-35.

[163] de Vos P, Lazarjani HA, Poncelet D, Faas MM (2014) Polymers in cell encapsulation from an enveloped cell perspective. Adv Drug Deliver Rev 67-68:15-34. doi:DOI 10.1016/j.addr.2013.11.005.

[164] Mohanty AK, Misra M, Drzal LT (2002) Sustainable bio-composites from renewable resources: Opportunities and challenges in the green materials world. J Polym Environ 10 (1-2):19-26. doi:Doi 10.1023/A:1021013921916.

[165] John MJ, Thomas S (2008) Biofibres and biocomposites. Carbohyd Polym 71 (3): 343-364. doi:DOI 10.1016/j.carbpol.2007.05.040.

[166] Faruk O, Bledzki AK, Fink HP, Sain M (2012) Biocomposites reinforced with natural fibers: 2000-2010. Prog Polym Sci 37 (11):1552-1596. doi:DOI 10.1016/j.progpolymsci.2012.04.003.

[167] Müssig J, Stevens C, editors (2010) Industrial applications of natural fibres: Structure, properties and technical applications. John Wiley & Sons.

[168] Bledzki AK, Gassan J (1999) Composites reinforced with cellulose based fibres. Prog Polym Sci 24 (2):221-274. doi:Doi 10.1016/S0079-6700(98)00018-5.

[169] Wambua P, Ivens J, Verpoest I (2003) Natural fibres: Can they replace glass in fibre reinforced plastics? Compos Sci Technol 63 (9):1259-1264. doi:Doi 10.1016/S0266-3538(03)00096-4.

[170] Reddy MM, Vivekanandhan S, Misra M, Bhatia SK, Mohanty AK (2013) Biobased plastics and bionanocomposites: Current status and future opportunities. Prog Polym Sci 38 (10-11):1653-1689. doi:DOI 10.1016/j.progpolymsci.2013.05.006.

[171] Mohanty AK, Misra M, Drzal LT (2001) Surface modifications of natural fibers and performance of the resulting biocomposites: An overview. Compos Interface 8 (5): 313-343. doi:Doi 10.1163/156855401753255422.

[172] George J, Sreekala MS, Thomas S (2001) A review on interface modification and characterization of natural fiber reinforced plastic composites. Polym Eng Sci 41 (9): 1471-1485. doi:Doi 10.1002/Pen.10846.

[173] Kalia S, Thakur K, Celli A, Kiechel M, Schauer CL (2013) Surface modification of plant fibers using environment friendly methods for their application in polymer composites, textile industry and antimicrobial activities: A review. Journal of Environmental Chemical Engineering 1:97-112.

[174] Matabola KP, De Vries AR, Moolman FS, Luyt AS (2009) Single polymer composites: A review. J Mater Sci 44 (23):6213-6222. doi:DOI 10.1007/s10853-009-3792-1.

[175] Kmetty A, Barany T, Karger-Kocsis J (2010) Self-reinforced polymeric materials: A review. Prog Polym Sci 35 (10):1288-1310. doi:DOI 10.1016/j.progpolymsci.2010.07.002.

[176] Karger-Kocsis J, Barany T (2014) Single-polymer composites (SPCs): Status and future trends. Compos Sci Technol 92:77-94. doi:DOI 10.1016/j.compscitech.2013.12.006.

[177] Nishino T, Matsuda I, Hirao K (2004) All-cellulose composite. Macromolecules 37 (20):7683-7687. doi:Doi 10.1021/Ma049300h.

[178] Qin C, Soykeabkaew N, Xiuyuan N, Peijs T (2008) The effect of fibre volume fraction and mercerization on the properties of all-cellulose composites. Carbohyd Polym 71 (3):458-467. doi:DOI 10.1016/j.carbpol.2007.06.019.

[179] Soykeabkaew N, Nishino T, Peijs T (2009) All-cellulose composites of regenerated cellulose fibres by surface selective dissolution. Compos Part a-Appl S 40 (4):321-328. doi:DOI 10.1016/j.compositesa.2008.10.021.

[180] Brown WF (1999) Vulcanized Fibre—An Old Material With A New Relevancy. Electrical Insulation Conference and Electrical Manufacturing and Coil Winding Conference Proceedings: 309-312.

[181] Brandenberger JE (1918) Composite cellulose film, US Patent 1266766.

[182] Cartier N, Escaffre P, Mathevet F, Chanzy H, Vuong R (1994) Structure and Recycling of Vegetable Parchment. Tappi J 77:95-101.

[183] Gindl W, Keckes J (2005) All-cellulose nanocomposite. Polymer 46 (23):10221-10225. doi:DOI 10.1016/j.polymer.2005.08.040

[184] Huber T, Bickerton S, Mussig J, Pang SS, Staiger MP (2012) Solvent infusion processing of all-cellulose composite materials. Carbohyd Polym 90 (1):730-733. doi:DOI 10.1016/j.carbpol.2012.05.047.

[185] Soykeabkaew N, Arimoto N, Nishino T, Peijs T (2008) All-cellulose composites by surface selective dissolution of aligned ligno-cellulosic fibres. Compos Sci Technol 68 (10-11):2201-2207. doi:DOI 10.1016/j.compscitech.2008.03.023.

[186] Haverhals LM, Reichert WM, De Long HC, Trulove PC (2010) Natural Fiber Welding. Macromol Mater Eng 295 (5):425-430. doi:DOI 10.1002/mame.201000005.

[187] Hull D, Clyne T (1996) An Introduction to Composite Materials. Cambridge University Press, Cambridge, UK, 2nd edition.

[188] Huber T, Mussig J, Curnow O, Pang SS, Bickerton S, Staiger MP (2012) A critical review of all-cellulose composites. J Mater Sci 47 (3):1171-1186. doi:DOI 10.1007/s10853-011-5774-3.

[189] Duchemin BJC, Mathew AP, Oksman K (2009) All-cellulose composites by partial dissolution in the ionic liquid 1-butyl-3-methylimidazolium chloride. Compos Part a-Appl S 40 (12):2031-2037. doi:DOI 10.1016/j.compositesa.2009.09.013.

[190] Duchemin BJC, Newman RH, Staiger MP (2009) Structure-property relationship of all-cellulose composites. Compos Sci Technol 69 (7-8):1225-1230. doi:DOI 10.1016/j.compscitech.2009.02.027.

[191] Abbott A, Bismarck A (2010) Self-reinforced cellulose nanocomposites. Cellulose 17 (4):779-791. doi:DOI 10.1007/s10570-010-9427-5.

[192] Pullawan T, Wilkinson AN, Eichhorn SJ (2013) Orientation and deformation of wet-stretched all-cellulose nanocomposites. J Mater Sci 48 (22):7847-7855. doi:DOI 10.1007/s10853-013-7404-8.

[193] Gindl W, Schoberl T, Keckes J (2006) Structure and properties of a pulp fibre-reinforced composite with regenerated cellulose matrix. Appl Phys a-Mater 83 (1):19-22. doi:DOI 10.1007/s00339-005-3451-6.

[194] Soykeabkaew N, Sian C, Gea S, Nishino T, Peijs T (2009) All-cellulose nanocomposites by surface selective dissolution of bacterial cellulose. Cellulose 16 (3):435-444. doi:DOI 10.1007/s10570-009-9285-1.

[195] Zhou Q, Malm E, Nilsson H (2009) Biomimetic design of cellulose-based nano- structured composites using bacterial cellulose. Polymer Preprints 50:7-8.

[196] Gindl-Altmutter W, Keckes J, Plackner J, Liebner F, Englund K, Laborie MP (2012) All-cellulose composites prepared from flax and lyocell fibres compared to epoxy-matrix composites. Compos Sci Technol 72 (11):1304-1309. doi:DOI 10.1016/j.compscitech.2012.05.011.

[197] Huber T, Pang SS, Staiger MP (2012) All-cellulose composite laminates. Compos Part a-Appl S 43 (10):1738-1745. doi:DOI 10.1016/j.compositesa.2012.04.017.

[198] Adusumali RB, Reifferscheid M, Weber H, Roeder T, Sixta H, Gindl W (2006) Mechanical properties of regenerated cellulose fibres for composites. Macromol Symp 244:119-125. doi:DOI 10.1002/masy.200651211.

[199] Röder T, Moosbauer J, Kliba G, Schlader S, Zuckerstatter G, Sixta H (2009) Comparative Characterisation of man-made regenerated cellulose fibres. Lenzinger Berichte 87:98-105.

[200] Chaudemanche C, Navard P (2011) Swelling and dissolution mechanisms of regenerated Lyocell cellulose fibers. Cellulose 18 (1):1-15. doi:DOI 10.1007/s10570-010-9460-4.

[201] Duchemin BJCZ, Newman RH, Staiger MP (2007) Phase transformations in microcrystalline cellulose due to partial dissolution. Cellulose 14 (4):311-320. doi:DOI 10.1007/s10570-007-9121-4.

[202] Nishino T, Takano K, Nakamae K (1995) Elastic-Modulus of the Crystalline Regions of Cellulose Polymorphs. J Polym Sci Pol Phys 33 (11):1647-1651. doi:DOI 10.1002/polb.1995.090331110.

[203] Pullawan T, Wilkinson A, Eichhorn S (2010) Discrimination of matrix-fibre interactions in all-cellulose nanocomposites. Compos Sci Technol 70:2325–2330.

[204] Pullawan T, Wilkinson AN, Zhang LNN, Eichhorn SJ (2014) Deformation micromechanics of all-cellulose nanocomposites: Comparing matrix and reinforcing components. Carbohyd Polym 100:31-39. doi:DOI 10.1016/j.carbpol.2012.12.066.

[205] Rosato DV, Rosato DV (1995) Injection molding handbook. Chapman & Hall, 2nd edition.

[206] Mazumdar SK (2001) Composites Manufacturing: Materials, Product, and Process Engineering, Chapter 6, CRC Press.

[207] Mazumdar SK (2001) Manufacturing techniques. Composites Manufacturing: Materials, Product, and Process Engineering, Chapter 6, CRC Press.

[208] Glancey JL (2010) Vacuum-Assisted Resin Transfer Molding. M. Schwartz, ed. Innovations in Materials Manufacturing, Fabrication, and Environmental Safety, Chapter 18, 531-544, CRC Press.

[209] Huber T, Bickerton S, Mussig J, Pang S, Staiger MP (2013) Flexural and impact properties of all-cellulose composite laminates. Compos Sci Technol 88:92-98. doi:DOI 10.1016/j.compscitech.2013.08.040.

[210] Gindl W, Martinschitz KJ, Boesecke P, Keckes J (2006) Changes in the molecular orientation and tensile properties of uniaxially drawn cellulose films. Biomacromolecules 7 (11):3146-3150. doi:Doi 10.1021/Bm060698u.

[211] Li DS, Liu ZY, Al-Haik M, Tehrani M, Murray F, Tannenbaum R, Garmestani H (2010) Magnetic alignment of cellulose nanowhiskers in an all-cellulose composite. Polym Bull 65 (6):635-642. doi:DOI 10.1007/s00289-010-0276-z.

[212] Pullawan T, Wilkinson AN, Eichhorn SJ (2012) Influence of Magnetic Field Alignment of Cellulose Whiskers on the Mechanics of All-Cellulose Nanocomposites. Biomacromolecules 13 (8):2528-2536. doi:Doi 10.1021/Bm300746r.

[213] Duchemin BJC, Staiger MP, Newman RH (2014) High-Temperature Viscoelastic Relaxation in All-Cellulose Composites. Macromol Symp 340 (1):52-58. doi:DOI 10.1002/masy.201300123.

[214] Manabe S, Iwata M, Kamide K (1986) Dynamic Mechanical Absorptions Observed for Regenerated Cellulose Solids in the Temperature-Range from 280-K to 600-K. Polym J 18 (1):1-14. doi:Doi 10.1295/Polymj.18.1.

[215] Kunioka M, Ninomiya F, Funabashi M (2006) Biodegradation of poly(lactic acid) powders proposed as reference test materials for international standard of biodegradation evaluation method. Journal of Polymer Degradation and Stability 91:1919-1928.

[216] Mathew AP, Oksman K, Sain M (2005) Mechanical properties of biodegradable composites from poly lactic acid (PLA) and microcrystalline cellulose (MCC). J Appl Polym Sci 97 (5):2014-2025. doi:Doi 10.1002/App.21779.

[217] Zhang L, Liu HQ, Zheng LS, Zhang JY, Du YM, Feng HQ (1996) Biodegradability of regenerated cellulose films in soil. Ind Eng Chem Res 35 (12):4682-4685. doi:Doi 10.1021/Ie950624r.

[218] Zhang L, Zhou JP, Huang J, Gong P, Zhou Q, Zheng LS, Du YM (1999) Biodegradability of regenerated cellulose films coated with polyurethane/natural polymers interpenetrating polymer networks. Ind Eng Chem Res 38 (11):4284-4289. doi:Doi 10.1021/Ie9900869.

[219] Kalka S, Huber T, Steinberg J, Baronian K, Mussig J, Staiger MP (2014) Biodegradability of all-cellulose composite laminates. Compos Part a-Appl S 59:37-44. doi:DOI 10.1016/j.compositesa.2013.12.012.

[220] Yousefi H, Nishino T, Shakeri A, Faezipour M, Ebrahimi G, Kotera M (2013) Water-repellent all-cellulose nanocomposite using silane coupling treatment. J Adhes Sci Technol 27 (12):1324-1334. doi:Doi 10.1080/01694243.2012.695954.

Chapter 2

An Assessment of Surface Properties and Moisture Uptake of Nonwoven Fabrics from Ginning By-products

Vince Edwards, Paul Sawhney, Alvin Bopp, Alfred French, Ryan Slopek, Michael Reynolds, Chuck Allen, Brian Condon and Joseph Montalvo

Additional information is available at the end of the chapter

http://dx.doi.org/10.5772/61329

Abstract

Greige (raw) cotton by-products resulting from cotton ginning and mill processes have long been bleached for use in absorbent nonwoven products. The potential to use greige cotton by-products as an economical source for absorbent nonwoven blends is explored. The nonwoven hydroentanglement of greige cotton lint with cotton gin motes and comber noils blends was analyzed for fiber surface polarity, swelling, and absorbance to assess properties with potential usefulness in absorbent nonwovens. The electrokinetic analysis of the fabric surface gives a composite picture of the relative hydrophilic/hydrophobic polarity absorbency and swelling properties. Nonwoven fabrics made with cleaned greige cotton lint separately blended with comber noils and ginning motes at 40:60 and 60:40 blend ratios demonstrated charge, swell, and percent moisture uptake profiles that are characteristic of the fabrics' crystalline/amorphous cellulosic content with some variance in swelling properties. However, cellulose crystallite size varied. X-ray diffraction patterns of the three different cotton constituents displayed similar crystalline cellulose compositions. An electrochemical double-layer analysis of charge based on a pH titration ($\zeta_{plateau}$) was employed to measure the relative fiber and fabric surface polarity which varied slightly between -21 and -29 mV. A relationship of fiber swelling ($\Delta\zeta$) and percent moisture content is apparent when greige cotton lint and other fibers are blended. The blended nonwoven materials possess absorbent properties characterized by similar moisture uptake (7.1-9.5 %) and fiber polarity, but some variation in swelling is based on the by-product additive and its percent content. The crystallinity, electrokinetic, and water binding properties of the nonwoven by-product materials are discussed in the context of the molecular features water, cellulose, and greige cotton components that enhance potential uses as absorbent nonwoven end-use products.

Keywords: nonwoven cotton, surface properties, cellulose, water of hydration, electrokinetic, by-products

INTECH
open science | open minds

1. Introduction

1.1. Cotton and cotton by-products in nonwovens

In recent years, the preference to use cotton fibers in nonwoven absorbent products has increased. Cotton fiber is naturally renewable and biodegradable. Cotton's characteristic soft hand, hypoallergenic properties, absorbency, and cellulosic composition have been historically utilized mostly in woven fabric products. Cotton's current use in nonwovens is estimated to be approximately 2 % (by volume/weight) of the total fiber consumption in nonwovens. Most of the cotton used at present in absorbent nonwovens is bleached cotton, including lint, gin motes, linters, comber noils, and the so-called other cotton textile processing wastes. However, the potential to use greige (nonbleached) cotton in nonwoven absorbent products has received increased attention based on innovations in cotton cleaning and nonwovens processes that open and expose the hydrophilic cellulosic component of greige cotton fiber to water absorption [1-3]. This affords an economical source of highly cleaned absorbent greige cotton nonwovens with the retention of properties inherent to the traditional cotton fabrics that generally require costly and eco-sensitive chemical scouring and bleaching processes.

Griege (raw) cotton gin motes are just one of several by-products (viz., cotton seeds, linters, motes, sticks, burrs, gin trash, among others) of the "cotton ginning process" that mechanically separates the (long) cotton fibers (greige lint) from the harvested seed cotton. The cotton ginning by-products are used in numerous applications [4-9]. The cotton gin motes typically have fibers up to ½ inch long and thus can be classified as textile fibers. However, limited use of cotton gin motes is made in traditional textiles made with spun yarns. Although cotton gin motes and textile comber noils have been used in cotton socks, their relatively lower cost and apparent differences in structural and surface properties make these cotton coproducts/fibers of interest to explore a broader nonwoven and high value end use. In addition, the processing innovations of modern nonwovens provide a facile conduit for efficiently blending these types of discounted by-products with greige cotton lint to explore new, value-added cotton-blend nonwoven products.

1.2. Absorbent applications

Highly cleaned greige cotton fiber retains most of its natural, native protective membrane or surface coating of waxes and pectin (native to the greige cotton fiber). In combination with surface-exposed cellulose from nonwoven hydroentanglement process conditions, unique fiber properties are retained when compared to scoured and bleached cotton. The amphiphilic surface character in nonwoven greige cotton, which is a combination of the polarity balance between the hydrophilic and the hydrophobic elements of the cotton material, is suitable for the application to the material layer components of incontinence absorbent products and wipes [10].

The potential for the topsheet application of nonwoven greige cotton for incontinence and wound dressing's materials is based on its suitability as it facilitates a mechanism of optimal polar gradients between the layers surrounding the absorbent core. During fluid transport, an optimal charge gradient between the layers from topsheet to backsheet helps facilitate better performance of wettable and absorbent materials and is associated with lower rewet, strike-through, and acquisition/distribution to the absorbent core [11]. Nonwoven greige cotton compares well with other commercial materials when analyzed for its performance as an incontinence layer surrounding the absorbent core. Moreover, when combined with synthetic fibers like polyester and polypropylene, greige cotton/synthetic blends possess an added scope of utilization in this regard [10]

1.3. Fabric surface properties and electrokinetic analysis

The zeta (ζ) potential is a measure of the charge and charge density on the surface of a particle or fiber and can be used to gauge the stability of a colloidal or the ability of a material to absorb liquid. The ζ potential can be either positive or negative depending on the surface chemistry. The ζ potential analysis is useful with absorbent materials. The fluid dynamics of the electro-chemical double-layer model enables measurement of functional properties similar to those that occur at the solid-liquid interface of incontinence materials [12]. The ζ potential decrease (a change in electrostatic potential at the shear plane of the electrochemical double layer) is caused by the swelling of the fibers and the outward movement of the aqueous shear plane where ions are in contact with the outer Helmholtz plane on the fiber surface [13]. Thus, a concomitant increase in the swelling of the fiber occurs for most fibers as the shear plane moves out to the more diffuse layer of ions, causing a decrease in ζ potential [14].

This chapter examines the electrokinetic properties of hydroentangled nonwoven materials made by blending clean greige cotton lint with greige cotton by-product fibers with a view to understanding the similarities the materials possess. The three constituent cotton fibers employed in the nonwovens were evaluated for their relative cellulosic crystallinity in relation to their moisture uptake properties. Some of the blends also included polyester fibers for comparison as were previously examined [10]. Two different measures of moisture uptake are also reported. One method [15] is simple, portable, easy to use, and involves an infrared lamp to dry the samples. All weight loss is attributed to moisture. The other method is the recent ASTM Karl Fischer titration method for water content developed for lint cotton, raw and processed. Water absorbency tests were also conducted. We report here the preparation, characterization, and electrokinetic analysis of cleaned greige cotton [16] in combination with gin motes and comber noils at two different blend ratios of the greige cotton and the by-products as a measure of the fabric polarity, swelling, and absorbent properties achievable with these fabric blends. The properties of the cotton materials are discussed in light of their water binding properties related to potential absorbent applications.

2. Experimental materials and methods

A commercially available bale of precleaned greige cotton was acquired from T. J. Beall, LLC. A quantity of cotton gin mote fibers was also obtained from T. J. Beall.

2.1. Hydroentanglement of fibrous webs into nonwoven fabric structures

The needle-punched webs of the different fiber blends were uniformly hydroentangled using a Fleissner MiniJet system (Figure 1). The system is equipped with one low water pressure jet head that wets the incoming feed web material on its top face, while two high water pressure jet heads alternatively impact the wetted substrate on either face. For all the fabrics, the low water pressure head was set to inject the water at 50 bars, and the two high water pressure heads were set at 125 bars. The fabric production speed was 5 m per minute. The resulting hydroentangled fabric was dried using a meter-wide, gas-fired drum dryer and wound onto a cardboard tube to form a compact fabric roll. The hydroentangling line was flushed and cleaned after each fabric production trial.

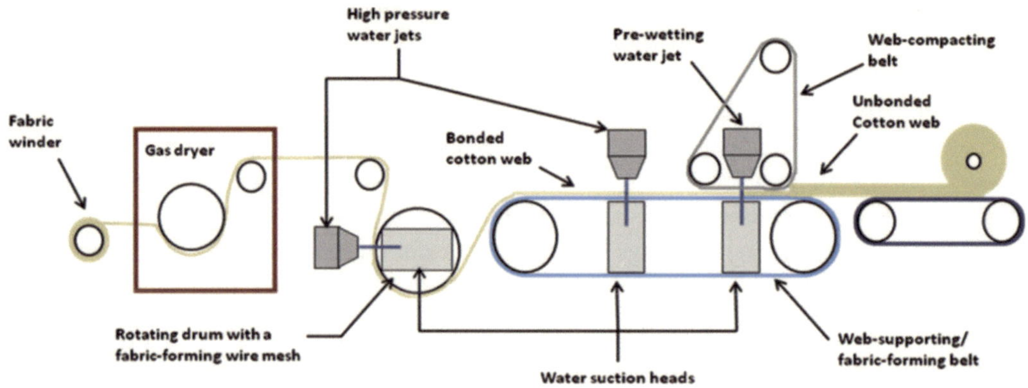

Figure 1. Diagram of a nonwovens hydroentanglement line. An outline schematic of the Fleissner MiniJet system used in the study.

2.2. Fabric absorbency measurements

The AATCC drop test measures the time it takes for one drop of water applied to a fabric held in an embroidery hoop to be absorbed (when the sheen disappears). The ATSM method uses a sample of fabric that is 76 mm wide and cut to a length that equals 5.0 ± 0.1 g. The sample is rolled into a cylindrical shape, upon itself, and placed in a basket of standardized weight and size. The basket is dropped from a height of 25 mm into a water bath, and the time it takes the sample and basket to sink is measured as sink time. After sinking, the sample is then allowed to remain submerged in water for 10 s. The basket is removed and allowed to drain for 10 s, and the sample is weighed to determine its water content. The weight of the water is reported as the absorptive capacity (grams of water held by 1 g of fabric).

The test methods used in the study are as follows:

- ASTM D6242—mass per unit area (weight) of nonwoven textile fabrics
- ATSM D5035—breaking force and elongation of textile fabrics (strip method)
- AATCC 79—absorbency of textiles (drop test)

- ASTM D1117 Section 5—absorbency time and absorptive capacity nonwovens
- ASTM D7785-12—standard test method for water in lint cotton by oven evaporation combined with volumetric Karl Fischer titration
- ASTM D2495-01—standard test method for moisture in cotton by oven drying
- ASTM D629-99—standard test method for moisture content for incontinence products
- AATCC 20A-2000—moisture content for incontinence products

2.3. Measurement of fabric polarity, charge, relative composition, swelling, and porosity

Streaming zeta potential experiments are carried out with an electrokinetic analyzer, which is manufactured in Ashland VA, USA, using the cylindrical cell developed for the measurement of fibrous samples. For each measurement, a fiber plug is placed between the Ag/AgCl hollow cylindrical electrodes of the cylindrical cell. The pH dependence of the zeta potential is investigated with the background electrolyte of 1 mM KCl solution. The swelling behavior of the incontinence products is measured using the Anton Paar analyzer with the cylindrical cell template. A 0.65-g sample is loaded into the cell and quickly rinsed with electrolyte solution. The flow rate is adjusted into the range of 60-100 mL/min by compression of the sample to remove trapped air. The pH of the sample is about 5.5 and was adjusted to 9.0 by adding 0.1 N NaOH solution.

2.4. Moisture uptake by ASTM D629-99

The moisture content of the cotton samples is measured using modified ASTM D629-99 and AATCC 20A-2000 methods. (The modification involved using an infrared lamp to dry the materials rather than a laboratory oven as called for in the standard methods.) The sample is conditioned overnight in a humidity chamber with the hygrometer reading at 70 % and a room temperature of 23°C. The moisture measurements are based on weight loss and are taken with an infrared moisture balance (Kett FD 240, manufactured by Kett Electric Laboratory in Tokyo, Japan). The balance is set for automatic wet-based moisture with a drying temperature of 110°C. Approximately, a 1-g sample was used for each measurement on the Kett FD 240.

2.5. Water content by Karl Fischer titration ASTM D7795-12

In order to determine the water content of cotton fibers via Karl Fischer titration (KFT), fiber samples must be conditioned to standard testing conditions, 21.1°C and 65 % relative humidity. The samples were weighed into 0.1-g specimens to four decimal places, placed in the Karl Fischer glass vials, and immediately crimped with septa caps. Each blend was formulated directly in the glass KFT vials by weight basis. All open vials were equilibrated to moisture equilibrium, sealed with a septum cap, and analyzed by standard KFT (ASTM D7785, 2012). The enclosed samples were then placed into mason jars that had been acclimated in the conditioned lab. The samples were then encapsulated in the jar right up until testing to maintain the environment. Prior to KFT testing cotton samples, blank vials were used for quality control measures. During testing, the vial is lowered into a 150°C single sample oven for approximately 5 min, while dry nitrogen (< 20 ppb water) is bubbled into the sample at 60 mL per minute. The water from the cotton sample from the vial is released and driven into the titration cell from which the percentage of moisture present is calculated from the volume of reagent consumed. The water content in the blank vial was 0.1 %. After analysis, a visual observation of the containers revealed all samples were white; there was no discoloration or scorching. Additionally, the polyester (PES) fibers had not melted.

2.6. Standard oven drying ASTM D2495-01

The typical method to measuring moisture or weight loss of cotton fibers involves using an oven where samples are heated to 105-110°C for 24 h. Empty sample bottles were weighed and then filled with 1 g samples and reweighed. Cottons were heated in a Yamato DKN 600 mechanical convection oven with a 150-L capacity and a mean flow rate of approximately 1.3 L/s. All weights were made in a standard conditioned laboratory.

2.7. X-ray diffraction

Wiley-milled samples of the comber noils, gin motes, and UltraClean cotton were pressed into pellets and placed in a powder diffractometer with Cu Kα radiation in reflection mode. The sample was secured in a paraffin base, which had a small effect on the comber noil pattern. For comparison, a pattern was simulated with the Mercury software [17] based on a modified crystal information file from the cellulose Iβ crystal structure [18]. The modification was a short increase in the *a*-axis of the unit cell to 7.906 from the original 7.784 value to better position the (200) peak because of the small differences in the cotton and tunicate unit cell [19]. A peak width at half maximum height of 1.85° 2θ was used. Additionally, the preferred orientation induced by pressing the sample pellet was compensated by a facility in the Mercury software, with a March-Dollase parameter of 1.2. The Scherrer formula was used to convert the peak width at half maximum (pwhm) to crystallite sizes perpendicular to the large (200) peak with a shape constant of 1.0. The Segal Crystallinity Index [20] was used to calculate the degree of relative crystallinity.

2.8. Environmental scanning electron microscope

A Philips XL-30 environmental scanning electron microscope (ESEM) was used to image the specimens, operating at 10-13 kV. The samples were mounted on standard Cambridge 1/200 SEM stubs using double-stick photo adhesive tabs. They were coated with 60/40 % gold/ palladium in a Technics' Hummer II sputter coater to a thickness of 20-30 nm.

3. Results and discussion

3.1. Preparation of the various fibrous materials for their respective nonwovens

The fibers utilized in this study were hand blended in ratios, as indicated in Table 1, before being converted into fibrous webs of nominal density (60-90 g/m^2). Fiber qualities of materials used in this study, including fiber length and properties illustrative of the fiber nomenclature, were consistent with those previously reported for greige cotton, gin motes, and comber noils [21]. UltraClean cotton, which is a form of greige cotton [2], was separately combined with the cotton gin motes and comber noils, whereupon the blends were carded, crosslapped, and subjected to light needle punching prior to their separate hydroentanglement at 50 bar wet-out water pressure and 125 bar hydroentangling water pressure. Figure 1 diagrams the process of hydroentanglement. This approach to greige cotton-based nonwoven production has previously been shown to increase absorbency while still retaining some of the cotton's native waxes and pectin [3]. Thus, depending on the hydroentangling process parameters and conditions [1-3], this approach increases hydrophobicity of the greige cotton nonwoven compared to an equivalent scoured and bleached cotton nonwoven product.

Sample No.	Fiber blend	Fabric weight (g/m²)
1	100% UC	55
2	100% PES	72
5	60% UC/40% GM	66.5
6	40% UC/60% GM	59.5
7	60% UC/40% CM	60.1
8	40% UC/60% CM	74.8

Fiber ID:

UC = UltraClean—unbleached precleaned cotton staple

PES = polyester staple fiber (1.5 denier, 3.8 cm (1.5″) long)

GM = cotton gin motes

CM = cotton textile comber noils

Table 1. Hydroentangled fabrics of different fiber blends

3.2. Fabric surface polarity

The moisture uptake (% MC) and electrokinetic results, including the fabric surface polarity ($\zeta_{plateau}$), swelling ($\Delta\zeta$), rate of swelling (k), isoelectric point (IEP), and related material density for the cotton/by-product blends, are shown in Table 2 for the cotton/by-product blends. The ζ potential is taken from a titration as previously shown [12-14] that assesses the surface charge based on pH and the planar portion along the x-axis of the plot is designated as the zeta plateau value ($\zeta_{plateau}$), which is a reflection of the relative hydrophilic versus hydrophobic character of the fabric. The ζ potentials reported here are consistent with the previously reported ζ potential titrations for both absorbent incontinence layers and greige cotton nonwoven incontinence topsheet and acquisition distribution layer prototypes [10,11].

3.3. Swelling and moisture uptake

Figure 2 shows the relationship between % moisture content and swelling behavior ($\Delta\zeta$) of the cotton blends. The swelling of the fiber blends results in the expansion of the electrochemical shear plane formed near the fibrous surfac [11,13] and is manifested as a decrease in the absolute value of ζ. The electrokinetic data presented in Figure 2 includes samples of the cotton blends investigated in this study. The percent moisture versus the Δζ for all ratios of cotton fiber reflects a correlation between the amount of moisture the material is prone to absorb and its degree of swelling. As the Δζ value (*x*-axis) of the cotton fabrics increases, the moisture uptake (% M) remains between 7.5 % and 9 %. KFT water content correlates with mean Δζ with an $R^2 = 0.88$. These two properties (moisture uptake and swelling) promote fluid transport. The increased swelling and relatively constant value for water uptake shown in Figure 2 can be contrasted with absorbency properties, as shown for the cotton blends in Table 3. As seen in Table 3, the ASTM sink time measurement (an indirect measurement of the material's swelling properties) principally correlates with material density in contrast with previous observations in that it correlates with swelling in greige cotton/polyester blends [10]. Swelling and rate of swelling of fabric blends are noticeably increased by blending cleaned greige cotton with by-products at a 60/40 ratio. On the other hand, the moisture uptake was the highest in the 100 % Gin Motes and the lowest in 60/40 UC/GM. This is consistent with the absorbency capacity (Table 3) being highest in the UC/GM blends. The gin motes have lower density and higher surface area and thus promote hydrophilic transport of water in the fabric. In addition, as discussed below, the smaller cellulose crystallite size of the gin motes is consistent with this and plays a role at a molecular level in the increased water absorption capacity observed in the cotton by-product nonwovens.

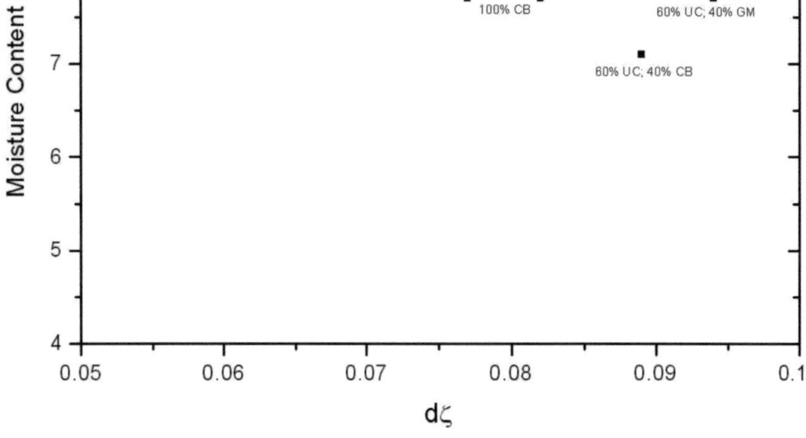

Figure 2. Plot of delta Zeta versus percent moisture content for the cotton blends of this study.

It is also notable that 60 % UC blended with 40 % GM, and CM results in the greatest swelling of the material. The hydrophobicity of the outer greige cotton fiber present in the UltraClean cotton may begin to contribute to the overall swelling of the material at 60 % UC present in the blend. There is little data on the presence of waxes and pectin in comber noils and gin motes, so a relative comparison of cotton cuticle contributions is not possible. However, recent contact angle measurements of greige cotton nonwovens have shown that approximately 32 % of the water droplet coverage is with wax-coated fiber [22]. Thus, the increased swelling due to increasing the ratio of UC may be a result of an additive contribution of waxes from the greige cotton, which are expected to contribute hydrophobicity to the fiber surface analogous to more hydrophobic fibers like polyester. Consistent with this observation, it is important to note that the greige cotton/polyester blend had the highest water absorption capacity. In addition, the similarity of the isoelectric points (IEPs) among the UltraClean cotton samples is consistent with the composition of the samples being cellulosic [13].

ID	Composition	KFT water content (%)	Moisture content (%)	IEP pH	Plateau potential (mV)	Swell test, k (min^{-1})	$\Delta\zeta$	Density (g/cm^3)
1	100% UC	7.46	8.2	2.3 (e)	-26	0.0044	0.065	0.147
2	100% PES	0.53	1.2	3.8	-23	0.0098 0.0130	0.133 0.197	0.118
3	60% UC/40% PES	4.77	4.6	2.6	-44	0.0140	0.102	0.110
4	100% CB	7.54	7.7	2.3	-29.0	0.014 0.0018	0.082 0.077	0.210
5	60% UC/40% CB	7.50	7.1	2.5	-23.0	0.0059	0.089	0.079
6	40% UC/60% CB	7.55	8.6	2.3	-22.0	0.0087	0.078	0.115
7	100% GM	7.58	9.5	2.3	-27.0	0.0150	0.077	0.114
8	60% UC/40% GM	7.54	7.7	2.5	-23.0	0.0093	0.094	0.103
9	40% UC/60% GM	7.50	9.0	2.3	-26.0	0.0091	0.063	0.131

Fiber codes: UC = UltraClean; PES = polyester; CB = comber noils; GM = gin motes.

KFT = Kraft Fisher Test; IEP = isoelectric point; plateau potential refers to the flat portion of the zeta potential titration curve; Swell test *k* refers to the rate of swelling; and $\Delta\zeta$ refers to the change in zeta potential.

Table 2. Electrokinetic data for the hydroentangled fabric samples made with the different fibers and their blends

3.4. Bound versus free water and crystallite size in cotton blends

It is an understatement to say that the nature of the binding of water to cotton plays a role in the swelling of the blended fabrics as are examined here. The microstructure of cotton fibers allows the penetration of water, in the case of the greige cotton nonwovens studied here. Water-accessible sites of cellulose are formed when cotton contacted by the high-pressure water jets during the nonwoven hydroentanglement process, which enhances the exposure of the primary and secondary cell wall of the fiber to an aqueous environment and results in an increase in cellulose-bound water. A subsequent loosening of the fiber cuticle resulting in the exposure of the cellulosic portion of the fiber is evidenced in the SEM image of nonwoven greige cotton shown in Figure 3. Thus, the hydroentanglement process promotes the disruption of the fiber cuticle that retains some wax and pectin while exposing cellulose fibrils and microfibrils to water penetration.

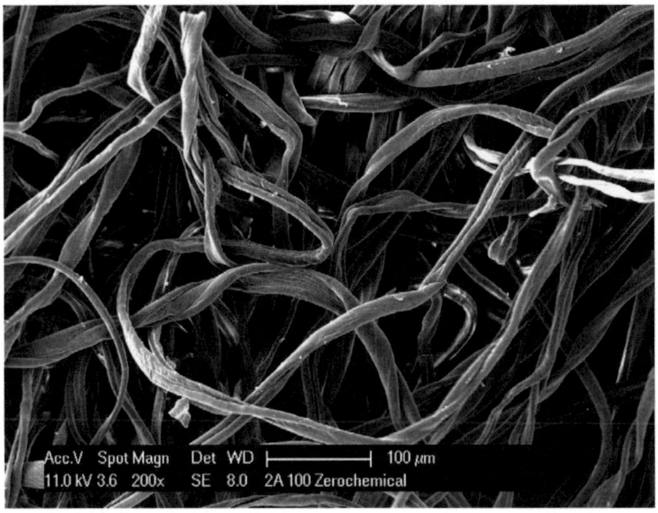

Figure 3. SEM image of hydroentangled greige cotton.

The 7-9.5 % moisture uptake observed for the cotton fabrics of this study and correlated in Figure 2 with swelling is consistent with previous studies that show moisture regain of dried cotton in this range [23]. It is also interesting to speculate how variation in cellulose crystallite size may affect binding of water to the cellulosics of this study. The X-ray diffraction patterns for the individual gin motes, comber noils, and greige cotton (UCC) are shown in Figure 4. The spectra all show the profile characteristic of cellulose I [18, 19]. Table 4 gives the percent crystallinity and cellulose crystallite size of the different types of cotton fibers compared with bleached cotton. The crystallinity index for each sample is as high as can be expected [20] for cotton and reflects nearly complete crystallinity. The calculated X-ray diffraction pattern is shown in Figure 3. The calculated pattern matches the observed patterns [28], especially that of the gin motes, fairly well despite the lack of any consideration of amorphous material in the calculation. The Segal Crystallinity Index values arise almost completely from the overlap of the wide observed peaks (20).

On the other hand, the crystallite size of cellulose in the gin motes and comber noils was larger than cellulose crystallites in greige cotton linters. It is important to note that bleached cotton also was found to have larger crystallite size, but the much higher density of the sample tested excludes it from comparison with the other samples. The relationship between crystallinity

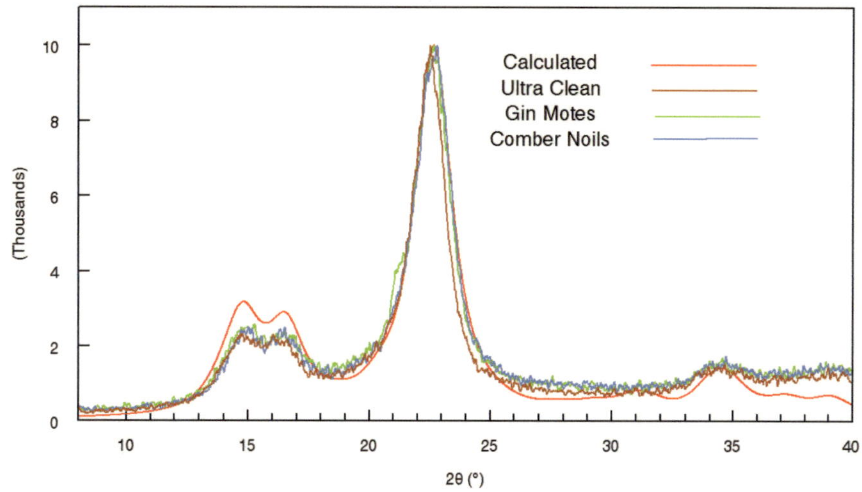

Figure 4. X-ray powder diffraction patterns of comber noil, gin motes, and UltraClean cotton. The comber noil pattern is affected by added intensity near 20° 2θ, but it does not seem to affect the crystallinity measurement. Also shown is a calculated pattern for the modified crystal structure of Nishiyama et al.[18]

and moisture uptake has received some attention in the literature over the years (24-26). Relevant to crystallinity, crystallite size, and moisture content, Nelson and O'Connor have previously noted that for two celluloses of the same crytallinity but different crystallite size, the cellulose with the smaller crystallite size will exhibit higher moisture regain (27). Observation of this phenomenon is based on more accessible water binding sites, i.e., hydroxyls on smaller crystallites, and is consistent with what is observed in this study (Table 4) in samples of similar density (Table 1) that exhibit differences in crystallite size, i.e., there is a parallel of crystallite size to absorption capacity in the cellulosic greige cotton, gin motes, and comber noils. Thus, as the crystallite size decreases, all things being equal, the absorption capacity of the fabrics consisting of 100 % greige cotton, gin motes, and comber noils increases. The calculated surface-to-volume ratio of smaller cellulose crystallites, as observed with the cotton by-products, is higher than the greige cotton, which infers more accessible hydroxyls for bound water. Thus, an increased capacity to hold water is observed as well in the relatively higher absorption capacity of the cotton by-product nonwovens, which have smaller cellulose crystallite size than the greige cotton (Table 4).

3.5. Moisture determinations

The water content results via Karl Fischer titration (KFT), following ASTM 7795, track the moisture results based on the Kett moisture determination balance that utilizes an infrared lamp (Table 3). The differences between KFT and Kett are due, in part, to the different nature of the two methods. The Kett measures moisture weight loss after drying under a large infrared heat lamp which, minimizes scorching.[15] On the other hand, scorching may contribute to oxidation, thus increasing weight loss (29). Because KFT is specific to water and is carried out under nitrogen, oxidation is eliminated. As a check, standard oven drying was used on the UC

	Carded	Carded	Carded	Carded
Fiber/fiber blend	WT (g/m2)	AATCC Drop Test (sec.)	ASTM Sink time (s)	ASTM Absorbency Capacity (g H_2O per g of fabric)
100% UC	54.6	<1	40.8	4.97
100% GM	51.9	2.4	32.1	6.88
100% CB	73.9	6.6	45.0	6.71
100% PES	68.6	>60	>300	0.14
60% UC/ 40% CB	60.1	<1	69.4	6.87
40% UC/ 60% CB	74.8	<1	79.0	6.12
60% UC/ 40% GM	66.5	5.4	26.4	7.25
40% UC/ 60% GM	59.5	<1	40.2	7.13
60% UC/ 40% PES	75.4	>60	113.8	7.05
40% UC/ 60% PES	66.6	>60	241.0	7.97
100% BC	103.6	<1	2.25	6.46

Fiber codes: UC = UltraClean; GM = gin motes; CB = comber noils; PES = polyester staple; BC = bleached cotton. (A sample of bleached cotton prepared by similar process was not available for the testing. This sample is provided for information and not comparison due to its higher density).

Table 3. Absorbency characteristics of the various hydroentangled fabrics

cotton and polyester fibers, generating a moisture content of 7.3 % and 0.43 %, respectively. The polyester water content value is consistent with that found in literature of fiber at standard testing conditions, around 0.4 % (30). Also, note that the KFT water content (Table 3) for polyester is 0.53 %, which is expected since the carrier gas is dry nitrogen. In addition, the small range of KFT values of the cellulosic blends is due to the specificity of KFT to water compared to the weight loss as measured with the Kett infrared heating method.[15] The tight range of crystallinities (Table 4) of the cellulosic is consistent with the KFT findings.

Sample	Crystallinity index	Crystallite size (Å)
UltraClean cotton	88.77	53.48
Comber noils	87.25	47.08
Gin motes	88.73	46.06
Bleached cotton	87.95%	65.30

Table 4. Crystallinity index and crystallite size of the different fibers

3.6. Cellulose-water binding considerations

The binding of water to cotton has been characterized in three states, including (1) strongly bound or nonfreezing water, (2) anisotropically constrained or perturbed water, and (3) unperturbed water, or water undergoing isotropic motion (24-26). Consideration of the phenomena of water binding to cellulosic fiber, from crystalline to fibrillar state, crystalline cellulose (crsytallites of 36 cellulose chains or more) has been characterized as low water binding (24-26, 31). However, ordered microfibrillar cellulose, which is composed of cellulose crystallites, possesses surface hydroxyls that present accessible water binding sites where penetrating water may form a monolayer (termed nonfreezing water) at a level of (0.1 g/g cotton) (24,31).

French et al. illustrated a cotton fiber moisture model based on the cotton fiber as a solid cylinder of cellulose with a central lumen (length 28 mm, diameter 15 μm) having crystallites at a size of 4 nm by 28 mm, a crystallite density of 1.63 g/cm^3, and a water monolayer of 0.25 nm thickness (25, 28). This model is contrasted with experimental models, including one based on TEM images of a water swollen cotton fiber to image water-accessible surfaces. In the cotton fiber water model at a moisture level of 5 %, a monolayer of water only covers 30 % of the surface, and this scenario accounts for one water molecule per cellobiose unit in the cellulose chains of the crystallite. This is somewhat consistent with a recent report on cotton water sorption based on sorption isotherms of cotton by Yakumin et al., showing a capacity of 41.5 g water/g of cellulose (32). They report that 37 % of the dry cotton cellulose is accessible to water binding at equilibrium moisture adsorption. These authors also point out that it is insufficient to use the cellulose crystallinity alone in calculating accessible water surfaces since water molecules can also enter defective regions in the crystallites formed during drying. This feature is also implicated in the incomplete removal of water being located in regions of the crystalline phase that are disorganized or "defective crystallites" upon drying of cotton. Nonetheless, the presence of a strongly bound monolayer of water on cotton is consistent historically with thermal calculations (90 cal/g), approximating that of the heat of fusion for ice and validating the hydrogen bonding forces of water to cellulose (31,33). From this state, further water sorption then assumes the character of a capillary-like condensation and has been characterized as free water, i.e., perturbed and unperturbed water. Thus, based on numerous past studies that characterize the role of water binding to moisture in cotton, the results of this study suggest that the cotton blends, which possess an average of 8 % moisture content, probably have most of the water strongly bound as nonfreezing water under ambient conditions. This is likely as well as it has previously been shown that the density of interfacial water (strongly bound water) on cellulose is increased when it is perturbed. This property improves the wettability of the cotton blends and may be seen as contributing synergistically to improving the swelling properties in the 60 % UC blends.

4. Conclusion

This study has shown that the ability of cotton gin mote fibers to modulate swelling and moisture uptake is beneficial in absorbent products. Depending on the end-use application, a hydroentangled nonwoven fabric made by using a blend of cotton mote fibers and greige (bleach-less) cleaned cotton lint gave optimal swelling and reasonably good moisture uptake. We compare properties of the nonwoven fabrics made with precleaned greige cotton that were blended in different proportions with cotton gin motes and comber noils. This approach opens up considerations for a more eco-friendly and economical use of cotton by-products, i.e., for wettable topsheet use and fluid transport layer in absorbent nonwovens. Here we contrast cotton by-product blends in greige cotton nonwovens with a similar previous study where polyester was blended with precleaned greige cotton lint (10). The results have shown that the cotton gin mote fibers, compared to the polyester fibers, yield improved moisture uptake while giving comparable swelling attributes. The cellulose-water interactions have been discussed in light of these findings. This moisture uptake/swelling property could assist toward promoting greater utilization of cotton and cotton motes in absorbent nonwoven products where end-use applications require comparable moisture uptake to cotton and similar swelling properties to polyester. Also, the water content by KFT seems to correlate with the ζ potential. Balanced material surface polarity, swelling, density, and moisture uptake is key to optimizing absorbent nonwovens for use in hygiene, incontinence, and even wound care applications, and the results of this study illustrate how these properties may be tuned in with cotton by products used in combination with greige cotton lint. Thus, it is demonstrated that the use of less expensive cotton gin motes—the by-product of cotton ginning process—in blends with cleaned greige cotton lint can potentially be useful and competitive for many nonwoven end-use products where absorbency or moisture uptake, swelling, biodegradability, and sustainability are desirable. This study demonstrates the versatility of nonwoven greige cotton when combined with cotton by-products as putative economical substitutes for synthetic fibers in absorbent applications. It further demonstrates the merit of focusing on material construction and analysis of fiber surface properties with novel by-product fibers at solid-liquid interfaces and the value of considering the molecular factors that influence properties of wettability and fluid transport as they exist in topsheet and layer components useful in absorbent prototypes, i.e., fibers that present green alternatives to petro-based approaches while demonstrating structure/function value. Future studies will focus on specific functional applications for incontinence hygiene and wound care nonwovens.

Author details

Vince Edwards[1*], Paul Sawhney[1], Alvin Bopp[2], Alfred French[1], Ryan Slopek[1], Michael Reynolds[1], Chuck Allen[1], Brian Condon[1] and Joseph Montalvo[1]

*Address all correspondence to: vince.edwards@ars.usda.gov

1 Southern Regional Research Center, US Department of Agriculture, USA

2 Chemistry Department, Southern University of New Orleans, USA

References

[1] Sawhney APS, Condon B, Reynolds M, Slopek R and Hui D. Advent of greige cotton non-wovens made using a hydro-entanglement process. *Textile Research Journal*. 2010; 80: 1540-9.

[2] Condon B, Gary L, Sawhney APS, et al. Properties of nonwoven fabrics made with UltraClean™ cotton. *World Journal of Engineering*. 2010; 7: 180-4.

[3] Sawhney P, Allen C, Reynolds M, Condon B and Slopek R. Effect of water pressure on absorbency of hydroentangled greige cotton non-woven fabrics. *Textile Research Journal*. 2012; 82: 21-6.

[4] Holt GA, Simonton JL, Beruvides MG and Canto A. Utilization of cotton gin by-products for manufacturing fuel pellets: an economic perspective. *Journal of the American Society of Agricultural Engineers*. 2004; 20: 423-30.

[5] Alagoz Z and Yilmaz EJ. Changes of organic content in two soils by cotton gin waste amendment, a by-product of agricultural industry. *Journal of Food, Agriculture and Environment*. 2009; 9: 250-2.

[6] Klasson KT, Wartelle LH, Lima IM, Marshall WE and Akin DE. Activated carbons from flax shive and cotton gin waste as environmental adsorbents for the chlorinated hydrocarbon trichloroethylene. *Bioresource Technology*. 2009; 100: 5045-50.

[7] Smith C. Adding value to (cotton gin) motes. *Cotton Farming*. 2007; 51: 14-24.

[8] Hughs SE, Wakelyn PJ and Green JK. Flammability of cotton gin trash/burrs. *Proceedings of the Beltwide Cotton Conferences*. 2007, p. 1974-6.

[9] Bajwa SG, Bajwa DS, Holt G, Coffelt T and Nakayama F. Properties of thermoplastic composites with cotton and guayule biomass residues as fiber fillers. *Industrial Crops and Products*. 2011; 33: 747-55.

[10] Edwards V, Condon B, Sawhney P, Reynolds M, Allen C, Nam S, et al. Electrokinetic analysis of hydroentangled greige cotton-synthetic fiber blends for absorbent technologies. *Textile Research Journal*.2013; 83(18):1949-60.

[11] Edwards JV, Prevost N, Condon B, Batiste S, Reynolds M, Allen C, et al. Electrokinetic properties of functional layers in absorbent incontinence nonwoven products. *Textile Research Journal*. 2012; 82(10):1001-3

[12] Grahame DC. The electrical double layer and the theory of electrocapillarity. *Chemical Reviews*. 1947; 41: 441-501.

[13] Ribitsch V and Stana-Kleinscheck K. Characterizing textile fiber surfaces with streaming potential measurements. *Textile Research Journal*. 1998; 68: 701-7.

[14] Grancaric AM, Tarbuk A and Pusic T. Electrokinetic properties of textile fabrics. *Coloration Technology*. 2005; 121: 221-7.

[15] USA K. Model FD 240 Infrared Moisture Determination Balance; 1997.

[16] Malkan SR and Wadsworth LC. Polymer-laid systems. In: Turbak A, (ed.). *Nonwovens: Theory, Process, Performance, and Testing*. Atlanta: TAPPI Press; 1993, p. 171-92.

[17] Macrae CF, Bruno IJ, Chisholm JA, et al. Mercury CSD 2.0—new features for the visualization and investigation of crystal structures. *Journal of Applied Crystallography*. 2008; 41: 466-70.

[18] Nishiyama Y, Langan P and Chanzy H. Crystal structure and hydrogen-bonding system in cellulose Iβ from synchrotron x-ray and neutron fiber diffraction. *Journal of the American Chemical Society*. 2002; 124: 9074-82.

[19] Nishiyama Y, Johnson G and French A. Diffraction from nonperiodic models of cellulose crystals. *Cellulose*. 2012; 19: 319-36.

[20] Segal L, Creely JJ, Martin AE and Conrad CM. An empirical method for estimating the degree of crystallinity of native cellulose using the x-ray diffractometer. *Textile Research Journal*. 1959; 29: 786-94.

[21] Sawhney P, Allen C, Reynolds M, Slopek R and Condon B. Whiteness and absorbency of hydroentangled cotton-based nonwoven fabrics of different constituent fibers and fiber blends. *World Journal of Engineering*. 2013; 10(2): 125-132.22.

[22] Buisson YL, Rajasekaran K, French AD, Conrad DC and Roy PS. Qualitative and quantitative evaluation of cotton fabric damage by tumble drying. *Textile Research Journal*. 2000; 70: 739-43.

[23] Taylor RE, French AD, Gamble GR, et al. 1H and 13C solid-state NMR of Gossypium barbadense (Pima) cotton. *Journal of Molecular Structure*. 2008; 878: 177-84.

[24] French AD, Goynes WR, Rousselle MA and Thibodeaux DP. Cotton fiber and moisture—some of the basics. *2004 Beltwide Cotton Conferences*. San Antonio, TX: Cotton Foundation. p. 2990-4.

[25] Venkatraman P, Ashbaugh H, Johnson GP and French AD. Simulation studies of the wetting of crystalline faces of cotton cellulose. *2010 Beltwide Cotton Conferences*. New Orleans, LA2010. p. 1577-80.

[26] Nelson ML and O'Connor RT. Relation of certain infrared bands to cellulose crystallinity and crystal lattic type, Part II. A new infrared ratio for estimation of crystallinity in celluloses I and II. *Journal of Applied Polymer Science*. 1964; 8: 1325-41.

[27] French A. Idealized powder diffraction patterns for cellulose polymorphs. *Cellulose*. 2014; 21(1):

[28] Fortier C, Montalvo J Jr, Von Hoven T, Easson M, Rodgers J and Condon B. Preliminary evidence of oxidation in standard oven drying of cotton: attenuated total reflec-

tance/fourier transform infrared spectroscopy, colorimetry, and particulate matter formation. *Textile Research Journal*. 2014; 84:157-73.

[29] Hedge RR, Dahiya A and Kamath MG. Polyester Fibers. 2004, http://www.engr.utk.edu/mse/Textiles/Polyester %20fiber.htm

[30] Rowland Stanley P. Cellulose: pores, internal surfaces, and the water interface. In: *Textile and Paper Chemistry and Technology*. American Chemical Society, 1977, pp. 20-45, DOI: 10.1021/bk-1977-0049.ch002

[31] Yakunin NA, Yakunina EN and Moryganov AP. Structural dependence of the features of the sorption behavior of cotton cellulose in reacting with water. *Fibre Chemistry*. 2008; 40: 529-32.

[32] Hermans, P.H. Contribution to the physic s of cellulose fibres. Journal of Polymer Science. 1947; 2(3): 354.

Chapter 3

Cellulose Grafting by Atom Transfer Radical Polymerization Method

Nevin Çankaya

Additional information is available at the end of the chapter

http://dx.doi.org/10.5772/61707

Abstract

Increased public awareness on environmental issues, fluctuations in raw material prices, and global difficulties in raw material supplies have made it necessary to find new, sustainable, environment-friendly, and inexpensive natural polymer sources. Cellulose and its derivatives, at this point, have become an important research matter once again, because of its easy accessibility, abundance, price, minimal effect on the environment, and new properties discovered with the help of technology.

This study explains cellulose and its modification, and gives information about ATRP, which is a controlled radical polymerization. Later, detailed information on grafting of cellulose by ATRP in both, the author's studies and other existing researches is provided.

Keywords: Cellulose, Grafting, Graft Copolymer, Atom Transfer Radical Polymerization (ATRP), Controlled Radical Polymerization (CRP)

1. Introduction

Cellulose is the most abundant natural polymer, which is used raw or substituted in a number of applications, for instance in paper, packaging, or lacquer technologies. Moreover, cellulose is biodegradable and renewable, which makes it preferable from an environment point of view. On the other hand, since its utilization would be limited in its natural form, cellulose has to be modified. This modification is mostly achieved by reaction of hydroxyl groups, yielding to cellulose esters or ethers. Besides, cellulose backbone can be grafted with synthetic polymers by either "grafting from" or "grafting onto" it, using various polymerization methods [1].

Thus, different materials with many different properties, such as elasticity, ion exchange ability, thermal stability, and mechanical properties can be obtained by grafting. Grafting is mostly realized via free-radical polymerization initiated with redox systems, mostly based on ceric or ferrous salts or sodium hydrogen sulfite systems together with peroxides. Using these methods, many new cellulose–backbone graft copolymers have been synthesized but key properties like number, density, length, and molecular mass distribution were out of control. In addition to the vinyl monomers, heterocyclic lactones can be grafted from cellulose or its derivatives by ring-opening polymerization (ROP), giving, in principle, biodegradable polymeric materials [2]. Also, nitroxide-mediated polymerization (NMP) [3] or reversible addition–fragmentation chain transfer process (RAFT) [4] have recently been applied into controlled grafting of cellulose with synthetic polymers. In recent years, a couple of papers reporting controlled cellulose grafting using ATRP have been published [5-6].

This chapter provides information about cellulose, ATRP and cellulose grafting by ATRP method. Later, some research conducted on cellulose grafting by ATRP are discussed.

2. Cellulose

Natural polymers are polymers that are biologically produced and have unique functional attributes. Polysaccharides (cellulose, starch, chitin, chitosan, dextran, inulin, levan etc.), proteins (collagen, gelatin, elastin, actin, etc.), and polynucleotides (DNA and RNA) are natural polymers. Increasing attention is being given to more complex carbohydrate polymers produced by bacteria and fungi, especially to polysaccharides such as xanthan, curdlan, pullulan, and hyaluronic acid.

Cellulose is regarded as the most abundant and renewable biopolymer in nature and the most common organic natural polymer, representing about 1.5×10^{12} tons of the annual biomass [7]. Together with its abundancy, cellulose has several other factors that make it a superior material, such as its renewability; high chemical stability; low price; high tenacity in the wet state; ability to yield to thin membranes; ease of the control of pore size, ranging from 1 to 100 nm in diameter; and controllable porosity [8].

2.1. Structure of cellulose

Natural and unprocessed cellulose in crystal form is called cellulose I (natural-raw cellulose). Crystal form of cellulose changes with external effects like chemical processes and temperature, which yields to various crystalline modifications depending on these processes. Cellulose II is obtained by mercerization of cellulose with concentrated alkaline solution, while cellulose III is obtained by inflating and decomposing of cellulose I or II with liquid ammonia (under -30°C). Cellulose IV is a high-temperature form and is prepared by heating cellulose I, II, or III in glycerol. When various crystalline forms of cellulose are heated up to 250°C, intramolecular hydrogen bonds loosen, which results in a different crystal structure. Cellulose V is obtained by processing cotton or paper with strong phosphoric acid or hydrochloric acid [9]. It is also one of the most promising raw materials for the modern industry, because of its low

http://dx.doi.org/10.5772/61707

cost and convertibility to various functional materials. Cellulose is a carbohydrate homopolymer consisting of β-D-glucopyranose units joined together by β-1,4-glycosidic linkages (Fig. 1). Unlike starch, glucose units in cellulose have long and unbranched chains caused by –CH_2OH groups alternating above and below the plane of rings. Lack of side chains lets cellulose molecules form organized structures. Cellulose has both crystallized (higher packing density) and amorphous (lower packing density) regions. Hydrogen bonds connect cellulose chains in the crystallized region, as well as van der Waals forces, which is also important in the lattice energy. Cellulose chains can either have parallel or antiparallel orientation, which is the source of the names of cellulose I and II, respectively. These two forms of cellulose have different structures. Other polymorphic forms are called cellulose III and IV. X-ray diffraction measurements prove that cotton cellulose contains about 60% crystalline cellulose I and 40% amorphous cellulose [10]. By delignification, removal of noncellulosic polysaccharides and low molecular mass cellulose components (called β- and γ-cellulose) the raw materials containing cellulose can be transformed to pure long-chain cellulose called α-cellulose. Pulp analysis of cellulose shows that it contains 44% carbon, 6.2% hydrogen, and 49% oxygen. Pure cellulose yields to approximately 95% D-glucose ($C_6H_{12}O_6$) when hydrolyzed [11].

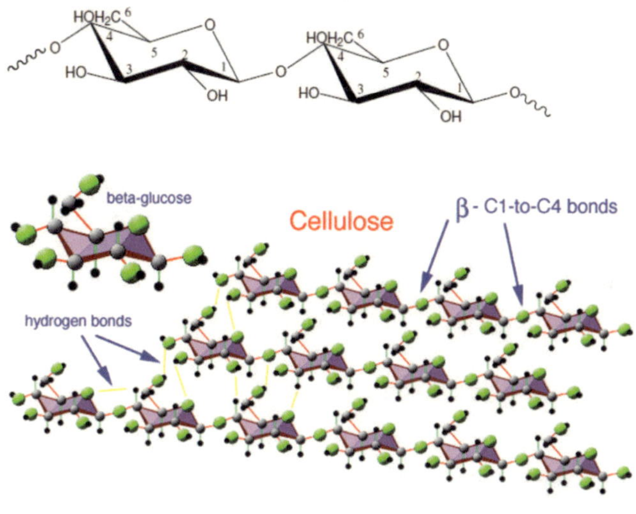

Figure 1. Cellobiose unit: two β-D-glucopyranose units joined together by β-1,4-glycosidic linkage [12]

Polymerization degree of natural cellulose is strongly related with cellulose source and separation/purification methods that are used. Cotton cellulose has the highest molecular mass which is around 800 000, while wood and other sources of cellulose have molecular masses around 160 000. Hydroxyl groups cause intense intra- and intermolecular hydrogen bonds in cellulose in solid form. These hydrogen bonds and high molecular mass make cellulose insoluble in classic solvents. Dissolving becomes possible when appropriate solvents (like Schweitzer solution –copper-II-hydroxide with concentrated ammonium hydroxide mixture) are used, or by modification of cellulose into derivatives like cellulose triacetate and cellulose nitrate [13-15].

When the distance between oxygen and hydrogen atoms in a cellulose molecule is less than 3 Å, these molecules interact with each other and form intra- or intermolecular hydrogen bonds (Fig. 2) [16].

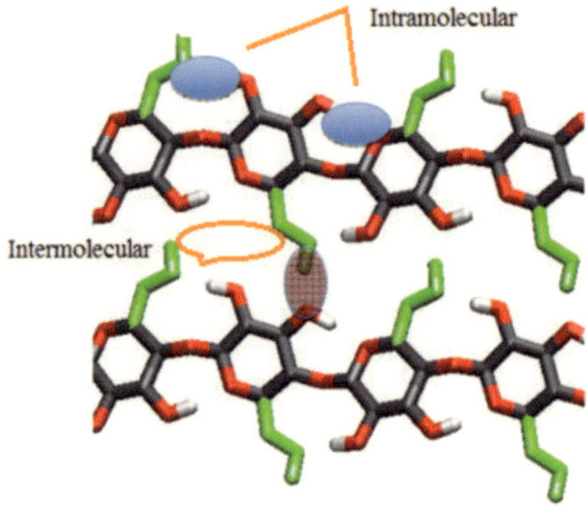

Figure 2. Intra and intermolecular hydrogen binding of cellulose

Equatorial orientation of the hydroxyl group together with the linear structure of cellulose results in both intermolecular and intramolecular hydrogen bonds. Sheet-like structure and crystalline form of the polymer are a result of intermolecular hydrogen bonds. Microfibrils formed by the cellulose molecules (Fig. 3) stack together and make up the fibrils, which gives the cellulose fibers. Hydrophilic property of the polymer is a result of hydroxyl groups. This makes it readily adsorb water [17].

2.2. Modification of cellulose and its derivatives

Its specific structure makes cellulose attractive. Cellulose consists of repetitive glucose units, which makes it more specific, architecturally diverse, reactive, and multifunctional. Some parameters that make cellulose a unique material to study are: isolation process itself, amount of hydrogen bonds, chain length and distribution, crystallinity, and functional group distribution in a repeating unit. They are also the factors that determine the reactions and properties of cellulose [7].

The first question that comes to mind: why should cellulose derivatives be established?

Compounds obtained by chemical changes of cellulose are called **cellulose derivatives**. While cellulose transforms into its derivatives, hydroxyl groups react and yield esters with organic and inorganic acids, ethers with some alcohols, alcoholates with bases and oxidation products with acids. They also react with halides, amines, and some complexes. Most important industrial derivatives of cellulose are cellulose esters and ethers. Producing alkali cellulose is a starting process to obtain these two products [18].

http://dx.doi.org/10.5772/61707

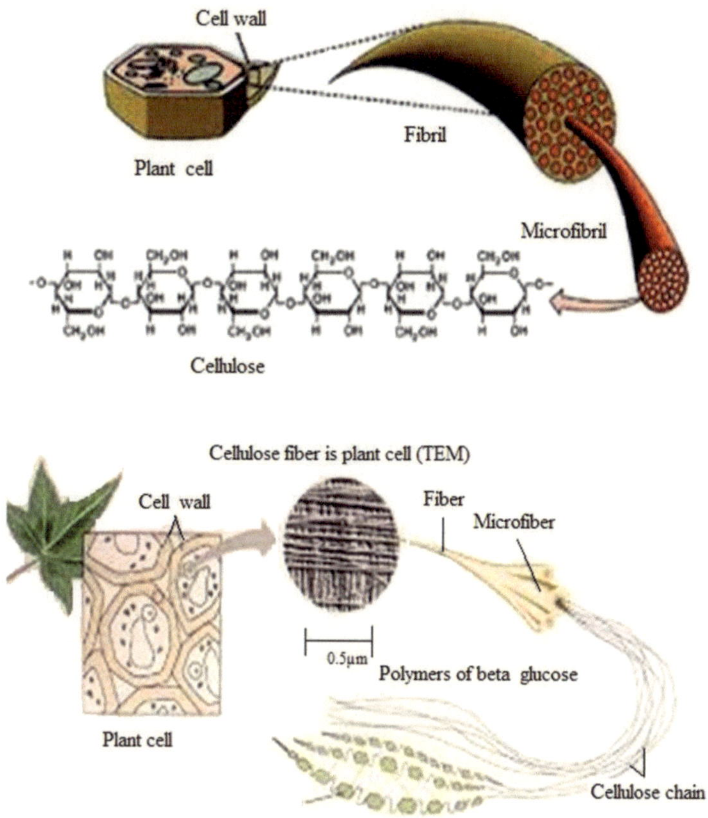

Figure 3. Structure of cellulose fibers and fibrils

Substitute groups in cellulose molecules are revealed when preparing cellulose derivatives. This process results in changes in physical properties, which makes cellulose derivatives industrially practical. This effect is revealed by both natural substitute groups and degree of substitution. Mechanical and physical properties of cellulose and cellulose derivatives differ with respect to average molecular mass. Increase in molecular mass also increases resistivity values; however, this effect becomes less significant after a certain degree [18, 19].

Although it has many useful properties, it is short of some properties that synthetic polymers have. Graft copolymerization is a significant way to alter the physical and chemical properties of cellulose like heat resistance, elasticity, resistance to abrasion and wear, ion-exchange capabilities, oil and water repellency, and antibacterial activity [20]. In the preparation of cellulose derivatives, making them soluble is often the decisive quality criterion, because cellulose is insoluble in organic solvents such as water, alcohol, acetone, benzene, chloroform etc. The usefulness of cellulose as a starting material for edible and biodegradable polymer is extended by chemical modification. Additionally, high solution viscosity is a desired property.

There are three ways to change the chemical properties of cellulose, as given in other sources [21]:

1. By preparing an ester or ether derivative of cellulose, **ethers** , e.g., ethyl cellulose (EC), methyl cellulose (MC), propyl cellulose (PC), hydroxypropyl cellulose (HPC), hydroxypropyl methyl cellulose (HPMC), hydroxylethyl cellulose (HEC), hydroxylethyl methyl cellulose (HEMC), methyl hydroxypropyl cellulose (MHPC), ethyl hydroxylethyl cellulose (EHEC), carboxymethyl cellulose (CMC), benzyl cellulose (BC); **esters** like cellulose acetate (CA) and cellulose xanthate, which are used to process cellulose into either fiber or film forms, during which the cellulose is regenerated by controlled hydrolysis; **acetals,** especially the cyclic acetal formed between secondary hydroxyl groups and butyraldehyde cellulose acetate, cellulose propionate, and cellulose acetate-butyrate [22].

2. By preparing a cross-linked derivative of cellulose.

3. By preparing a graft copolymer of cellulose, namely, a branched derivative of cellulose (This study reports about cellulose grafting by ATRP).

The most significant cellulosic applications are in the paper, wood products, textiles, film, and fiber industries. However, recently it has also attracted significant interest as a source of biofuel production [23]. Some usages of soluble cellulose are given in Table 1 [24].

Cellulose derivatives	**Areas of utilization**
Ethers	
MC-EC (methylcellulose- ethylcellulose)	Foods, dyes, drugs, detergents, cosmetics, textiles, lacquers, finishing, inks, glues, electrical insulators, fire extinguishers, electrical appliances, water retainers, stabilizers, borehole liquids, etc.
CMC-NaCMC (carboxymethylcellulose-sodiumcarboxymethylcellulose) (E-466)	Foods, cosmetics, textiles, drugs, paper, detergents, glues, etc.
HEC (hydroxyethlycellulose)	Dyes, chemicals, liquid detergents, rubber, oil wells, polymerization emulsions, etc.
HPC-HPMC (hydroxypropylcellulose-hydroxypropylmethylcellulose)	Foods, drugs, papers, plastics, ceramics, gels (as water retainers), eye lenses, stabilizers, etc.
Esters	
Acetates (organic esters: cellulose acetate, cellulose triacetate, cellulose propionate, cellulose acetate propionate, propionate, cellulose acetate butyrate, etc.	Textiles, films, plates, sheaths, coils, isolation, and nonconductive parts
Filaments	Garments, threads, furniture, packaging, etc.
Plastics	Various
Tows (linen and cannabis fibers)	Textiles, yarns, cigarette filters, etc.
Nitrates (Nitrocellulose) (inorganic ester)	Automotive, textiles, dyes, lacquers, polishes, films, explosives, cement, plates, etc.
Nitriles	Food and other packaging

Table 1. Usage of soluble cellulose

http://dx.doi.org/10.5772/61707

2.3. Nanocellulose

Mechanical properties of natural materials like elasticity or endurance could differ from nanoscale to macroscale. First-generation cellulosic engineering products like cotton and wood have been used in various fields depending on their structural sizes. However, some vital properties like functionality, durability, and homogeneity could not have been obtained in following generation cellulose-based products by using conventional cellulosic materials. Although conventional forestry products are still being widely used, these products have failed to meet the requirements of high-performance materials. Deep research interest in forestry products and by-products triggered by the necessity of sustainability has gone into a further phase after the invention of cellulose nanoparticles. Since cellulose extracted at nanoscale did not have compositional errors at hierarchical dimension like first-generation cellulose, this new method has paved the way to the development of next-generation cellulose based on composite materials. Although wood pulp is commonly used, nanocellulose can be produced from any cellulose material. Fibrils are isolated from wood-based fibers by using high-pressure homogenizer. Amorphous part in natural cellulose is hydrolyzed by acid hydrolyzation and crystal part at nanoscale is expelled by centrifuge (Fig. 4) [14, 25].

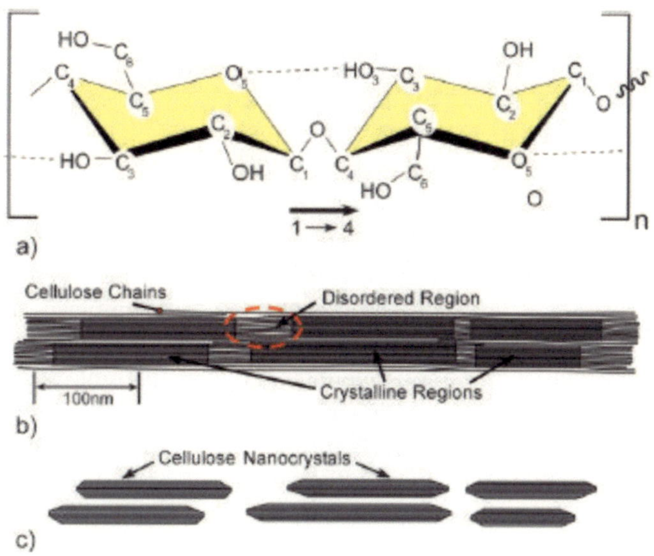

Figure 4. Schematics of (a) repeating single cellulose chain unit with the directionality of the -4 linkage and intra chain hydrogen bonding (dotted line), (b) idealized cellulose microfibril illustrating a suggested configuration of the crystalline and amorphous regions, and (c) cellulose nanocrystals with disordered regions dissolved in acid

3. Atom Transfer Radical Polymerization (ATRP)

Ostu, et al. (1982) invented of the term "Living/controlled radical polymerization" in their work on the iniferter mechanism [26]. Various studies have proven that Controlled Radical

Polymerization (CRP) is the most versatile method for surface-initiated polymerization (SIP) on model surfaces, and it includes methods such as Atom Transfer Radical Polymerization (ATRP), nitroxide-mediated and iniferter-based (INItiators-transFER-terminaTER agent) polymerizations (NMP) and Reversible Addition–Fragmentation chain Transfer (Radical Addition Fragmentation Transfer) (RAFT) polymerization [27]. The polymerization is obtained by implication of techniques generally based on the reversible homolytic cleavage of (a) a dormant chain end group into the corresponding polymeric radical (which induces chain propagation) and (b) a stable, persistent radical that cannot undergo addition to monomer [27, 28]. The radical-dormant adduct equilibrium can be reached by the application of thermal energy, light, or the addition of a catalyst as a function of the particular method considered [27].

ATRP, which is quite versatile in yielding to polymers with low polydispersity and controlled molecular weight, is probably the most widely studied CRP [29]. Moreover, ATRP has proven to be useful in synthesizing graft copolymers with well-defined structures utilizing a variety of monomers. The application of ATRP to cellulose could prove to be attractive when preparing novel cellulose derivatives with well-defined side-chain structures. Besides, reports on ATRP and the synthesis of cellulosic graft copolymers show that the method is promising [30-32].

3.1. Mechanism of ATRP

In 1995, two different research groups separately reported very similar controlled radical polymerization techniques, the ATRP, both based on catalytic systems used for the atom transfer radical addition reaction (ATRA), or in other words, Kharasch reaction, a method of forming carbon-carbon bonds between organic halides and alkenes known to be productive. The other system, reported by Matyjaszewski, *et al.*, is the polymerization of styrene under catalyzation of CuCl/ 2, 2′-bipyridyl (bpy) with 1-phenylethyl chloride as an initiator. Since the outset of reports, many reports have been published on ATRP of styrene, acrylates, methacrylates, and acrylonitrile by different transition metal complexes. ATRP proves to be versatile when compared to other controlled radical polymerization methods, providing control in the polymerization of various monomers under different reaction conditions and enabling to prepare polymers with a wide range of architectures. Some examples are blocks, grafts, gradient copolymers, stars, combs, branched, and hyper-branched. ATRP is one of the fastest growing subjects in chemistry, with the number of publications approximately doubling each year [26].

The mechanism of ATRP can be seen in Scheme 1. In or der to generate a growing radical or an active species ($P_n^{\cdot}$), a transition metal complex (M^n-Y/L) undergoes one-electron oxidation together with abstraction of a halogen (X) from an initiator or a dormant species (P_n-X). The active species initiate monomers (CH_2=CH_2R_1) to yield new growing radicals. This reaction may continue until the release of halogen atoms from the oxidized metal back to form dormant species. This turns the polymer chain ends from a dormant to a propagating and active state. In this process, the oxidized metal complexes (X-M^{n+1}-Y/L) are utilized as persistent radicals and reduce the rippling concentration of growing radicals in order to terminate by generation of a steady low concentration of active radical chain ends with short life span. This mechanism has the advantage of keeping concentration of radical intermediates in the reaction medium

http://dx.doi.org/10.5772/61707

low before adding new monomers, because of fast but reversible transformation into nonactive species. Speedy initiation and the fast reversible deactivation provide uniform growth in all chains, which in turn allows achieving narrow polydispersities. Besides, it leads to suppression of termination reactions. An ATRP reaction is considered to be successful if it has a small contribution of terminated chains, and a uniform growth of all the chains, managed by fast initiation and rapid reversible deactivation. This process, shown in Scheme 1, takes place with a rate constant of activation ($k_{a)}$ and deactivation ($k_{d)}$. The growth of polymer chains comes by the addition of the intermediate radicals to monomers in a similar manner to a constant of propagation ($k_{p)}$. Termination reactions (k_t) would also occur through radical coupling and disproportionation in ATRP in a great amount [5, 24, 29, 31, 32-39].

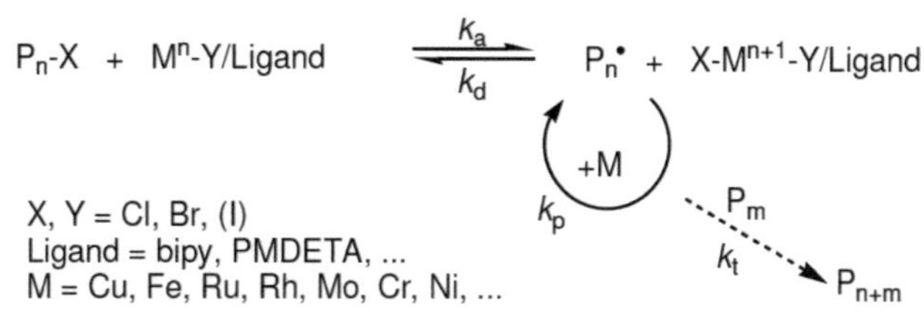

Sheme 1. Mechanism of metal-catalyzed ATRP

3.2. Ligands and initiators commonly used in ATRP

Various transition metals, usually in the form of salts, have been used in ATRP together with various complexing ligands. Copper is the most common of them, due to its low cost and versatility. Ligand's duty is to solubilize the metal ion, and this affects the reduction potential of the transition metal ion too. Although alkyl iodides have been used too, alkyl bromides and chlorides are more common initiators [40]. Some common ligands and alkyl halides are 2,2'-bipyridine and ligands bound nitrogen. For example: N,N,N$^{/}$,N$^{//}$,N$^{///}$–pentamethyl diethylene triamine (PMDETA), tetramethyl ethylene diamine (TMEDA), 1,14,7,10,10- hexamethyl triethylene tetramine (HMTETA), hexahexyl triethylenetetramine (HHTETA) [41], tris [2-diamethylamino ethylamine (Me_6-TREN)], alkyl pyridyl methanimine [34]. Some general copper containing ligands are given in Fig. 5 [39].

3.3. Catalysts commonly used in ATRP

The catalyst is arguably the most important component in ATRP, since it determines the position of the atom transfer equilibrium and the dynamics of exchange between the dormant and active species. A transition metal catalyst needs to fulfill some requirements in order to be considered as efficient. These are: the metal having at least two oxidation states separated by one electron; the metal center having reasonable affinity toward a halogen; and the coordination sphere around the metal being expandable upon oxidation to accommodate a (pseudo)-halogen; and the ligand being relatively strongly. Moreover, the dynamics and position of the ATRP equilibrium is expected to be appropriate for the particular system [31].

Structure		Name	Reference
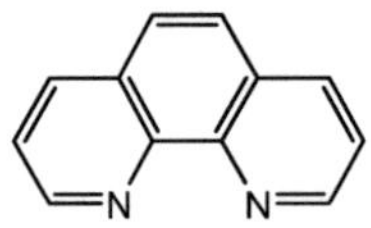		1,10-Phenonthroline (*o*-phen)	42
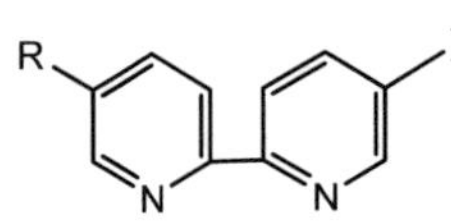	R=H	2-2'-bipyridine (bpy)	43
	$R=C_7H_{15}$	4,4'-di-n-heptyl-2,2'-bipyridine(dHbpy)	43
	$R=CH(CH_2CH_2CH_2CH_3)_2$	4,4'-d-(5-nonyl)-2,2'-bipyridine (dNbpy)	43
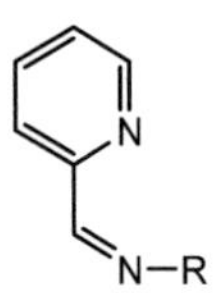	$R=C_3H_7$	N-(n-propyl)pyridyl methanimine (NPPMI)	44
	$R=C_8H_{17}$	N-(n-octyl)pyridyl methanimine (NOPMI)	44
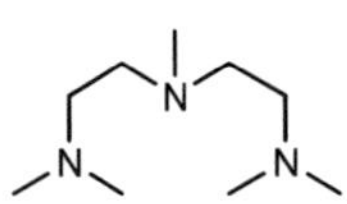		N,N,N',N",N'"-pentamethyl diethylenetriamine (PMDETA)	45
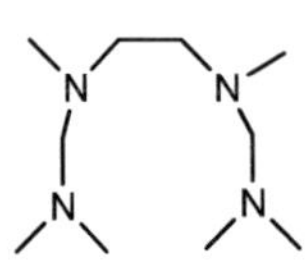		1,1,4,7,10,10-hexamethyl triethylenetetramine (HMTETA)	45
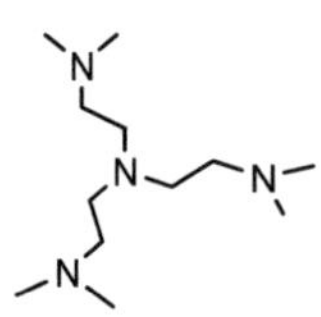		tris[(2-dimethylamino)ethyl] amine (Me_6TREN)	46
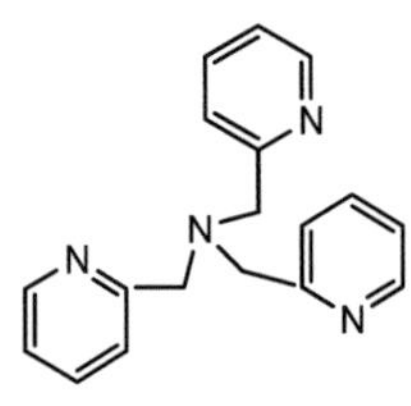		tris[(2-pyridyl)methyl] amine (TPMA)	47
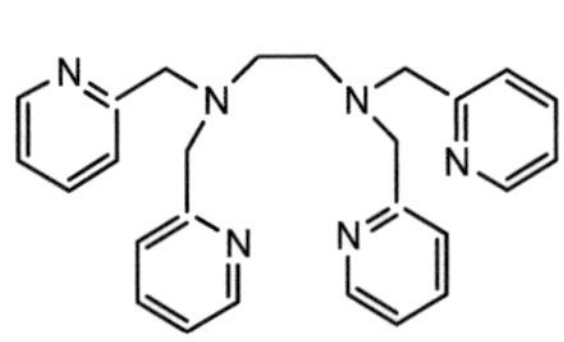		N,N,N',N'-tetrakis (2-pyridylmethyl) ethylenediamine (TPEN)	48

Figure 5. Examples of some ligands used in copper mediated ATRP

3.4. Halide exchange effect in ATRP

Replacing a chloride catalyst with a bromide catalyst is an effective way to achieve initiation rates higher than propagation. This results in halide exchange giving predominantly Cl-ended polymer and bromide catalyst by a mechanism (shown in Fig. 6). Utilizing a bromide catalyst with a soluble chloride salt would yield the same effect too. The ongoing radical effect brings two activation- deactivation equilibria involving P· with equilibrium constants K_1 and K_2, K_2 being greater than K_1 since $K_{ex} >> 1$. The halide exchange is not great with the initiator when compared with the polymer as a result of the lower concentration of R· than that would have been established by the regular persistent radical effect because of the capture by the monomer, so the initiator remains a bromide, while the polymer becomes a chloride predominantly. Initiation becomes faster than propagation as a result of the weaker C–Br bond in the initiator compared to the stronger C–Cl bond in the polymer. On the other hand, although majority of the polymers are Cl- ended, most of the deactivation events occur by the faster process of Br atom transfer, inasmuch as $k_{da2} > k_{da1}$, resulting in narrow molecular weight distribution polymers. Nevertheless, its slower frequency of activation results in P–Cl build up. Since P–Cl being more stable than P–Br, the end group loss due to side reactions at higher conversions becomes much less [39].

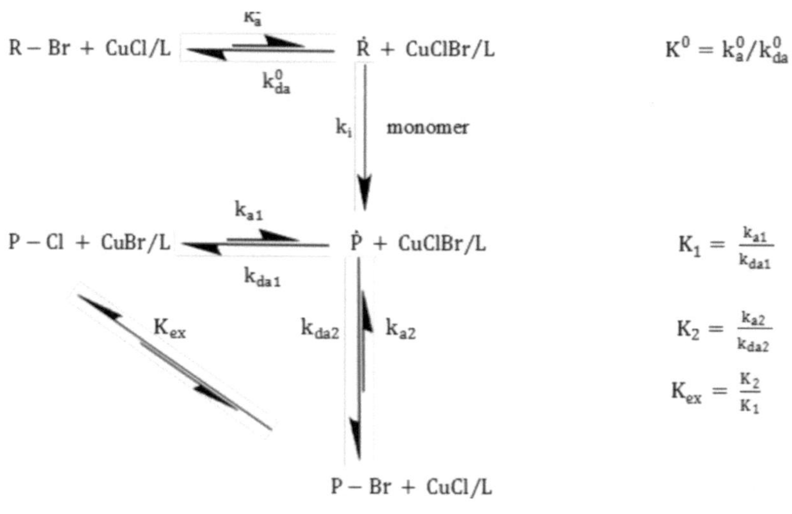

Figure 6. Halide exchange using bromide initiator and chloride catalyst

4. Some studies about cellulose grafting by atom transfer radical polymerization method

Graft copolymers of cellulose are obtained by binding synthetic polymers to cellulose – a natural polymer – with covalent bonds. This process – called grafting – changes behaviors and properties of cellulose. Numerous researchers have studied this subject. Below are some researches on grafting by ATRP method. Other than grafting studies on cellulose, there are other natural polymers grafting studies.

Coşkun M. et al grafted styrene, methyl methacrylate, methacrylamide and acrylomorpholine on pulverized-raw cellulose by using ATRP method. They prepared cellulose chloroacetate as macroinitiator by reacting chloroacetyl chloride with OH group of primary alcohol in pulverized cellulose. CuBr and 1,2-dipiperidinoethane was used as catalyst and transient metal compound. The reactions were followed by FT-IR and mass increase in cellulose chloroacetate. Dye affinity and dye absorption values of cellulose, macroinitiator, and graft polymers were determined by acidic alizarin yellow and alkaline bromocresole green, while humidity, water, and dye absorption capacities were measured by thermal stability [35].

Our research group grafted cellulose by using two different methods, through free-radical graft copolymerization and atom transfer radical polymerization (ATRP). N-(4-nitrophenyl) acrylamide (4NPA), and N-cyclohexylacrylamide (NCA) monomers were synthesized by the researchers, while some commercial monomers were used for this grafting. In the first method, primary OH groups in powder cellulose were first transformed to ester groups with methacryloyl chloride and cellulose methacrylate was obtained. This cellulose methacrylate was grafted with the synthesized monomers of N-(4-nitrophenyl) acrylamide (4NPA) and N-cyclohexylacrylamide (NCA); 4-vinylpyridine (4VP), acrylamide (AM), methacrylamide (MAM), diacetone acrylamide (DAAM), and methylmethacrylate (MMA) monomers, in the presence of 2,2'-azobis (isobutyronitrile) in acetonitrile as solvent. In the second method, with the ATRP, primary OH groups of powder-raw cellulose were firstly reacted with chloroacetyl chloride and cellulose chloroacetate was obtained, where cellulose chloroacetate was used as macroinitiator. 4NPA, NCA, 4VP, AM, and DAAM monomers reacted in the presence of copper (I) chloride and 2,2'-bipyridine as ligands at 130°C in the N,N-Dimethylformamide (DMF), where cellulose chloroacetate was used as macroinitiator. In the third method, Poly [Cell-g-(N-(4-nitrophenyl) acrylamide-co-methyl methacrylate)] {Poly [Cell-g-(4NPA-co-MMA)]} was prepared in the presence of CuCI/2,2'-bipyridine. In the fourth method, Poly [Cell-g-(N-cyclohexylacrylamide)-co-methyl methacrylate)] {Poly [Cell-g-(NCA-co-MMA)]} was prepared similarly. The monomer reactivity ratios of both graft copolymerization reactions were calculated by using Fineman-Ross (F–R), Inverted Fineman-Ross (I–F–R), Yezrielev-Brokhina-Roskin (Y–B–R), Kelen-Tüdos (K–T), and Extended Kelen-Tüdos (E–K–T) methods. All of the graft copolymers were characterized by elemental analysis, IR, TGA techniques. Water and metal ion (by Atomic Absorption Spectrum) uptake properties of the graft copolymers obtained with the first method were determined. Electrical conductivity of the graft copolymers by ATRP obtained with the second method was determined. The results are discussed below [24].

Cankaya et al. synthesized the graft copolymers of cellulose methacrylate by free- radical polymerization. Cellulose methacrylate was prepared by esterification of primary -OH group on raw-powder cellulose with methacryloyl chloride with a 21.3% yield by mole. The vinyl monomers such as 4NPA, NCA, 4VP, AM, MAM, DAAM, and MMA were grafted into the cellulose methacrylate via free-radical polymerization using 2,2'-azobis (isobutyronitrile) as an initiator in acetonitrile. The graft copolymers were characterized by FT-IR spectra, elemental analysis, and thermal analysis. Thermal stabilities of the graft copolymers were determined by TGA method and thermal stability of the copolymers decreased with grafting. Water-uptake capacities increased the grafting and its metal ion sorption tendency (Ni^{2+}, Co^{2+}, Cu^{2+}, Cd^{2+}, Pb^{2+}, Fe^{3+} and Cr^{3+}) improved with the grafting. –The study was conducted by using 8 graft copolymers and 7 metals and no comparison between graft copolymers and metals was possible, which is an expected result. It could be concluded that +3 valence metals were held by graft copolymers more than +2 valence metals, and cellulose methacrylate holds metals more than cellulose [49].

Cankaya et al. synthesized graft copolymers of cellulose chloroacetate by ATRP. Cellulose chloroacetate was prepared by esterification of primary alcoholic OH groups on raw-powdered cellulose. Cellulose was first esterified with chloroacetyl chloride, yielding cellulose chloroacetate which behaves as a macroinitiator because of chloroacetyl groups on it (Scheme 2). The graft copolymers of cellulose with some monomers such as NCA, 4VP, and DAAM were prepared by means of these groups using the Cu(I)/2,2′-bipyridine complex as a catalyst in DMF at 130°C by ATRP. Reactions of grafting on cellulose are indicated in Scheme 3. Formed homopolymers were removed from graft copolymers. The electrical conductivity of the graft copolymers was measured as a function of temperature. The electrical conductivity of the copolymers increased with increase in temperature, and this indicates that the studied copolymers exhibit semiconducting behavior. The electrical and optical properties of all the copolymers changed with grafted monomers. The cellulose exhibits the highest conductivity and lowest optical absorption edge with 4VP monomer [32].

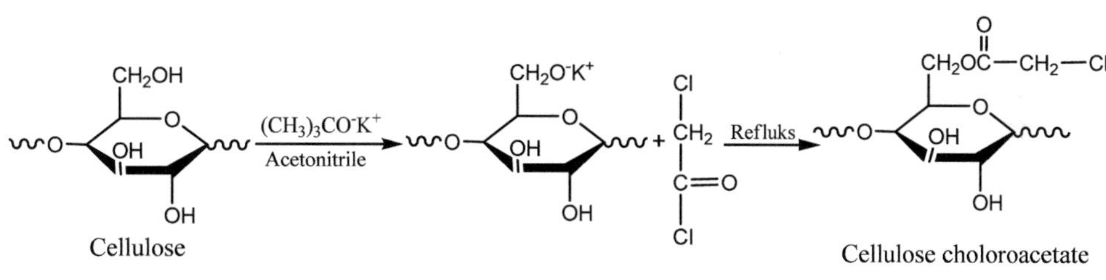

Sheme 2. Synthesis of cellulose chloroacetate macroinitiator

Cankaya et al. prepared graft copolymers with 4NPA and MMA of cellulose chloroacetate, using cellulose chloroacetate/Cu(I)/2,2′-bipyridine complex as a catalyst in DMF at 130°C, by ATRP. Cellulose chloroacetate behaved as a macroinitiator because of chloroacetyl groups on it (Scheme 4). Poly [Cell-g-(N-(4-nitrophenyl) acrylamide-co-methyl methacrylate)] {Poly [Cell-g-(4NPA-co-MMA)]} graft copolymers were characterized by elemental analysis, FT-IR spectra, and thermal analysis. FT-IR spectra of Cell-*g*-(4NPA-*co*-MMA) graft copolymer showed that as 4NPA content decreased, 1670 cm^{-1} amide carbonyl peak sharpness decreased and as MMA content increased, 1745 cm^{-1} ester sharpness increased. Also, 1745 cm^{-1} ester band in the cellulose chloroacetate macroinitiator was observed for each graft copolymer (Fig.7). Thermal stabilities of the graft copolymers were determined by TGA method and thermal

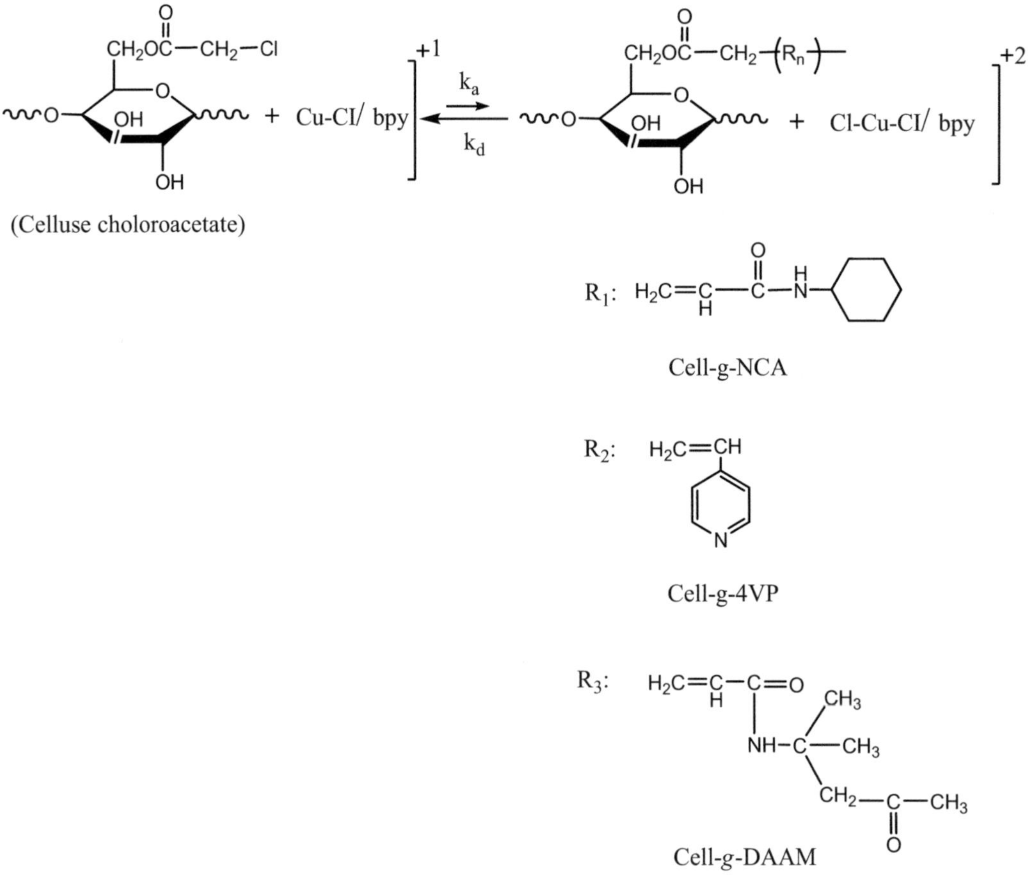

Sheme 3. The grafting of cellulose chloroacetate with some monomers by ATRP

stability of the copolymers increased with the increase in amount of MMA, while it decreased with the increase in amount of 4NPA (Fig.8). In order to investigate the effect of 4NPA with MMA monomers interactions on grafting, the graft copolymerization was also studied using different feed compositions ranging from 0.15 to 0.85. This study was undertaken in order to determine of the monomer reactivity ratios in the grafting of cellulose with 4NPA and MMA by ATRP, with linear methods such as the F–R, Inverted F–R, Y–B–R, K–T, and Extended K–T methods. The reactivity ratios of 4NPA and MMA in the graft copolymerization found was r_1=0.017–0.116 and r_2=1.209–1.472, with various methods (Table 2). According to these methods, the r_1 and r_2 values were 0.011 and 1.263; $r_1.r_2$= 0.014, respectively. The values calculated by all methods were found to be equal to each other. Two monomer mixtures on cellulose had a tendency to form alternative copolymers because the $r_1.r_2$ was close to zero. It could be concluded that 4NPA monomer with amide group and larger molecule size was less reactive on grafting onto cellulose with respect to MMA monomer with ester group and smaller molecule size [30].

http://dx.doi.org/10.5772/61707

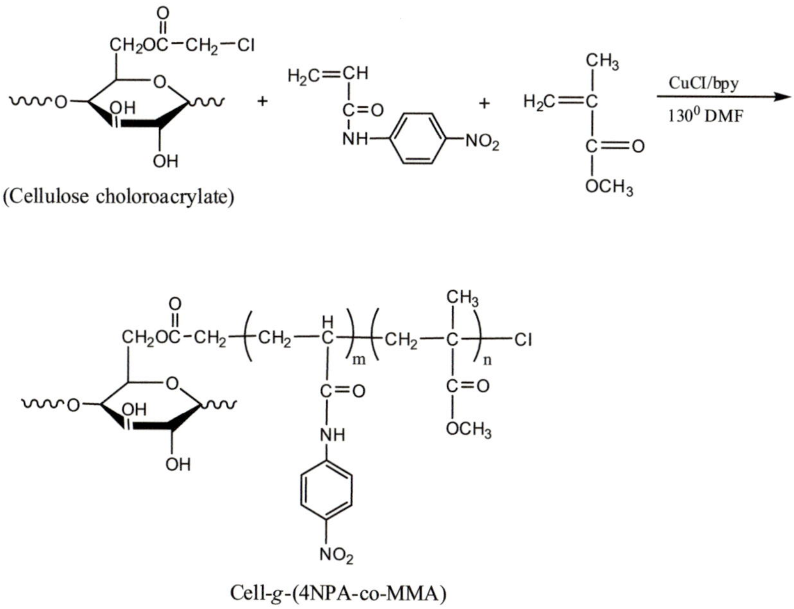

Sheme 4. Synthesis of the serials Cell-*g*-(4NPA-*co*-MMA)

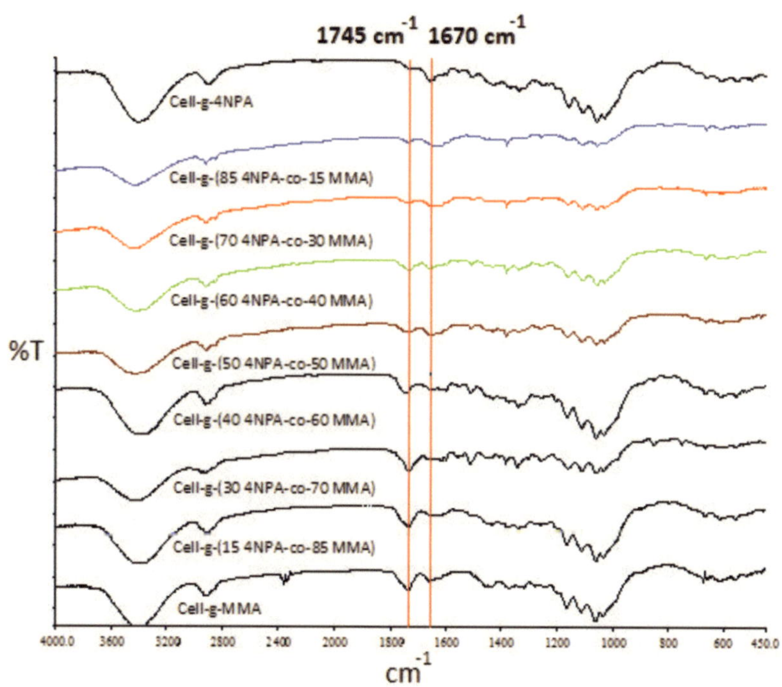

Figure 7. FT-IR spectra of the serials Poly [Cell-g-(4NPA-co-MMA)] graft copolymers

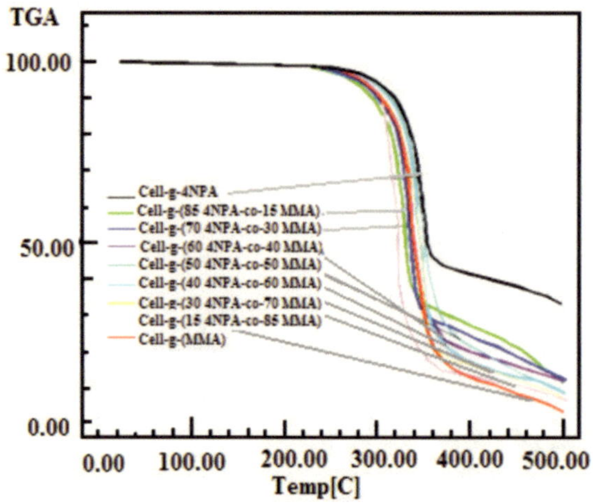

Figure 8. Thermal Gravimetric Analysis curves of the serials Poly [Cell-g-(4NPA-co-MMA)] graft copolymers

Methods	r_{4NPA}	r_{MMA}	$r_1.r_2$
FR	0.096	1.365	0.130
IFR	0.011	1.263	0.014
YBR	0.017	1.209	0.021
KT	0.116	1.455	0.169
EKT	0.109	1.472	0.160

Table 2. Comparison of monomer reactivity ratios for Poly [Cell-g-(4NPA-co-MMA)] graft copolymers with five methods

The same method was utilized to graft NCA and MMA monomers on cellulose. For this purpose, poly [Cell-g-(N-cyclohexylacrylamide-co-methyl methacrylate)] {Poly [Cell-g-(NCA-co-MMA)]} graft copolymers were synthesized (Scheme 5). Cellulose graft copolymers were characterized by elemental analysis, FT-IR spectra, and thermogravimetric analysis. The FT-IR spectra of the graft copolymers are presented in Fig. 9. In the FT-IR spectra, the ester band of cellulose chloroacetate macro initiator at 1745 cm^{-1} was observed for each graft copolymer. Also, the FT-IR spectra of Cell-*g*-(NCA-*co*-MMA) showed that as the NCA content decreased in the graft copolymer, the amide carbonyl peak sharpness (band at 1670 cm^{-1}) decreased too, and the 1745 cm^{-1} ester sharpness increased with MMA content. (Fig.9). Thermal stabilities were compared considering the thermogravimetric curves. Residual percentage in the graft copolymers decreased at 500 °C, while the MMA amount decreased (Fig.10). Also, the monomer reactivity ratios in the grafting of cellulose with NCA and MMA were calculated by using 7 different feed compositions ranging from 0.15 to 0.85 by ATRP. Reactivity ratios were determined by linear methods, such as the F–R, Inverted F–R, Y–B–R, K–T, and Extended

K–T methods, which were r_1 = 0.004-0.128 for NCA cellulose graft copolymer and r_2 = 0.657-0.907 for MMA cellulose graft copolymer (Table 3). Two monomer mixtures on cellulose had a tendency to form an alternating copolymer, because the value of r_1, r_2, and $r_1.r_2$ approached zero. It was concluded that MMA monomer with ester group and smaller molecular size was more reactive on grafting onto cellulose with respect to NCA molecule with amide group and larger molecular size [1].

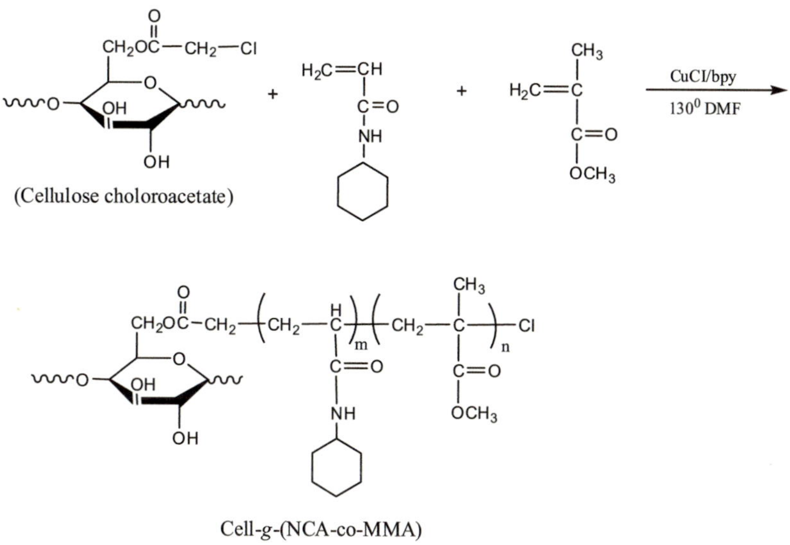

Sheme 5. Synthesis of the serials Cell-g-(NCA-co-MMA)

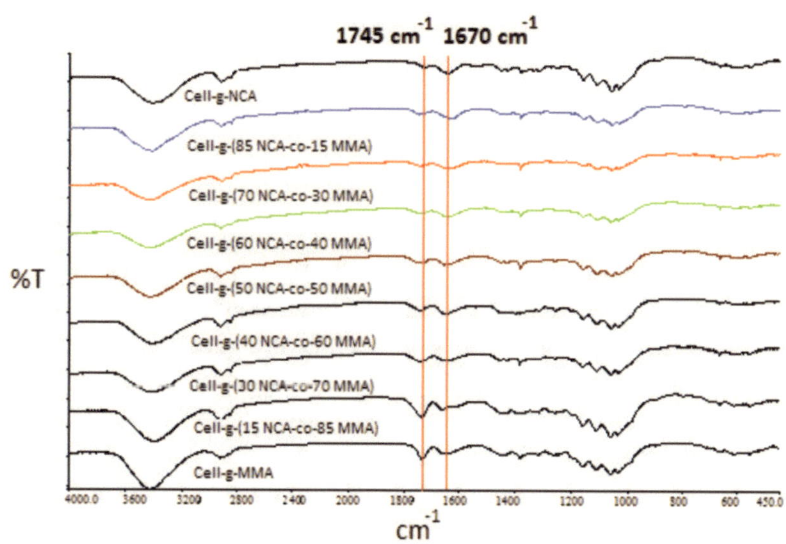

Figure 9. FT-IR spectral area of the serials Poly [Cell-g-(NCA-co-MMA)] graft copolymers

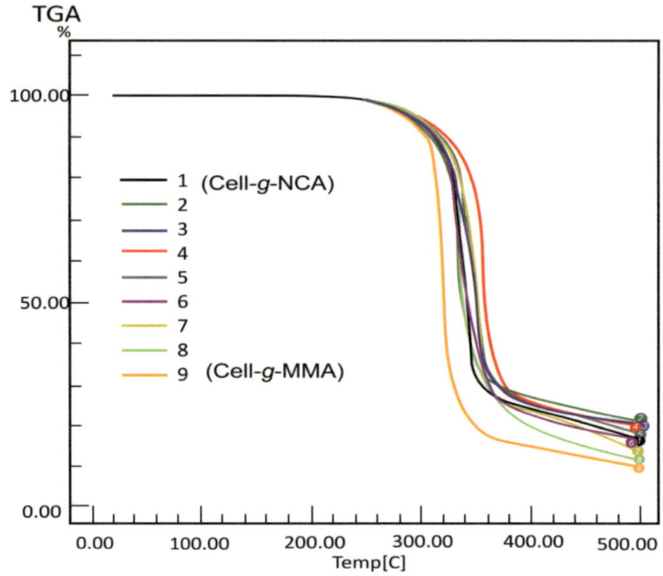

Figure 10. TGA curves of the serials Poly [Cell-g-(NCA-co-MMA)] graft copolymers

Methods	r_{NCA}	r_{MMA}	$r_1.r_2$
FR	0.022	0.696	0.015
IFR	0.128	0.907	0.116
YBR	0.004	0.657	0.003
KT	0.066	0.832	0.055
EKT	0.065	0.837	0.054
Average	0.057	0.786	0.049

Table 3. Comparison of monomer reactivity ratios for Poly [Cell-g-(4NPA-co-MMA)] graft copolymers by different methods

Billy M. et al. obtained more flexible and hydrophilic graft copolymers with various hydrophilic/hydrophobic balances by implanting methyldiethylene glycolmethacrylate (MDEGMA) on cellulose acetate with 2-bromoisobutyrilbromine initiator in cyclopentanone solvent with ATRP method. It was observed that MDEGMA copolymerization is in line with Hanns Fischer's kinetic modeling with radical persistency. Then ATRP grafting of two different cellulose acetate with different ATRP initiator group numbers were studied with respect to macroinitiator, which yielded two graft copolymers with different nanostructures, one being a short graft copolymer, and the other, a longer graft copolymer with the same mass ratio. Their morphologies were analyzed by X-ray diffraction method and it was found that phase separation depended on the number and length of poly (MDEGMA) grafts. As a result, cellulose acetate copolymers yielded to powerful films that could be utilized in membrane applications [50].

Chun-xiang L. et al. obtained cellulose graft poly (methylmethacrylate) copolymers in 1-allyl-3-methylimidazolium chloride (BMIMCl) which is an ionic liquid, using 2,2/-bipyridine and CuBr as catalysts. Cellulose chloroacetate as macroinitiator was synthesized by direct acetylation of cellulose with chloroacetyl chloride in BMIMCl in the absence of catalyst. Copolymerization was done in ionic liquid BMIMCl in the absence of homopolymer by-product. When an ionic liquid was used, polymers separated from the catalysts easily. Grafted polymers were characterized by ^{1}H-NMR, GPC (gel permeation chromatography), and AFM (atomic force microscope) methods. According to the findings from AFM, cellulose-grafted copolymers in the solution merged into a sphere-like shape [51].

Glaied O. et al. grafted cationic poly [2-(methacryloyloxy)ethyl]-trimethylammonium chloride (PMeDMA) on cellulose fibers in aqueous solution by using ATRP. Binding 2-bromoisobutyryl bromide – which is an efficient initiator – on hydroxyl groups on cellulose surface, cellulose macroinitiator, was synthesized. Obtained fiber/ PMeDMA complex was characterized by infrared, SEM, and XPS (X-ray photoelectron spectroscopy), which yielded some evidences that showed the polymerization was on the surface of the cellulose. Different initiators were added and withdrawn in order to better define the polymers and it was proven that polymerization was under control in this heterogeneous media by SEC (size exclusion chromatography). It was concluded that cellulose modified by cationic PMeDMA grafting had better mechanical properties [52].

Meng T. et al. synthesized methylmethacrylate and styrene graft copolymers by direct acetylation of cellulose with ATRP method in 1-alyl-3-methylimidazoliumchloride (AMIMCl) – which is an ionic liquid at room temperature – in the absence of catalysts and chemical preservative by using 2-bromoisobutyrylate as macroinitiator. Synthesized cellulose graft copolymers were characterized by using FT-IR, ^{1}H-NMR, ^{13}C-NMR, and GPC methods. It was concluded by TEM (transmission electron microscopy) and static and dynamic laser scattering measurements that cellulose graft copolymers in the solution could aggregate into a sphere-like polymer structure [53].

Yan Q. et al. grafted cellulose-g-poly (N,Ndimethlyamino ethylmetacrylate)-g-poly (ε-caprolactone) (EC-g-PDMAEMA-g-PCL) by using ring opening and ATRP methods. Unlike others, (EC-g-PDMAEMA-g-PCL) brush copolymers show unique physicochemical properties and have multifunctional structure. These bio adaptive copolymers formed mycelia in water and these mycelia merged more with altered pH value. Therefore, it was concluded that these mycelia could be used as perfect nanocarriers for controlled drug release. It was also concluded that crystal structure and crystal morphology of the copolymers could be controlled by altering the length of side chains [54].

Yi J. et al. synthesized temperature-dependent cellulose nanocrystals-g-(N,N-dimethylamino ethylmethacrylate) (CNC-g-PDMAEMA) with ATRP method. Graft copolymers were characterized by FT-IR, TGA, GPC methods. AFM characterization showed that the width of the nanocrystals was 10–40 nm, while their length was 100–400 nm. The liquid-crystal ratio of the graft copolymers was studied by polarized optical microscopy. It was observed that graft copolymers showed lyotropic property and change of copolymer chains with respect to temperature studied [55].

Xu Q et al. synthesized a new amphoteric polymer by grafting azo polymers to cellulose nanocrystals that showed liquid crystal behavior both in solvent and during heating process. This new crystal prepared by hydrolyzing filter paper with acid was defined by AFM. Liquid crystal polymers poly {6-[4-(4-methoxyphenylazo) phenoxy] hexyl methacrylate} (PMMAZO) were successfully grafted to cellulose nanocrystals by ATRP. The structure and thermal properties of PMMAZO-grafted nanocrystals were investigated by FT-IR and TGA methods; phase structure and transition were studied by DSC and polarized optical microscopy (POM) methods. Experiment results showed that PMMAZO-grafted cellulose nanocrystals showed both thermotropic and lyotropic properties of liquid crystal formation [56].

Bhut B.V. et al. grafted poly (2-dimethylaminoethylmetacrylate) on cellulose successfully by ATRP method. Surface topography, porous membrane morphology as a result of modification, and changes in chemical functions was characterized by ATR-FTIR AFM and SEM. In order to evaluate protein binding capacity of the graft copolymer Bovine serum albumin was utilized and it was concluded that graft has increased in time [57].

Yan L. and Tao W. synthesized graft copolymer of cellulose with N,N-dimethyl acrylamide (DMA) in homogeneous media by ATRP method. First, cellulose was dissolved in N,N-dimethylacetoamide (DMAc)/LiCl system and 2-bromoisobutyloylbromide (B*i*BBr)-Cell-B*i*B macroinitiator was formed. Then DMA was polymerized with cellulose chain under existence of Cell-B*i*B at dimethyl sulfoxide. FT-IR, NMR, and GPC measurements showed that Cell-PDMA graft copolymer was formed. Protein adsorption studies proved that modified cellulose membranes prepared as Cell-PDMA had good protein adsorption resistivity [58].

Sui X. et al. synthesized cellulose-*g*-poly (*N,N*-dimethylamino-2-ethylmetacrylate) (cellulose-*g*-PDMAEMA) graft copolymer with ATRP method. Macroinitiator was obtained by direct acetylation of cellulose with 2-bromopropionyl bromide in 1-alyl-3-methylimidazoliumchloride, which is an ionic liquid at room conditions. Copolymers were obtained by undertaking DMAEMA into ATRP process in the presence of CuBr/PMDETA (pentamethyldiethylenetriamine) –as catalyst in DMF solvent without homopolymer by-product. Grafted copolymers were characterized by ^{1}H-NMR, FT-IR, and TGA, and the results proved that PDMAEMA bonded to cellulose with covalent bond. Moreover, cellulose grafted PDMAEMA in aqua was studied under various concentration, temperature, and pH values with UV, TEM, AFM, and DLS (Dynamic light scattering) methods, and was observed that copolymers showed properties expected from PDMAEMA to temperature and pH. It is thought that the method used in this study could be used in preparation of polysaccharides in various biomaterials [59].

Kang H. et al. synthesized ethyl cellulose-g-poly (2-hydroxyethylmetacrylate) (EC-g-PHEMA) in methanol by ATRP method and characterized with GPC and ^{1}H-NMR. Kinetic study showed that polymerization was controllable and (EC-g-PHEMA) aggregated in aqua as mycelia. Morphology of mycelia was determined by DLS and TEM and its formation was discussed [60].

http://dx.doi.org/10.5772/61707

Kang H. et al. synthesized ethyl cellulose-g-poly (acrylic acid) (EC-g-PAA) and ethyl cellulose-g-*tert*-butylacrylate (EC-g-P*t*BA) copolymers by ATRP method. Using 2-bromoisobutyrilbromide as initiator, hydroxyl groups in ethyl cellulose were esterified and in existence of ethyl cellulose as macroinitiator EC-g-PAA and EC-g-P*t*BA graft copolymers were synthesized and were characterized by FT-IR, ^{1}H-NMR, and GPC methods. Molecular mass of the copolymers increased with polymerization while polydispersity decreased. This study showed that polymerization had kinetics of first degree [61].

Vlcek P. et al. acetylated cellulose diacetate (CDA) with 2-bromoisobutyryl bromide or dichloroacetyl chloride and obtained a new macroinitiator with functionality for ATRP method for different reaction conditions and synthesized new graft copolymers of styrene (St), MMA and butyl acrylate (BuA) with it. Poly (CDA-g-St) and poly (CDA-g-MMA) were used as macroinitiator of BuA polymer and consequently poly [CDA-g-(St-b-BuA)] and poly [CDA-g-(MMA-b-BuA)] diblock graft polymers were obtained, showing that different graft copolymer macroinitiators and different monomers could change graft polymers' length and density [62].

Shen D. et al. grafted polystyrene copolymer on ethyl cellulose (EC) by ATRP method and characterized mycelium properties of the graft copolymer by using DLS, AFM and TEM. Increased concentration resulted in increased mycelia. TEM and AFM images revealed that mycelia had spherical shapes and nucleus-crust structure. All of the macroinitiator was used; molecular mass of the graft copolymers increased and polydispersity decreased. Kinetic study showed that polymerization was of first degree [63].

Shen D. et al. synthesized graft copolymers of ethyl cellulose with polystyrene and polymethylmethacrylate by ATRP method. OH groups in ethyl cellulose reacted with 2-bromoisobutyrilbromide, which is known as an effective ATRP initiator. Functionalized ethyl cellulose has been synthesized as ATRP initiator in toluene, where CuBr/N,N,N′,N′′,N′′- pentamethyldiethylenetriamine was used as catalyst. Molecular mass of the macroinitiators copolymer increased, and polydispersity decreased. Kinetic studies showed that polymerization took place in the first minutes. Graft copolymers were characterized by LLS (laser light scattering) and approved by AFM [64].

Shen D. et al. grafted polymethylmethacrylate (PMMA) on cellulose diacetate by using ATRP method and characterized by ^{1}H-NMR and GPC. OH groups in CDA reacted with 2-bromoisobutyryl bromide, which is known to be an effective initiator for ATRP and was used as functional CDA macroinitiator in MMA graft copolymer. Polymerization was performed under 70°C and in N,N,N′,N′′,N′′- pentamethyldiethylenetriamine/CuBr/1,4-dioxane system. All of the macroinitiator was used, molecular mass increased, and polydispersity decreased. Kinetic studies showed that polymerization was of first degree [65].

Wang et al. [66] reported that pH responsive poly [ethyl cellulose-grafted-(2-diethylamino) ethyl methacrylate] (EC-g-PDEAEMA) synthesized by ATRP can be used in the pH-responsive release of rifampicin (RIF). In addition, another graft product obtained by ATRP, which is graft copolymer of ethyl cellulose with azobenzene-containing polymethacrylates [67], has been reported to be used in some applications such as sensors and optical materials.

5. Conclusion

Cellulose graft copolymers can be obtained either in homogeneous or heterogeneous media. Most widely used methods for graft confirmation are FT-IR, NMR, TGA, SEM, TEM, XRD, and XRF. Monomer grafted on cellulose can give a hydrophilic or hydrophobic character to the new copolymer, which can pave the way to various applications. Reactions in homogeneous media usually show better controllability and yield to higher number of grafts per cellulose chain. CRP methods, such as ATRP, make it possible to obtain pre-designed copolymers, which make grafting a promising research area.

6. Abbreviations list

CRP: Controlled Radical Polymerization

ATRP: Atom Transfer Radical Polymerization

NMP: Nitroxide-Mediated Polymerization

RAFT: Reversible Addition–Fragmentation Chain Transfer Process

ATRA: Atom Transfer Radical Addition Reaction

SIP: Surface-Initiated Polymerization

ROP: Ring-Opening Polymerization

TEM: Transmission Electron Microscopy

AFM: Atomic Force Microscope

XPS: X-Ray Photoelectron Spectroscopy

SEC: Size Exclusion Chromatography

TGA: Thermal Gravimetric Analysis

DLS : Dynamic Light Scattering

GPC: Gel Permeation Chromatography

GPC: Gel Permeation Chromatography

LLS: Laser Light Scattering

$C_6H_{12}O_6$: D-glucose

EC: Ethyl Cellulose

MC: Methyl Cellulose

PC: Propyl Cellulose

http://dx.doi.org/10.5772/61707

BC: Benzyl Cellulose

HPC: Hydroxypropyl Cellulose

HPMC: Hydroxypropyl Methyl Cellulose

HEC: Hydroxylethyl Cellulose

HEMC: Hydroxylethyl Methyl Cellulose

MHPC: Methyl Hydroxypropyl Cellulose

EHEC: Ethyl Hydroxylethyl Cellulose

CMC: Carboxymethyl Cellulose

CA: Cellulose Acetate

P_n: An Active Species

M^n-Y/L: Transition Metal Complex

X: Halogen

P_n-X: Initiator or a Dormant Species

CH_2=CH_2R_1: Active Species Initiate Monomers

X-M^{n+1}-Y/L: Oxidized Metal Complexes

k_a: Rate Constant of Activation Amit

k_d: Rate Constant of Deactivation

k_p: A Constant of Propagation

k_t: Termination Reactions

PMDETA: N,N,$N^{/}$,$N^{//}$,$N^{///}$–Pentamethyl Diethylene Triamine

TMEDA: Tetramethyl Ethylene Diamine

HMTETA: 1,14,7,10,10-Hexamethyl Triethylene Tetramine

Me_6-TREN: Tris [2-diamethylamino ethylamine]

o-phen: 1,10-phenanthroline

bpy: 2,2′-bipyridine

dNbpy: 4,4′-di-(5-nonyl)-2,2′-bipyridine

NPPMI: N-(n-propyl)pyridylmethanimine

NOPMI: N-(n-octyl)pyridylmethanimine

PMDETA: N, N, N′, N′′, N′′′-pentamethyldiethylenetriamine

HMTETA: 1,1,4,7,10,10-hexamethyltriethylenetetramine

TPMA: Tris[(2-pyridyl)methyl]amine

TPEN: N, N, N′,N′-tetrakis(2-pyridylmethyl)ethylenediamine

4NPA: N-(4-Nitrophenyl) acrylamide

NCA: N-Cyclohexylacrylamide

4VP: 4-Vinylpyridine

AM: Acrylamide

MAM: Methacrylamide

DAAM: Diacetone acrylamide

MMA: Methylmethacrylate

F–R: Fineman-Ross method

K–T: Kelen-Tüdos method

E–K–T: Extended Kelen-Tüdos method

I–F–R: Inverted Fineman-Ross method

Y–B–R: Yezrielev-Brokhina-Roskin method

DMF: N,N-Dimethylformamide

DMF: N,N-Dimethylformamide

Acknowledgements

The author thanks M. M. Temüz who was the author's PhD thesis advisor in Fırat University.

Author details

Nevin Çankaya*

Address all correspondence to: nevin.cankaya@usak.edu.tr; nevincankaya@hotmail.com

Uşak University, Faculty of Science and Art, Department of Chemistry, Uşak, Turkey

References

[1] Cankaya N., Temüz M. M. Monomer reactivity ratios of cellulose grafted with N-cyclohexylacrylamide and methyl methacrylate by atom transfer radical polymerization. *Cellulose Chem Technol*. 2014; 48(3-4):209-215.

[2] Teramoto Y., Ama S., Higeshiro T., Nishio Y. Cellulose Acetate-graft-Poly(hydroxyalkanoate)s: Synthesis and dependence of the thermal properties on copolymer composition. *Macromol Chem Physics*. 2004, 205(14):1904-1915. DOI: 10.1002/macp.200400160

[3] Daly W. H., Evenson T. S., Iacono S. T., Walker-Jones R. Recent developments in cellulose grafting chemistry utilizing Barton ester intermediates and nitroxide mediation. *Macromol Symp*. 2001;174(1):155. DOI: 10.1002/1521-3900(200109)174:1

[4] Perrier S., Takolpuckdee P., Westwood J., Lewis D. M. Versatile chain transfer agents for reversible addition fragmentation chain transfer (RAFT) polymerization to synthesize functional polymeric architectures. *Macromolecules*. 2004; 37:2709-2717.

[5] Siegwart D. J., Oh J. K., Matyjaszewski K. ATRP in the design of functional materials for biomedical applications. *Progr Polym Sci*. 2012; 37:18-37.

[6] Cai X., Riedl B., Bouaziz M., Lignocellulosic composites with grafted polystyrene interfaces. *Compos. Interf*. 2005;12(1):25-39.

[7] Poletto M., Pistor V., Zattera A. J. Structural Characteristics and Thermal Properties of Native Cellulose. Chichester: *Cellulose–Fundamental Aspects-Intech*; 2013. p. 45-68 http://dx.doi.org/10.5772/50452.

[8] Barbucci R. editor. Integrated Biomaterials Science. Chichester: Kluwer Academic Publishers; 2002. 1037 p. ebook: 0-306-47583-9 print: 0-306-46678-3

[9] Gümüşkaya E. Selülozun Kristal Yapısı. Kafkas University Artvin Forest Faculty. 2005; 6 (1-2): 69-78.

[10] Wojnarovits L., Foldvary C. M., Takacs E. Radiation-induced grafting of cellulose for adsorption of hazardous water pollutants: A review. *Rad Physics Chem*. 2010; 79:848-862.

[11] Brendel O., Iannetta P.P.M., Stewart D. A rapid and simple method to isolate pure alpha-cellulose. Phytochem. Anal. 2000; 11:7-10.

[12] http://www.brooklyn.cuny.edu/bc/ahp/LAD/C4c/C4c_polysaccharides.html

[13] Saçak M., editor. Lif Teknolojisi. Chichester: Ankara Üniversitesi Fen Fakültesi Döner Sermaye işletmesi Yayınları. 1994. 144p., http://kitaplar.ankara.edu.tr/detail.php?id=702

[14] http://www.solverkimya.com/site/makaleler/endustriyel-urunler-makaleleri/seluloz-ve-kimyasal-yapisi.html

[15] http://www.tksd.org.tr/doc/kimyasal_forum/Kimyasal_Forum_EYLUL_2013.pdf

[16] http://www.tappi.org/content/events/10nano/papers/12.5.pdf

[17] Carlmark A. Atom Transfer Radical Polymerization from Multifunctional Substrates [thesis]. Department of Polymer Technology, Royal Institute of Technology, Stockholm; 39p. 2002.

[18] Kırcı H., Ateş S., Akgül M. Selüloz türevleri ve kullanım yerleri. Fen ve Mühendislik Dergisi. 2001; 4(2):119-130.

[19] Ott E., Spurlin H.M., Grafflin M.W, editors. *Cellulose and Cellulose Derivatives Part II*, Chichester: Wiley-Interscience Publishers; 1954.1075 p.

[20] Carlmark A, Malmström E. E. ATRP Grafting from cellulose fibers to create block-copolymer grafts. *Biomacromolecules*. 2003; 4:1740-1745.

[21] Gürdağ G., Sarmad S. Cellulose Graft Copolymers: Synthesis, Properties and Applications. In: Kalia S., Sabaa M.W. editors. *Polysaccharide Based Graft Copolymers*. Chichester: Springer-Verlag; 2013. p. 15-57. DOI: 978-3-642-36565-2.

[22] Ghanbarzadeh B., Almasi H. Biodegradable polymers. In: Ghanbarzadeh and Almasi, editors. *Biodegradation-Life of Science*. Chichester: InTech, 2013. p:141-186 http://dx.doi.org/10.5772/56230

[23] La Mantia F.P., Morreale M. Green composites: A brief review. Some of the location of use of cellulose derivatives is given in the table follows. *Composites: Part A*. 2011; 42:579-588.

[24] Cankaya N. Selüloz Üzerinde Aşılama Çalışmaları [thesis]. Elazığ: Firat University; 132p. 2011.

[25] Moon R. J., Martini A., Nairn J., Simonsenf J., Youngblood J. Cellulose nanomaterials review: structure, properties and nanocomposites. *Chem Soc Rev*. 2011; 40:3941-3994.

[26] Mishra V., Kumar R. Living radical polymerization: A review. *J Sci Res-Banaras Hindu Univ Varanasi*. 2012;56:141-176, ISSN: 0447-9483

[27] Benetti E. M. Molecular engineering of designer surfaces by controlled radical polymerizations: Brushes, hedges and hybrid grafts [thesis]. Twente: Twente University; 164p. 2009.

[28] Maynor B. W., Li Y., Liu J. Au "Ink" for AFM "Dip-Pen" nanolithography. Langmuir. 2001;17: 2575-2578.

[29] Cankaya N., Demirelli K. Atom transfer radical copolymerization of phenoxycarbonylmethyl methacrylate with methylmethacrylate: Their monomer reactivity ratios and thermal stabilities. *J Chem Soc Pak*. 2011; 33(6):884.

[30] Cankaya N., Temüz M. M. Characterization And Monomer Reactivity Ratios of Grafted Cellulose with N-(4-Nitrophenyl)acrylamide and methyl methacrylate by atom transfer radical polymerization. *Cellulose Chem Technol.* 2012; 46(9):551.

[31] Matyjaszewski K., Xia J. H. Atom transfer radical polymerization. *Chem Rev.* 2001; 101:2921–2990.

[32] Cankaya N., Temüz M. M., Yakuphanoglu F. Grafting and semiconducting properties of N-Cyclohexylacrylamide, 4-Vinylpyridine, Diacetone Acrylamide and Diallylamine onto cellulose by atom transfer radical polymerization. *FEBS Lett.* In press.

[33] Yin M. Synthesis and Controlled Radical Polymerization of Multifunctional Monomers [thesis]. Dresden: Dresden University; 200p. 2004.

[34] Cankaya N. editor. Aril gruplu metakrilat monomerinin atom transfer radikal polimerizasyonu. Chichester: Türkiye Alim Kitaplari, moreBooks; 2015. 80 p. ISBN: 978-3-639-67248-0

[35] Coskun M., Temüz M. M. Grafting studies onto cellulose by atom-transfer radical polymerization. *Polym Int.* 2005; 54:342.

[36] Matyjaszewski K., Tsarevsky N. V., Braunecker W. A., Dong H., Huang J., Jakubowski W., Kwak Y., Nicolay R., Tang W., Yoon J. A. Role of Cu-0 in controlled/"living" radical polymerization. *Macromolecules.* 2007; 40:7795-7806.

[37] Lin C. Y., Coote M. L., Gennaro A., Matyjaszewski K. Ab initio evaluation of the thermodynamic and electrochemical properties of alkyl halides and radicals and their mechanistic implications for Atom Transfer Radical Polymerization. *J Am Chem Soc.* 2008; 130:12762–12774.

[38] Tsarevsky N. V., Matyjaszewski K. "Green" Atom Transfer Radical Polymerization: from process design to preparation of well-defined environmentally friendly polymeric materials. *Chem Rev.* 2007; 107:2270-2299.

[39] Mandal B. M. editor. *Fundamental of Polymerization.* Chichester: World Scientific; 2013. 445 p. DOI:978-981-4322-46-1

[40] Bergenudd H. Understanding the Mechanisms Behind Atom Transfer Radical Polymerization-Exploring the Limit of Control [thesis]. Royal Institute of Technology: University of Warwick; 74p. 2011.

[41] Acar M. H., Bıçak N. Synthesis of hexylated triethylenetetramine: New ligand for homogeneous Atom Transfer Radical Polymerization. *Polym Sci Part A: Polym Chem.* 2003; 41(11):1677-1680. DOI:10.1002/pola.10716

[42] Bencini A., Lippolis V. 1,10-Phenanthroline: A versatile building block for the construction of ligands for various purposes. *Coord Chem Rev.* 2010; 254(17-18): 2096-2180.

[43] Matyjaszewski K., Pattern T. E., Xia J. Controlled/"living" radical polymerization. Kinetics of the homogeneous atom transfer radical polymerization of styrene. *J Am Chem Soc*. 1997; 119:674-680.

[44] Haddleton D. M., Jasieczek C. B., Hannon M. J., Shorter A. J. Atom transfer radical polymerization of methyl methacrylate initiated by alkyl bromide and 2-Pyridinecarbaldehyde Imine Copper(I) complexes. *Macromolecules*. 1997; 30:2190-2193.

[45] Xia J., Matyjaszewski K. Controlled/"living" radical polymerization. Atom transfer radical polymerization using multidentate amine ligands. *Macromolecules*. 1997;30:7697-7700.

[46] Xia J., Gaynor S.G., Matyjaszewski K. Controlled/"living" radical polymerization. Atom transfer radical polymerization of acrylates at ambient temperature. *Macromolecules*. 1998; 31:5958-5959.

[47] Xia J., Matyjaszewski K. Controlled/"living" radical polymerization. Atom transfer radical polymerization catalyzed by Copper(I) and Picolylamine complexes. *Macromolecules*. 1999; 32:2434-2437.

[48] Tang H., Arulsamy N., Radosz M., Shen Y., Tsarevsky N. V., Braunecker W. A., Tang W., Matyjaszewski K. Highly active copper-based catalyst for atom transfer radical polymerization. *J Am Chem Soc*. 2006; 128:16277-16285.

[49] Cankaya N., Temüz M. M. Grafting of some monomers onto cellulose and determination of metal and water uptake properties. *Cellulose Chem. Technol.* 2015; 49(2): 135-141. DOI: 0576-9787

[50] Billy M., Ranzani da Costa A., Lochon P., Clément R., Dresch M., Etienne S., Hiver J. M., David L., Jonquières A. Cellulose acetate graft copolymers with nano-structured architectures:Synthesis and characterization. *Eur Polym J*. 2010; 46:944-957.

[51] Chun-xiang L., Huai-yu Z., Ming-hua L., Shi-yu F., Jia-jun Z. Preparation of cellulose graft poly(methyl methacrylate) copolymers by atom transfer radical polymerization in an ionic liquid. *Carbohy Polym*. 2009; 78:432–438.

[52] Glaied O., Dube M., Chabot B., Daneault C. Synthesis of cationic polymer-grafted cellulose by aqueous ATRP. *J Coll Interf Sci*. 2009; 333:145-151.

[53] Meng T., Gao X., Zhang J., Yuan J., Zhang Y., He J. Graft copolymers prepared by atom transfer radical polymerization (ATRP) from cellulose. *Polymer*. 2009; 50:447-454.

[54] Yan Q., Yuan J., Zhang F., Sui X., Xie X., Yin Y., Wang S., Wei Y. Cellulose-based dual graft molecular brushes as potential drug nanocarriers: Stimulus-responsive micelles, self-assembled phase transition behavior, and tunable crystalline morphologies. *Biomacromolecules*. 2009; 10:2033-2042.

[55] Yi J., Xu Q., Zhang X., Zhang H. Temperature-induced chiral nematic phase changes of suspensions of Poly(N,N-Dimethylaminoethyl Methacrylate)-grafted cellulose nanocrystals. *Cellulose*. 2009; 16:989-997.

[56] Xu Q., Yi J., Zhang X., Zhang H. A novel amphotropic polymer based on cellulose nanocrystals grafted with azo polymers. *Eur Polym*. 2008; 44:2830-2837.

[57] Bhut B. V., Wickramasinghe S. R., Husson S. M. Preparation of high-capacity, weak anion-exchange membranes for protein separations using surface-initiated atom transfer radical polymerization. *Membrane Sci*. 2008; 325:176-183.

[58] Yan L., Tao W. Graft copolymerization of N,N-Dimethylacrylamide to cellulose in homogeneous media using Atom Transfer Radical Polymerization for hemocompatibility. *Biomed Sci Engin*. 2008; 1:37-43.

[59] Sui X., Yuan J., Zhou M., Zhang J., Yang H., Yuan W., Wei Y., Pan C. Synthesis of cellulose-graft-Poly(N,N-dimethylamino-2-ethylmethacrylate) copolymers via homogeneous ATRP and their aggregates in aqueous media. *Biomacromolecules*. 2008; 9:2615-2620.

[60] Kang H., Liu W., Liu R., Huang Y. A novel, amphiphilic ethyl cellulose grafting copolymer with Poly(2-HydroxyethylMethacrylate) side chains and its micellization. *Macromol Chem Phys*. 2008; 209:424-430.

[61] Kang H., Liu W., He B., Shen D., Ma L., Huang Y. Synthesis of amphiphilic ethyl cellulose grafting poly(acrylic acid) copolymers and their self-assembly orphologies in water. *Polymer*. 2006; 47:7927-7934.

[62] Vlcek P., Janata M., Latalova P., Kriz J., Cadova E., Toman L. Controlled grafting of cellulose diacetate. *Polymer*. 2006; 47:2587-2595.

[63] Shen D., Yu H., Huang Y. Synthesis of graft copolymer of ethyl cellulose through living polymerization and its self-assembly. *Cellulose*. 2006; 13:235-244.

[64] Shen D., Yu H., Huang Y. Densely grafting copolymers of ethyl cellulose through atom transfer radical polymerization. *Polym Sci: Part A: Polym Chem*. 2005; 43, 4099-4108.

[65] Shen D., Huang Y. The Synthesis of CDA-G-PMMA Copolymers through atom transfer radical polymerization. *Polymer*. 2004; 45:7091-7097.

[66] Wang D., Tan J., Kang H., Ma L., Jin X., Liu R., Huang Y. Synthesis, self-assembly and drug release behaviors of pH-responsive copolymers ethyl cellulose-graft-PDEAEMA through ATRP. *Carbohy Polymer*. 2011; 84:195-202.

[67] Tang X., Gao L., Fan X., Zhou Q. Controlled grafting of ethyl cellulose with azobenzene-containing polymethacrylates via atom transfer radical polymerization. *J Polym Sci: Part A: Polym Chem*. 2007; 45:1653-1660.

Chapter 4

Chemical Functionalization of Cellulosic Materials — Main Reactions and Applications in the Contaminants Removal of Aqueous Medium

Roosevelt D.S. Bezerra, Paulo R.S. Teixeira, Ana S.N.M. Teixeira, Carla Eiras, Josy A. Osajima and Edson C. Silva Filho

Additional information is available at the end of the chapter

http://dx.doi.org/10.5772/61431

Abstract

The cellulose is the most abundant biopolymer in the world and presents a higher chemical variability for presence of several hydroxyl groups. These hydroxyl groups allow surface modification of biomaterials, with insertion of several chemical groups which change cellulose characteristics. This natural biopolymer and its derivatives have been used a lot as adsorbent, from several contaminants of aqueous medium due to biocompatibility, chemical degradability, and variability. Therefore, this chapter has the objective to review the literature about several cellulose surfaces or cellulosic material (incorporation of carboxymethyl, phosphorus, carboxyl, amines, and sulfur), presenting the main characteristics of reactions and showing its adsorption in application of aqueous medium (metals, dyes, and drugs), locating the main interactions between biomaterial/contaminant.

Keywords: cellulose, chemical modification, interations

1. Introduction

The cellulose (Figure 1) is the most abundant polysaccharide on the earth, being the main structural component of plants and seaweed cell walls. The cellulose is formed from the repeated units of D-glucose, which are linked by glycosidic linkages β(1 → 4). This natural polysaccharide has become the more used material due to its physical and structural properties and its biocompatibility. These properties arise from multiple interactions of hydrogen, which result in semicrystalline polymer, containing highly structured crystalline regions, and also, in materials with high tensile strength [1].

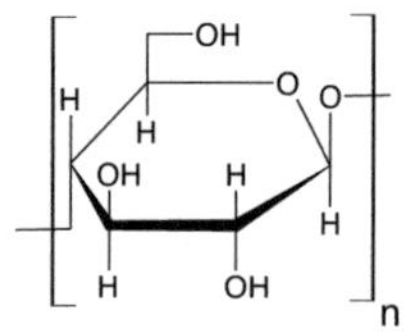

Figure 1. Molecular structure of cellulose.

The cellulose is mainly obtained from four resources: forest, agriculture, industrial, animal waste. The biomass which is obtained from all sources has three main components: cellulose, hemicellulose, and lignin, with its percentage depending highly on the obtained source. Therefore, the biomass has been extracted and processed in order to separate different components and to isolate cellulose [1].

The cellulose has lot of hydroxyl groups which can bond to several functional groups through variability of chemical modifications [2]. These chemical modifications provoke the formation of covalent bond through interaction between the modifier agent and the interactive center of solid surface, where insertion of organic molecules, in the surface of solid support, gives it advantageous and additional properties that differ from the original support [3].

The cellulose derivatives modified by several chemical reactions have various application possibilities in different fields. Among the various possibilities, one can be highlighted the usefulness of cellulosic derivatives in contaminants removal of aqueous medium. For example, the modified cellulose with diethylenetriamine was applied in the adsorption of Cu (II) and Pb (II) [4]. The cellulose modification with aminoethanethiol was more efficient with adsorption of red reactive dye RB [3] than pure cellulose. The phosphated cellulose presented higher adsorption capacity of ranitidine drug [6] than the pure cellulose [7]. Both the phosphated bacterial cellulose as well as the bacterial cellulose containing quaternary salt were efficient in protein adsorption.

Therefore, this chapter's goal is to perform a review of the literature about the main chemical reactions on cellulosic material surfaces and its applications in contaminants removal of aqueous medium.

2. Modifications and applications in contaminants removal

2.1. Carboxymethylation

The sodium carboxymethylcellulose (CMC) is a polyelectrolyte formed when the chloroacetic acid, or its sodium salt, reacts with alkaline cellulose. The CMC is a copolymer of two units: β-D-glucose and salt β-D-glucopyranose 2-*O*-(carboxymethyl)-monosodium, not spread randomly around macromolecules which are linked by β-1,4-glycosidic linkages [8].

The CMC is used in a lot of industrial fields, such as in food industry as thickener or viscosity modifier for stabilizing emulsion, in the petroleum industry as ingredient of drilling mud.

Moreover, it is a component of a lot of products, such as detergents, toothpaste, water-based paints, textile sizing, and several paper products. Lately, it has been applied in green synthesis study and stabilization of silver nanoparticles [9].

The CMC synthesis is divided in two stages: (I) alkalinization and etherization (II). In stage (I), the cellulose is dispersed in sodium hydroxide solution and ethanol. In stage (II), the sodium chloroacetic is added to solution; the mixture is hectic at 353 K, according to Figure 2 [9,10,11].

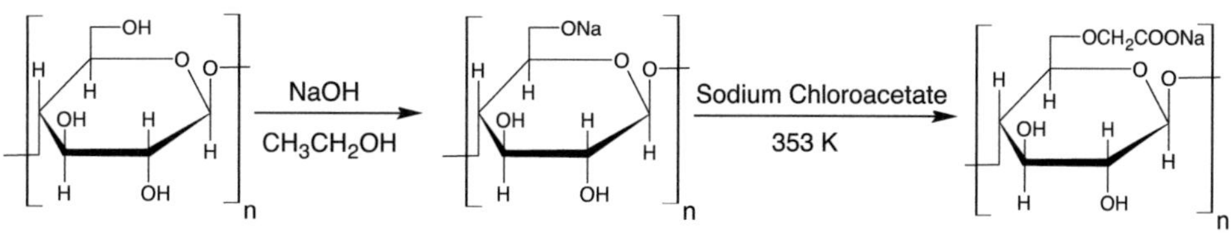

Figure 2. Preparation scheme of the CMC.

After the introduction of carboxymethyl group, the cellulose presents an anionic feature and there occurs an increase in its water solubility [12]. This carboxymethylation changes completely the electric charge of cellulose surface, turning it in ion exchanger. And the increase of carboxymethyl groups also promotes the increase of zeta potential value on CMC on to pure cellulose [10]. After carboxymethylation, these new characteristics acquired by celluloses are very important in the application in contaminants removal of aqueous medium, because the adsorption process is extremely affected by interactions which can occur between adsorbate and adsorbent, and they can be affected by pH of the medium [5,6,7].

For example, in the removal study of the methylene blue cationic dye (MB) (Figure 3 (a)), the pH is an important factor in the adsorption process of CMC. In acid conditions, the carboxymethyl group of CMC is protonated, and there occurs an ion exchange ($-CH_2COONa$ turning into $-CH_2COOH$), and thus adsorption of MB is a disadvantage. In alkaline conditions, the carboxymethyl group is totally deprotonated ($-CH_2COO^-$), furthering the MB dye adsorption. These results indicate which adsorption mechanism of the MB in CMC is due to ion exchange reactions, according to Figure 3 (b) [11].

The adsorption of the metal strontium (II) (Sr^{2+}) in CMC proved efficient and independent of pH. The ion exchange is the main reason for adsorption of ions Sr(II) in carboxymethyl groups of the CMC. In acidic pH, i.e., pH less than 2.0, the excess ions will compete to the Sr (II) ions, for match to $-COO^-$, as shown in Eq.1. Thus, as the pH of the medium increase, there will be an increase in the polymer hydrolysis, and hence it will grow the available sites for interaction to the strontium ions (II) [13].

$$\left[n\left(\text{-COO}\right)\text{Sr}\right]^{(2-n)+} + \text{nH}^{+} \leftrightarrow \text{nCOOH} + \text{Sr}^{2+}\left(n=1,2\right) \quad (1)$$

The adsorption of Sr (II) in CMC is highly influenced by ionic force from the medium. As the concentration of KCl increases, the ionic force of the solution grows, and the ion adsorption Sr

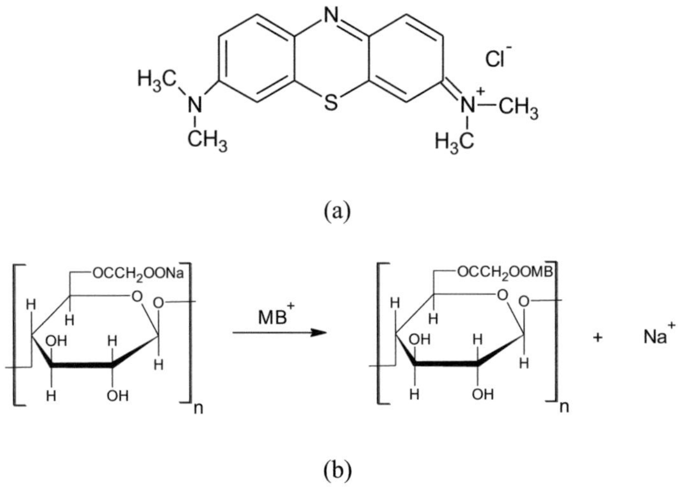

Figure 3. (a) Molecular structure of methylene blue dye. (b) Adsorption mechanism of the MB in CMC is due to ion exchange.

(II) in CMC decreases. The ionic force effect can be explained by two factors: (I) despite the presence of KCl in adsorption system which can improve the electrostatic attraction among $-COO^-$ groups present in the CMC and the Sr (II) ions, the K^+ ions destroy the ionic bond among $-COO^-$ groups and the Sr (II) ions, similar to H^+ (Eq.2) ions, (II) the affinity growth between KCl and water, after adding KCl, decreases the metal mobility, which provokes the decrease in adsorption [13].

$$\left[n(\text{-COO})\text{Sr}\right]^{(2-n)+} + nK^+ \leftrightarrow n\text{COOK} + Sr^{2+}\ (n = 1,2) \tag{2}$$

2.2. Phosphating

The chemical incorporation of phosphate in cellulose structure changes its properties meaningfully, since the synthetized biomaterial starts showing the characteristics of the phosphate group. This phosphated biomaterial is used in textile industry such as flames delayer to cellulose based. Also, it is used as a biomaterial for the disease treatment about calcium ions transport [6].

Due to importance of the phosphated group, a lot of methods to cellulose phosphating have been developed, either by free hydroxyl groups in positions C2, C3, or C6 or by shift from the ester or ether group already in the cellulose. The latest groups are usually used in homogeneous reaction using toxic reagents and mainly organic solvents which promote partial or total break of cellulose fibers. The synthesis of phosphate cellulose derivatives have been performed in several solvents [dimethylsulfoxide (DMSO)-methylamine, DMAc-LiCl, SO_3-triethylamine, formic acid, trifluoroacetic acid, N,N-dimethylformamide (DMF)-N_2O_2, paraformaldehyde, DMF, molten or aqueous urea, NaOH] with various phosphorus-containing compounds

http://dx.doi.org/10.5772/61431

(phosphoric acid and phosphinic acid, phosphorus oxyacids, phosphorus pentoxide, amido-phosphates, alkyl or aryl derivatives of phosphorous acid) [14].

The phosphoric acid (H_3PO_4) is the main phosphate precursor used in cellulose phosphating reaction. The reaction using phosphoric acid, such as phosphate precursor, can be performed by two manners (heterogeneously and homogeneously), and depending on the way followed, the reaction will produce different phosphated products [14, 15].

The cellulose reaction with phosphoric acid in aqueous medium (heterogeneously) is performed through adding H_3PO_4 (85%) in a system which has only cellulose. The system temperature is raised to 373 K and maintained at this temperature for 30 min, as shown in Figure 4. In this reaction, the phosphoric acid is linked to cellulose chain under the form phosphorous acid group, by ester linkage, through reactions from cellulose hydroxyl group. The product formed in this reaction also established a tautomeric equilibrium, due to the presence of hydrogen atom linked directly to phosphorus, as can be viewed on Figure 4 (a and b) [14].

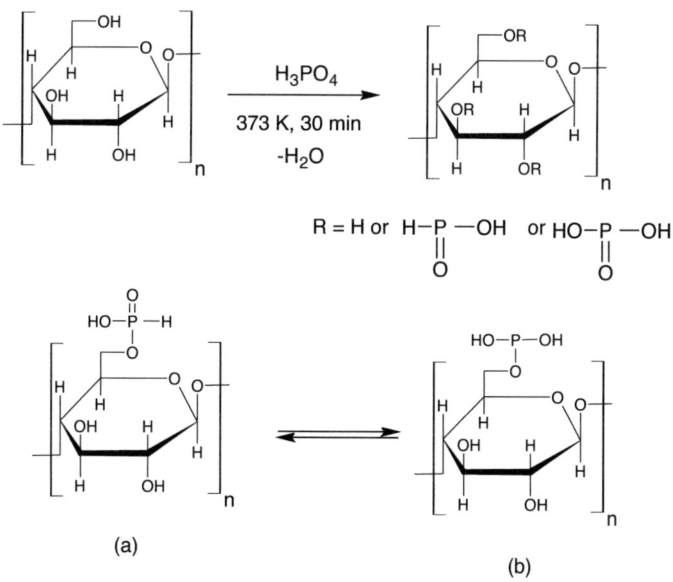

Figure 4. Reaction scheme of cellulose phosphorylation with H_3PO_4 in aqueous medium, and possible tautomeric products (a and b) of phosphated cellulose structure.

The cellulose reaction as phosphoric acid in molten urea (homogeneously) is initially performed merging urea at 413 K. Then, it adds cellulose the suspension, water, and phosphoric acid. The reaction takes place for 30 min at 423 K, as shown in Figure 5. In this reaction, in addition to the listed structures (Figure 4 (a and b)), the phosphoric acid can form ester products disubstituted and trisubstituted from cellulose [14].

The cellulose phosphorylation using phosphoric acid and urea can also be performed using DMF as solvent. Initially, in this reaction, it puts cellulose in contact with urea. After 1 hour,

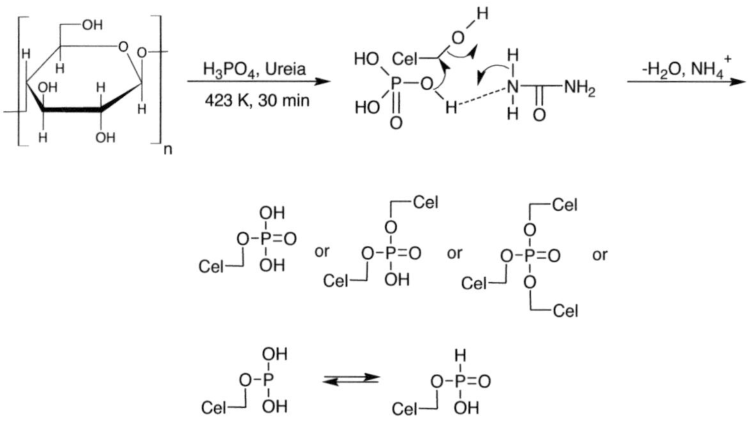

Figure 5. Reaction scheme of cellulose phosphorylation H_3PO_4 in molten urea and the possible gained products.

it adds phosphoric acid and the mixture are stirred for 4 h at 403 K. The reaction products are monophosphated cellulose and water, as shown in Figure 6 [16,17].

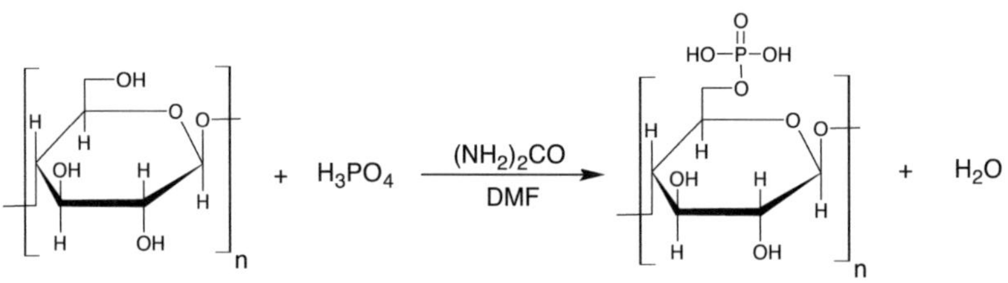

Figure 6. Reaction scheme of cellulose phosphorylation H_3PO_4 in urea using DMF as the solvent.

The cellulose phosphorylation increases the capacity of cellulose ionic exchange, because the introduction of phosphated group in the structure promotes the growth of the active groups. This characteristic was noticed by addition of cellulosic solvents [rice straw (37.4% of cellulose, 44.9% of hemi-cellulose, 4.9% of lignin and 13.1% of ashes) and phosphate rice straw] in contact with a 1 mol L^{-1} de NaCl solution. The ionic exchange reaction occurs by substitution of H^+ of the adsorbents for Na^+ which is in solution, generating HCl in solution, according to Figure 7 [18].

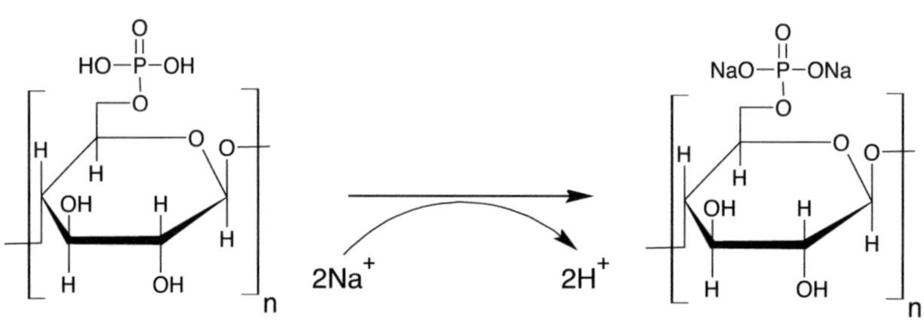

Figure 7. Ionic exchange solution from the phosphated solution.

http://dx.doi.org/10.5772/61431

The ionic exchange capacity is very important in the application in adsorption of heavy metals from aqueous medium. Surveys have showed which bacterial cellulose has no adsorption capacity of lanthanides metals (La^{3+}, Sm^{3+} e Ho^{3+}), since phosphated bacterial cellulose presents adsorption capacity of these metals. This adsorption process is influenced by the pH of medium. This pH dependence in adsorption indicates the pH mechanism is cation exchange. In addition, the bacterial phosphated cellulose has high adsorption capacity of lanthanides ions "stiff acid", based on the high affinity of the phosphoric acid group "stiff base" by the principle of base acid stiff acid (HSAB) [16].

The phosphated cellulose also proved efficient in the removal of drug ranitidine (Figure 8). Surveys have shown that after phosphating reaction, a growth of negative charge of cellulose occurred, due to the presence of phosphate group. It made the drug ranitidine reaction higher on modified cellulose surface than on pure cellulose surface. The ranitidine adsorption mechanism on the phosphated cellulose surface was through electrostatic interaction, where the negative charge of phosphate groups interacts with positive charge on the nitrogen atom of drug. As the same way, the OH^- ions can interact with the positive ions of drug, stopping drug interaction with phosphated cellulose, as reported above [6,7].

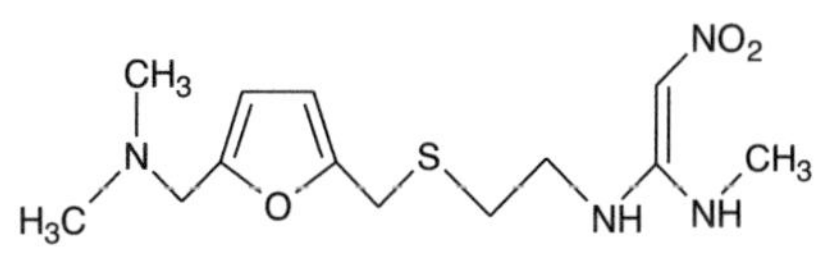

Figure 8. Molecular structure of ranitidine.

2.3. Acylation

The acylation with cyclic organic anhydrides of five members is a well-established reaction, which at first it involves all three hydroxyl groups of the cellulose unity (acts as nucleophiles), for getting a steady ester and reactive groups from carboxyl acids. In the practical use, these acylation processes are used in covering, cosmetics, in the pharmaceutical and food industries, as membranes and antibody filters, enzyme, protein, ion separation, etc. [19].

The cellulose derivatives which have free carboxyl groups are prepared by cellulose reaction or cellulosic material with succinic, maleic, or phthalic in the solvent absence. This process consists of heating cellulose with the amount of each anhydride up to anhydride melting fusion. For each case, the cellulose/anhydride proportion is 1/10, the mixture is stirred for 20 min and the reaction is interrupted by addition of DMA [20]. The full scheme of reaction is shown in Figure 9.

The three reactions shown above have covalent incorporation of carboxyl groups. This incorporation can be confirmed by the growth of the amount of carbon in the cellulose [20,21]. The cellulose reaction with maleic anhydride presents a high incorporation level, due to absence of solvent in this synthesis, which is well-proportioned by high temperature and by easy availability of reagents. The maleic anhydride presents incorporation, which after

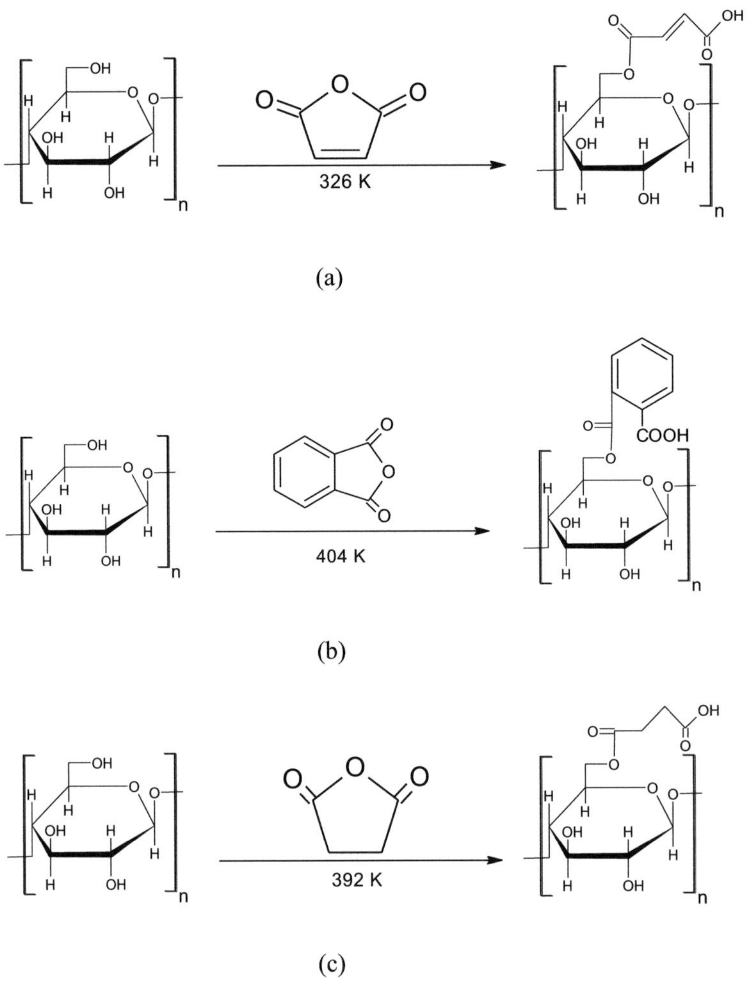

Figure 9. Cellulose reaction scheme or cellulose material with anhydride (a) maleic, (b) phthalic, and (c) succinic.

opening the ring forms two isomers of the cellulose: the cellulose maleate (cis) and cellulose fumarate (trans). This possibility to form both isomers is due to the presence of withdrawing groups near to unsaturation, which can affect isomerization according to the potential effect for removing electrons, and also due to impediment caused by these groups. In addition, the reaction α,β-unsaturation, with hydroxyl C(2), C(3), and (C6), through Michael's addition can produce other composts as sub products [21].

The cellulose reaction with phthalic anhydride also has a high level of anhydride incorporation by the same reasons quoted above in the cellulose reaction with maleic anhydride. The reaction using this aromatic anhydride can create some attractive interactions involving the aromatic rings of the anhydride, aromatic–aromatic interactions (Ar–Ar). These interactions can be strong enough to disturb the β-(1 → 4) bonds [22]. The cellulose reaction with succinic anhydride is similar to cellulose reaction with maleic anhydride, the only difference being no formation of cis–trans isomer, due to the absence of unsaturation in succinic anhydride.

These cellulose derivatives are enough in the removal of several contaminants from the aqueous medium. For example, the cellulose derivatives modified with maleic and phthalic anhydride were applied in adsorption of green malachite dye (Figure 10 (a)), and were shown to be more effective than pure cellulose. The dye adsorption was influenced by initial pH of medium. The addition of green malachite dye, on the surface of two cellulosic derivatives, increased as the pH increased until pH 6, where adsorption was established. This happened because in low pH, the carboxyl groups, of modified biopolymers, can be protonated, due to high proton concentration, decreasing the amount of adsorbed cationic dye. Also, this survey has shown the higher amount of carboxyl groups in cellulose, higher is the adsorption of green malachite dye [23]. The cellulose derivative modified with maleic anhydride showed effective in adsorption of divalent metals (Co^{2+} and Ni^{2+}), where the adsorption mechanism in this system is complexation among basic centers of carboxyl groups and divalent cations, as shown in Figure 10(b). For facilitating this complexation, and any lower minimization of pH, the protons of carboxyl groups were substituted by sodium. Figure 10 (b) is shows a counter-ion for neutralizing free cationic charge, where the cation can interact with a carboxyl group or two distinct basic centers [21].

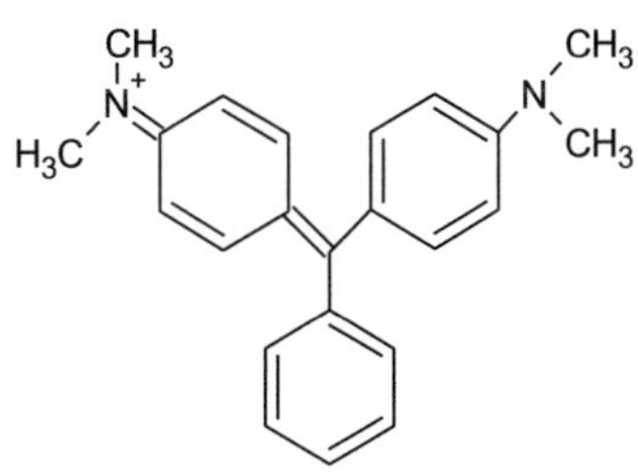

(a)

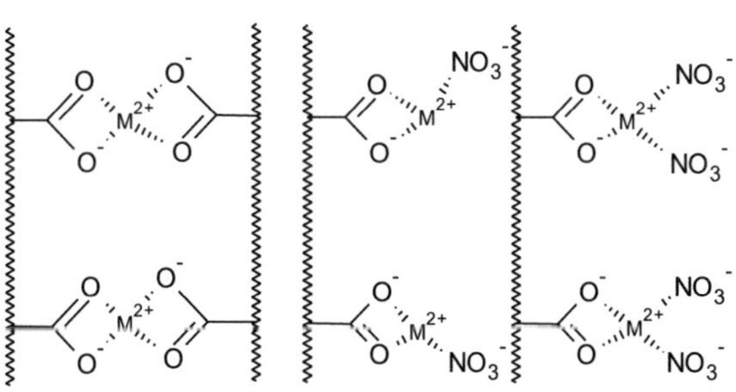

(b)

Figure 10. (a) Molecular structure of green malachite cationic dye. (b) Possible structures for formation of divalent cation complex and modified cellulose with maleic anhydride.

The cellulose derivatives modified with succinic and phthalic anhydrides proved efficient in adsorption of metal Cu^{2+}. In these cases, the adsorption is favored by growth of pH, where the sorption is favored in solutions with high pH to the point of zero charge (pH_{pzc}) for each modified material (modified cellulose with phthalic anhydride pH_{pzc}= 5.4 and with succinic anhydride pH_{pzc}= 5.7). Under acid conditions, the chemically modified biopolymers can be protonated, due to high concentrations of H^+ ions, decreasing the amount of adsorbed cation. On the other hand, the growth of pH deprotonates carboxyl group, a condition which favors chelating capacity, and consequently, the capacity of adsorbed Cu^{2+} increases [20]. The cotton fibers and other cellulosic derivatives have an improvement about 40% in the adsorption of blue methylene cationic dye (Figure 3(a)) after the chemical modification with succinic anhydride. The presence of carboxyl groups in biomaterials means significant rise in the adsorption, which can result from electrostatic interactions among these groups and dye reactive groups [24].

2.4. Amination

The molecule incorporation which has basic centers (primarily nitrogen, oxygen, and sulfur) in the cellulose structure raises its contaminants' adsorption capacity [3]. Thus, the nitrogen incorporation in cellulose or in the cellulosic materials is an important chemical modification which allows the insertion of active sites in this biomaterial. These sites can interact in aqueous medium with contaminants molecules, in heterogeneous system in which the interactive processes are defined by solid/liquid interface [25].

An example of incorporation of basic centers in cellulose structure is the reaction with ethylene-1,2-diamine. The first stage of reaction consists in allogeneic derivative synthesis. In the halogenation, chlorine addition is preferable, since it is the most effective allogeneic, whose preferential sequence is: chlorine > bromine, iodine > fluorine [25]. The first step of reaction series consists in the cellulose chlorination, a favored process by nucleophilic attack from thionyl chloride on hydroxyl group in the biopolymer backbone, resulting in chlorine atom pending in polymer structure. This replaced atom is more reactive than the original hydroxyl group, where a total substitution of hydroxyl in C6 by chlorine can happen. It happens due to hydroxyl group which is in C6 being more reactive than other hydroxyl groups, following order C6>> C3≈C2, as shown in Figure 11 [26,27].

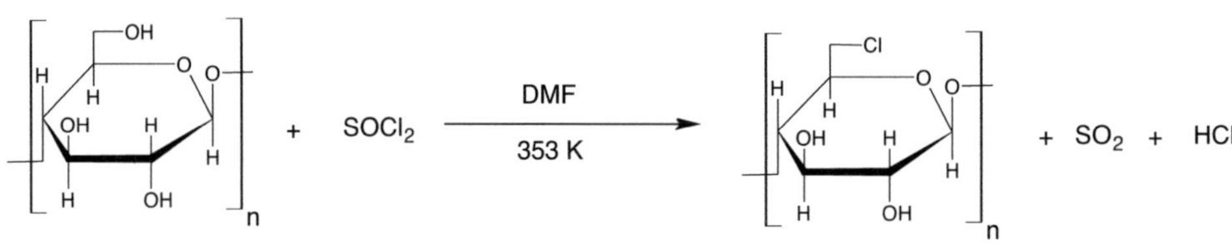

Figure 11. Reaction of cellulose chlorination.

The second step of reaction consists in chlorinated cellulose reaction to react with ethylene-1,2-diamine under reflux for 3 h in the absence of solvent, according to Figure 12 [28]. When the

http://dx.doi.org/10.5772/61431

ethylene-1,2-diamine reacts with chlorinated cellulose and replaced chlorine, the inter- and intramolecular reactions present in biomaterial which are very much responsible for its organization and crystalline arrangement become amorphous. In addition, the molecule incorporation of ethylene-1,2-diamine provides the growth of the amount of carbon in the structure, the increasing C7 and C8 [29].

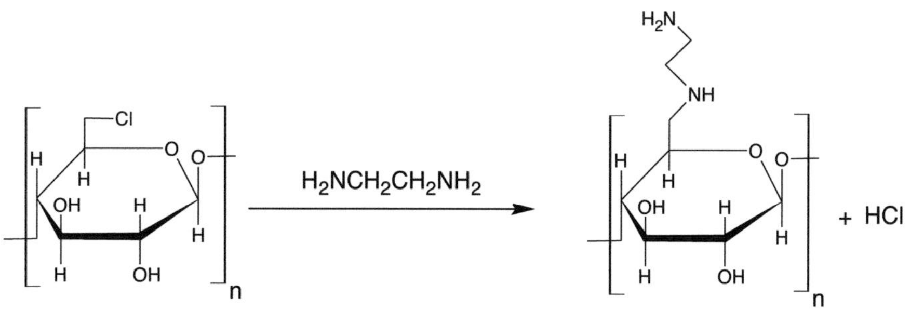

Figure 12. Reaction of chlorinated cellulose with ethylene-1,2-diamine.

Another important reaction of amination is organic molecule incorporation of 2- aminomethylpyridine. As the latest example, initially cellulose is chlorinated and, then it reacts with 2-aminomethylpyridine under reflux for 4 h [25], according to Figure 13. The amount of incorporated nitrogen in cellulose by this reaction is low, when compared to similar patterns. This fact can be explained by steric impediment which is caused by molecule hydrophobicity [30].

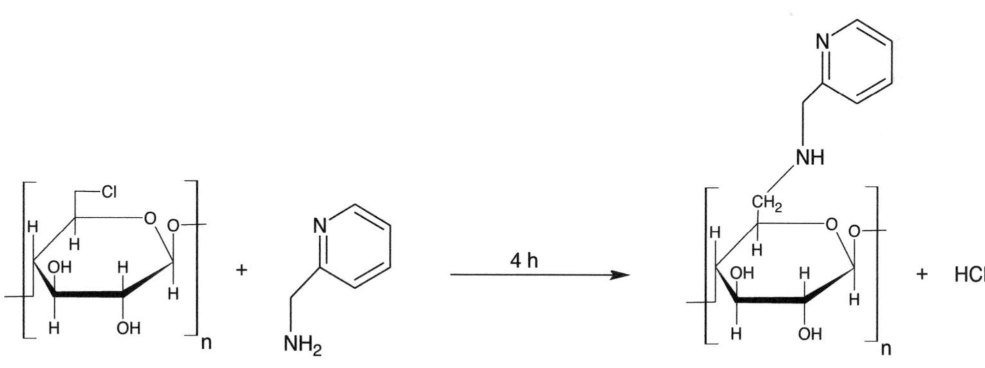

Figure 13. Reaction of chlorinated cellulose with organic 2-aminomethylpyridine.

These cellulosic derivatives containing basic centers (in this case nitrogen) are extremely used in the contaminants removal of the aqueous medium. For example, the modified cellulose 2-aminomethylpyridine is proved to be effective in divalent cations removal (Cu^{2+}, Co^{2+}, Ni^{2+}, and Zn^{2+}). In this case, the adsorption mechanism is based on the complexation process of a cation in two basic centers, which indicates that cellulose chain acts as a bidentate chelating agent. The scheme of metal complexation is shown in Figure 14 (a), where the available nitrogen centers are coordinated to divalent cations, with charge being counterbalanced for counter-ions. The modified cellulose with ethylene-1,2-diamine also proved to be efficient in divalent cations removal quoted above. In this case, the adsorption order was: $Co^{2+} > Cu^{2+} > Zn^{2+} > Ni^{2+}$. Its adsorption mechanism is similar to that quoted above, where there is transfer of cations from solutions to available basic centers in ethylene-1,2-diamine anchored in the cellulose, by cations complexation through available amine groups, as shown in Figure 14 (b) [28].

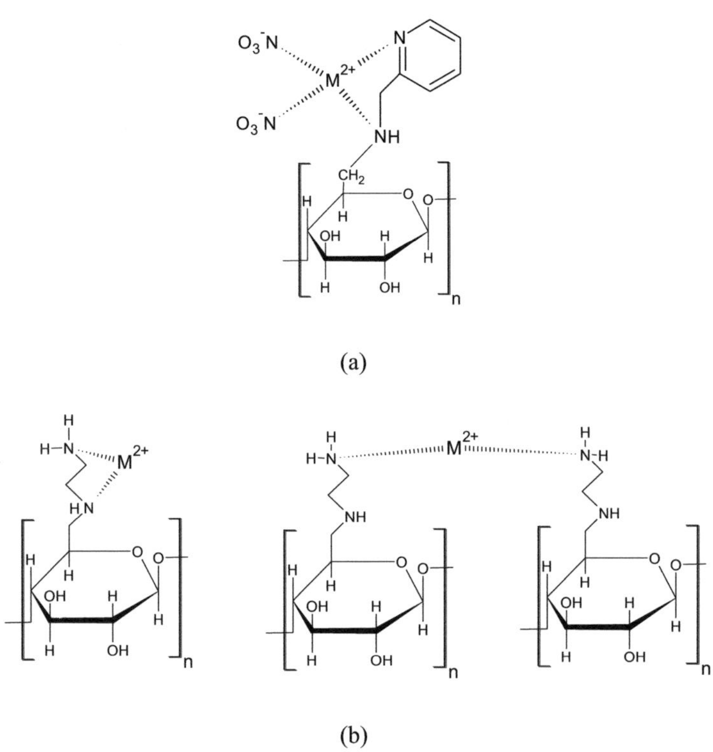

Figure 14. (a) Proposed scheme of complexation for divalent cations with a modified cellulose with 2-aminomethylpyridine. (b) Proposed scheme of complexation for divalent cations with the modified cellulose with ethylene-1,2-diamine.

Another amination example is the introduction of ethylene-1,2-diamine in the cellulose structure through a rusty cellulosic intermediate. The first stage consists in nanocrystalline cellulose synthesis (NCC) through hydrolysis with sulfuric acid. In the second stage, the carbon C2 and C3 of the cellulose is oxidised with sodium periodate ($NaIO_4$), in the absence of light at 313 K, forming nanocrystalline dialdehyde from cellulose. After this, ethylene-1,2-diamine is added, the mixture is continuously stirred for 6 h at 303 K. Finally, with imine intermediate reduction through addition of $NaBH_4$ at room temperature, as shown in Figure 15 [31], the zeta potential of the final product of this reaction is formed, which is highly influenced by pH. In lower pH, the zeta potential was positive due to amine protonation ($NCC\text{–}NH_3^+$). By the pH growth, the zeta potential decreases and it becomes negative in alkaline region, resulting in amine group deprotonation (NCC–NH_2) and dissociation of sulfate groups on surface [31].

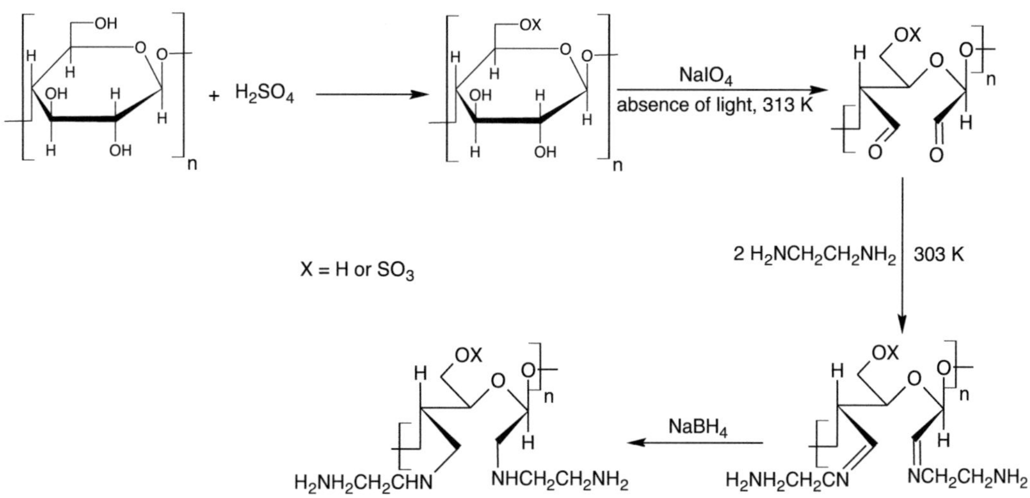

Figure 15. Oxidation with sodium periodate and reaction of nanocrystalline cellulose amination.

These cellulosic derivatives containing basic centers were efficient in dye removal. For example, the modified cellulose with ethylene-1,2-diamine, having as intermediate an oxidized cellulose derivative, was efficient in adsorption of the red anionic acid dyes GP, congo red 4 BS, and pale yellow reactive K-4G (Figure 16). The pH affected adsorption of the three dyes on the structure of modified cellulosic biomaterial with ethylene-1,2-diamine. The isoelectric point (pH_{pzc}) of this material is about 8, which indicates its surface at $pH < 8$ is positively charged, whereas at $pH > 8$ is negatively charged. Hence, in acid conditions the interaction occurs between the protonated amine group of the cellulosic material and the active anionic site of dye, which favors the adsorption. Whereas in alkaline regions, the surface charge of biomaterial becomes negative, and, consequently, it narrows interaction with dye anionic active site, due to electrostatic repulsion [31].

2.5. Sulfonation

The cellulose functionalization with sulfur incorporation is more used in contaminants removal of the aqueous medium. Several reactions are studied and their products are applied in contaminants adsorption. Among reactions that have been studied is the cellulose oxidation forming a dialdehyde, reacting with sodium bisulfite and creating a sulfonated cellulosic material [32]. In this reaction, initially the cellulosic material oxidation occurs with sodium periodate ($NaIO_4$), as shown in Figure 15. Then, the cellulosic material 2,3-dialdehyde formed is dispersed in water and it is treated with sodium bisulfite ($NaHSO_3$), creating a sulfonated cellulosic material, according to Figure 17. Afterward, sulfonation and oxidation of the surface morphology of cellulosic material are modified. In this case, the cellulose nanospheres are

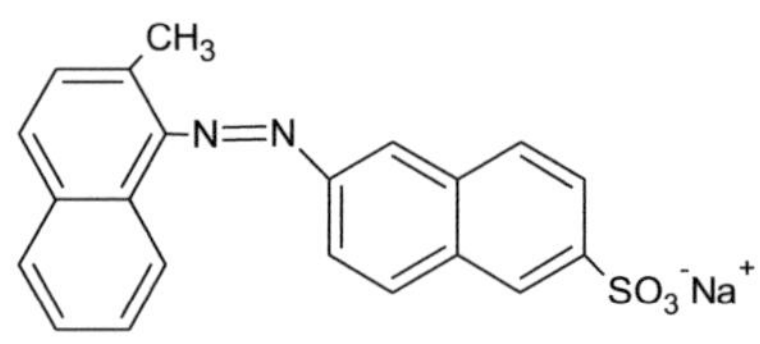

(a)

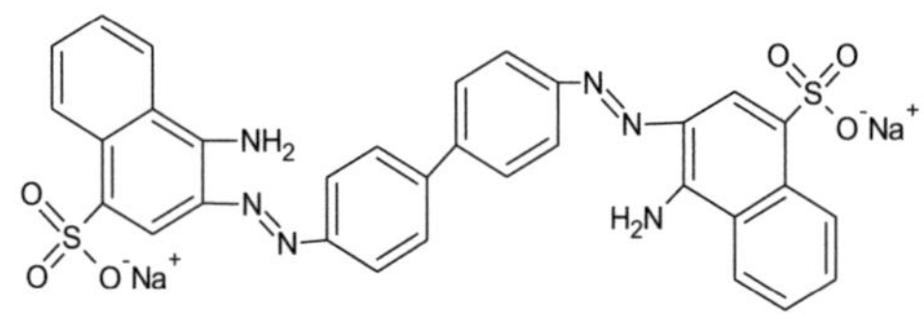

(b)

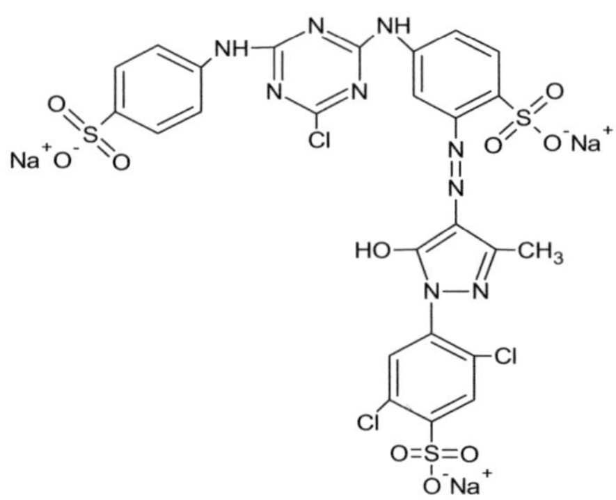

(c)

Figure 16. Chemical structure of the red acid dyes GR (a), congo red 4BS (b), and pale yellow reactive K-4G (c).

gradually deformed, losing their spherical forms with oxidation/sulfonation growth. However, these changes in the nanosphere molds were stretched after sulfonation, with formation of rods aggregates. In addition, the presence of sulfonic group has provided an increasing value of water retention in the cellulosic material structure [32].

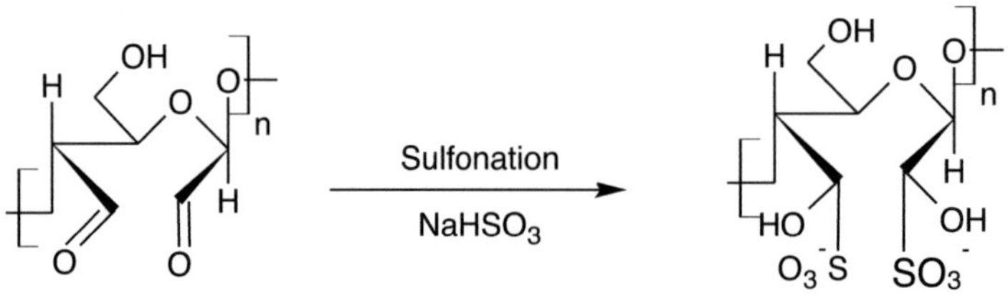

Figure 17. Sulfonation of cellulosic material.

The sulfonated cellulosic derivative (obtained from wheat straw) modified similar to the latest reaction proved efficient in heavy metal lead removal (Pb^{2+}). The sulfonic group introduction in the cellulosic biomaterial structure provokes the growth of electron density of the biomaterial, increasing its affinity for metal ions. In lower levels of pH, the sulfonic groups are protonated, which results in bad ions adsorption Pb^{2+}. Since the increase in pH provides adsorption growth, because sulfonic groups are deprotonated, and consequently in its ionic form, the soft acids form higher covalent complex than ionic with binders containing sulfur. Thus, in the beginning of adsorption, ions can bond with two binders to form complexes, soon, in lower concentration, the ions (all Pb^{2+}) can interact with two linkage locals, having a higher adsorption. Another possible mechanism is ionic exchange with adjacent carboxyl group, but as happens with other materials, there is possibility both mechanisms can participate together during the adsorption, since there are several different active groups on adsorbent surface [33].

Another example of sulfur molecule incorporation is chemical modification of cellulosic biomaterials, derivative from mesocarp and epicarp from babaçu coconut, with ethylenesulfide. In this reaction, the cellulosic biomaterials are put in contact with ethylenesulfide, for 3 h at 323 K, as shown in Figure 18 (a). The amount of incorporated sulfur in these materials was lower than the amount of incorporated sulfur in chitosan, which occurs due to amino reactive center free of chitosan, being higher than hydroxyl groups of carbon 6, of these cellulosic biomaterials [34,35]. Another methodology used to incorporate ethylenesulfide in cellulose chain is by reaction with available amine group, which is from the reaction ethylene-1,2-diamine (Figure 12). Since this amine group has been introduced by reaction with chlorine cellulose (Figure 11) for increasing the cellulose surface reactivity, in this reaction, after amination and chlorination, the cellulose containing amine groups are placed in contact with ethylenesulfide for 4 h at 328 K, according to Figure 18 (b) [19].

The cellulosic biomaterials derivative from mesocarp and epicarp from babaçu coconut proved efficient in adsorption of copper divalent cation (Cu^{2+}). This adsorption process was influenced by pH, where the increase in pH promoted the increasing ions adsorption, with maximum adsorption at pH 6 for two biomaterials. This occurs because the surface of bio-adsorbents changes its polarization with the value of pH of solution and with $pH_{(pzc)}$ of solid. The $pH_{(pzc)}$ is the parameter that indicates the pH value in which a solid exhibit electrically neutral surface so the number of positive charges equals the number of negative charges. The $pH_{(pzc)}$ of two modified cellulosic biomaterials is 5.6, then in pH lower of this value the biomaterial surface is positively charged, which disfavors the adsorption of cation ions Cu^{2+}. Whereas in pH values above $pH_{(pzc)}$, the biomaterials surface is negatively charged, and consequently fostering cation adsorption occurs [34].

The modified cellulose with ethylenesulfide, according to Figure 18 (b), also proved to be efficient in removal of several divalent cations (Pb^{2+}, Cd^{2+}, Ni^{2+}, Co^{2+}, Cu^{2+}, and Zn^{2+}). This modified biomaterial presents higher capacity in extracting cations from aqueous solution, due to presence of nitrogen and sulfur acids, which are highly reactive adsorption sites which can coordinate with metal cations. The cations adsorption order was: $Pb^{2+} > Cd^{2+} > Ni^{2+} > Co^{2+} > Cu^{2+} > Zn^{2+}$. The copper is less preferred by modified cellulose than nickel and cobalt, because they have favorable complexation equilibrium constants [29].

Another reaction of sulfur incorporation in cellulose structure is its reaction with aminoethanethiol, using the cellulose chlorination with intermediate reaction. At first, chlorine cellulose

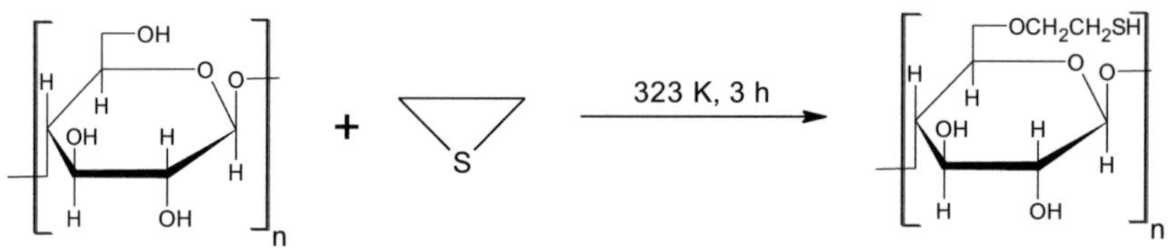

(a)

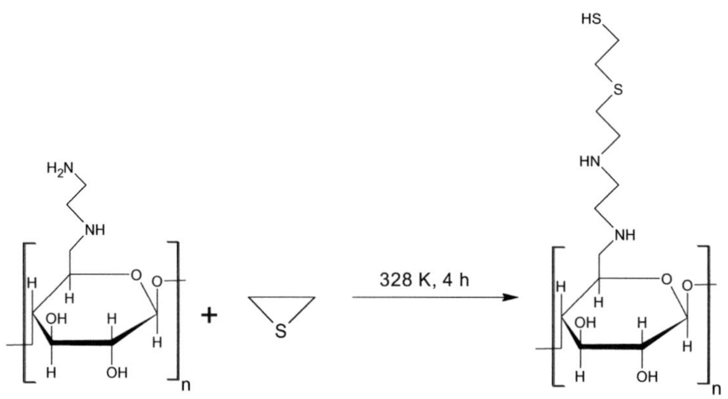

(b)

Figure 18. (a) Cellulosic biomaterials reaction with ethylenesulfide. (b) Cellulose reaction, containing amine groups with ethylenesulfide.

is made, as shown in Figure 11. Then, the chlorine cellulose is placed in contact with aminoethanethiol at 363 K, after the triethylamine is used for removing excess of HCl, according to Figure 19. The molecule introduction of aminoethanethiol in the cellulose structure provokes changes in inter- and intramolecular interactions in biomaterial; these interactions are responsible by bigger part of biomaterial organization, by biomaterial crystalline arrangement, and, consequently, if some disturbance occurs due to molecule introduction in biomaterial, it becomes amorphous, such as with molecule introduction of aminoethanethiol [3].

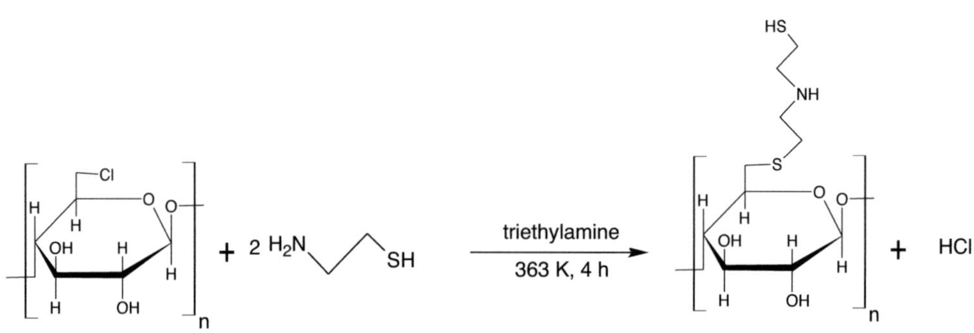

Figure 19. Reaction of incorporation of aminoethanethiol group in the chlorinated cellulose structure.

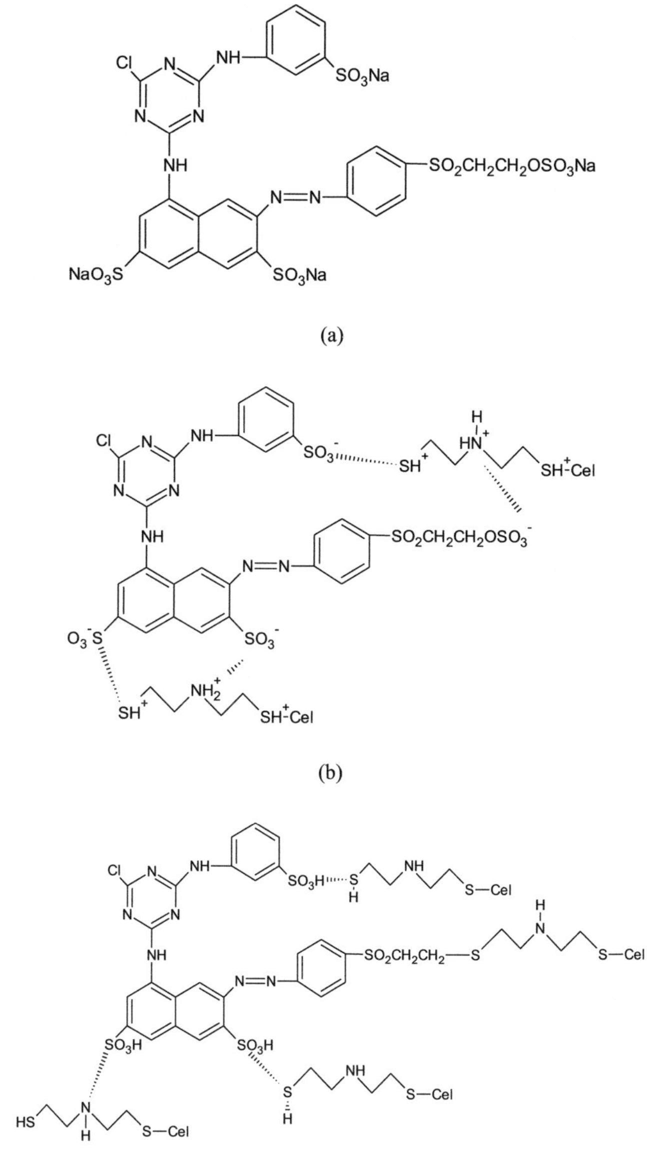

Figure 20. Molecular structure of reactive red anionic dye RB (a) and scheme of modified cellulose interaction with anionic dye in acid medium (b) and basic medium (c).

This modified biomaterial with aminoethanethiol group proved efficient in removal of reactive red anionic dye (Figure 20 (a)). In this survey, it is noted that the adsorption of dye is maximal at pH 8 and 9, which indicate two predominant mechanisms in adsorption: one in acid medium and other in basic medium. In acid medium, the adsorption is favored by electron interaction. The $pH_{(pzc)}$ of modified material is nearest to 6, then, below this pH, the biomaterial surface is positively charged, due to protonation of nitrogen and sulfur acids. These biomaterial positive sites interact with negative ions in the dye favoring the adsorption (Figure 20 (b)). Since in basic medium there will be no protonation on biomaterial surface, adsorption is favored by hydrogen interaction and/or covalent interactions. The first stage of the mechanism involves the group transformation of β-sulfatoethylsulfonic of dye $SO_2CH_2CH_2OSO_3Na$, in a vinylsulfonic group, $-SO_2CH{=}CH_2$. In alkaline conditions, the group divides forming sulfate waste. Then, the vinylsulfonic group interacts with the modified cellulose surface through a covalent bond. In addition, it can have formation of hydrogen bond between sulfur/nitrogen atoms of modified cellulose and the hydrogen sulfonic dye group (Figure 20 (c)) [3].

3. Conclusion

The chemical modification (incorporation of carboxymethyl, phosphorus, carboxyl, amines, and sulfur) of cellulosic materials is very important for the application in the removal of contaminants from aqueous medium. After the chemical modification, cellulosic materials exhibit new chemical properties that are more advantageous than the starting materials. These new chemical properties allow an increase in interaction between the modified cellulosic materials and the contaminants (metals, dyes, and drugs) during adsorption processes in aqueous medium and these interactions are strongly influenced by solution pH. Therefore, the cellulosic materials modified with various chemical groups are promising for application in removing contaminants from aqueous medium of the environment.

Acknowledgements

The authors thank the Council for Scientific and Technological Development (CNPq), Foundation for Supporting Research of Piauí (FAPEPI), and Coordination Support in Higher Education (CAPES) for financial support. To the Federal University of Piauí (UFPI) and Federal Institute of Piauí (IFPI), to provide work research conditions.

Author details

Roosevelt D.S. Bezerra[1*], Paulo R.S. Teixeira[1], Ana S.N.M. Teixeira[1], Carla Eiras[2], Josy A. Osajima[2] and Edson C. Silva Filho[2]

*Address all correspondence to: rooseveltdsb@ifpi.edu.br

1 Federal Institute of Education, Science and Technology of Piaui _ Campus Teresina Central, Teresina, Brazil

2 Federal University of Piaui _ LIMAV, Teresina, Brazil

References

[1] Isik M, Sardon H, Mecerreyes D. Ionic liquids and cellulose: dissolution, chemical modification and preparation of new cellulosic materials. Int J Mol Sci 2014; 15(7): 11922 – 49. doi: 10.3390/ijms150711922

[2] Tian Y, Wu M, Liu R, Wang D, Liu X, Liu W, Ma L, Li T, Huang Y. Modified native cellulose fibers: a novel efficient adsorbent for both fluoride and arsenic. J Hazard Mater 2011; 185(1): 93 – 100. doi: 10.1016/j.jhazmat.2010.09.001.

[3] Silva L S, Lima L C B, Silva F C, Matos J M E, Santos M R M C, Júnior L S S, Sousa K S, da Silva Filho E C. Dye anionic sorption in aqueous solution onto a cellulose surface chemically modified with aminoethanethiol. Chem Eng J 2013; 218: 89 – 98. doi: 10.1016/j.cej.2012.11.118.

[4] Lin Q, Zheng Y, Wang G, Shi X, Zhang T, Yu J, Sun J. Protein adsorption behaviors of carboxymethylated bacterial cellulose membranes. Int J Biol Macromol 2015; 73: 264 – 9. doi: 10.1016/j.ijbiomac.2014.11.011

[5] Silva L S, Lima L C B, Ferreira F J L, Silva M S, Osajima J A, Bezerra R D S, da Silva Filho E C. Sorption of the anionic reactive red RB dye in cellulose: assessment of kinetic, thermodynamic, and equilibrium data. Open Chem 2015; 13: 801 – 12. doi: 10.1515/chem-2015-0079.

[6] Bezerra R D S, Silva M M F, Morais A I S, Osajima J A, Santos M R M C, Airoldi C, da Silva Filho E C. Phosphated cellulose as an efficient biomaterial for aqueous drug ranitidine removal. Materials 2014; 7(12): 7907 – 24. doi: 10.3390/ma7127907

[7] Bezerra R D S, Silva M M F, Morais A I S, Osajima J A, Santos M R M C, Airoldi C, da Silva Filho E C. Natural cellulose for ranitidine drug removal from solutions. J Environ Chem Eng 2014; 2: 605 – 11. doi: 10.1016/j.jece.2013.10.016.

[8] Yasar F, Togrul H, Arslan N. Flow properties of cellulose and carboxymethyl cellulose from orange peel. J Food Eng 2007; 81: 187 – 99. doi: 10.1016/j.jfoodeng.2006.10.022.

[9] Zhang G, Zhang L, Deng H, Sun P. Preparation and characterization of sodium carboxymethyl cellulose from cotton stalk using microwave heating. J Chem Technol Biotechnol 2011; 86: 584 – 9. doi: 10.1002/jctb.2556.

[10] Lin Q, Zheng Y, Wang G, Shi X, Zhang T,Yu J, Sun j. Protein adsorption behaviors of carboxymethylated bacterial cellulose membranes. Int Biol Macromol 2015; 73: 264 – 9. doi:10.1016/j.ijbiomac.2014.11.011.

[11] Yan H, Zhang W, Kan X, Dong L, Jiang Z, Li H, Yang H, Cheng R. Sorption of methylene blue by carboxymethyl cellulose and reuse process in a secondary sorption. Colloids Surf A 2011; 380: 143 – 51. doi:10.1016/j.colsurfa.2011.02.045.

[12] Kargl R, Mohan T, Bracic M, Kulterer M, Doliska A, Stana-kleinschek K, Ribitsch V. Adsorption of carboxymethyl cellulose on polymer surface: evidence of a specific interaction with cellulose. Langmuir 2012; 28: 11440 – 7. doi: 10.1021/la302110a.

[13] Wang M, Xu L, Peng J, Zhai M, Li J, Wei G. Adsorption and desorption of Sr (II) ions in the gels based on polysaccharide derivates. J Hazard Mater 2009; 171: 820 – 6. doi: 10.1016/j.jhazmat.2009.06.071.

[14] Kokol V, Bozic M, Vogrincic R, Mathew AP. Characterisation and properties of homo- and heterogeneously phosphorylated nanocellulose. Carbohydr Polym 2015; 125: 301 – 13. doi: 10.1016/j.carbpol.2015.02.056.

[15] Suflet D M, Chitanu G C, Popa V I. Phosphorylation of polysaccharides: new results on synthesis and characterisation of phosphosrylated cellulose. React Funct Polym 2006; 66: 1240 – 9. doi: 10.1016/j.reactfunctpolym.2006.03.006.

[16] Oshima T, Kondo K, Ohto K, Inoue K, Baba Y. Preparation of phosphorylated bacterial cellulose as an adsorbent for metal ions. Reac Funct Polym 2008; 68: 376 – 83. doi: 10.1016/j.reactfunctpolym.2007.07.046.

[17] Oshima T, Taguchi S, Ohe K, Baba Y. Phosphorylated bacterial cellulose for adsorption of proteins. Carbohydr Polym 2011; 83: 953 – 8. doi:10.1016/j.carbpol.2010.09.005.

[18] Rungrodnimitchai S. Modification of rice straw for heavy metal ion adsorbents by microwave heating. Macromol Symp 2014; 295: 100 – 6. doi: 10.1002/masy.200900146.

[19] de Melo J C P, da Silva Filho E C, Santana S A A, Airoldi C. Synthesized cellulose/succinic anhydride as an ion exchanger. Calorimetry of divalent cations in aqueous suspension. Thermochim Acta 2011; 523: 29 – 34. doi: doi:10.1016/j.tca.2011.06.007.

[20] Vieira A P, Santana S A A, Bezerra C W B, Silva H A S, de Melo J C P, da Silva Filho E C. Airoldi C. Copper sorption from aqueous solutions and sugar cane spirits by chemically modified babaçu coconut (*Orbignya speciosa*) mesocarp. Chem Eng J 2010; 161: 99 – 105. doi: 10.1016/j.cej.2010.04.036.

[21] de Melo J C P, da Silva Filho E C, Santana S A A, Airoldi C. Maleic anhydride incorporated onto cellulose and thermodynamics of cation-exchange process at the solid/liquid interface. Colloids Surf A 2009; 346: 138 – 45. doi: 10.1016/j.colsurfa.2009.06.006.

[22] de Melo J C P, da Silva Filho E C, Santana S A A, Airoldi C. Exploring the favorable ion-exchange ability of phthalytated cellulose biopolymer using thermodynamic data. Carbohyd Res 2010; 345: 1914 – 21. doi: 10.1016/j.carres.2010.06.012.

[23] Zhou Y, Min Y, Qiao H, Huang Q, Wang E, Ma T. Improved removal of malachite green from aqueous solution using chemically modified cellulose by anhydride. Int Biol Macromol 2015; 74: 271 – 7. doi: 10.1016/j.ijbiomac.2014.12.020.

[24] Kassale A, Barouni K, Bazzaoui M, Martins J I, Albourine A. Methylene Blue adsorption by cotton grafted with succinic anhydride. Prot Met Phys Chem 2015; 51: 382 – 9. doi: 10.1134/S2070205115030120.

[25] da Silva Filho E C, Júnior L S S, Silva M M F, Fonseca M G, Santana S A A, Airoldi C. Surface cellulose modification with 2-aminomethypyridine for copper, cobalt, nickel and zinc removal from aqueous solution. Mat Res 2013; 16: 79 – 87. doi: 10.1590/S1516-14392012005000147.

[26] da Silva Filho E C, de Melo J C P, da Fonseca M G, Airoldi C. Cation removal using cellulose chemically modified by a Schiff base procedure applying green principles. J Colloid Interface Sci 2009; 340: 8 – 15. doi: 10.1016/j.jcis.2009.08.012.

[27] da Silva Filho E C, Santana S A A, de Melo J C P, Oliveira F J V E, Airoldi C. X-ray diffraction and thermogravimetry data of cellulose, chlorodeoxycellulose and amino-deoxycellulose. J Therm Anal Calorim 2010; 100: 315 – 321.doi: 10.1007/s10973-009-0270-6.

[28] da Silva Filho E C, de Melo J C P, Airoldi C. Preparation of ethylenediamine-anchored cellulose and determination of thermochemical data for interaction between cations and basic centers at the solid/liquid interface. Carbophydr Res 2006; 341: 2842 – 50. doi: 10.1016/j.carres.2006.09.004.

[29] da Silva Filho E C, Lima L C B, Silva F C, Sousa K S, Fonseca M G, Santana S A A. Immobilization of ethylene sulfide in aminated cellulose for removal the divalent cations. Carbohydr Polym 2013; 92: 1203 – 10. doi: 10.1016/j.carbpol.2012.10.031.

[30] da Silva Filho E C, Júnior L S S, Santos M R M C, Fonseca M G, Sousa K S, Santana S A A, Airoldi C. Thermochemistry of interaction between cellulose modified with 2-aminomethylpyridine and divalent cations. J Therm Anal Calorim 2013; 114: 423 – 9. doi: 10.1007/s10973-012-2917-y.

[31] Jin L, Li W, Xu Q, Sun Qiucun. Amino-functionalized nanocrystalline cellulose as an adsorbent for anionic dyes. Cellulose 2015; doi: 10.1007/s10570-015-0649-4.

[32] Zhang J, Jiang N, Dang Z, Elder T.J, Ragauskas A.J. Oxidation and sulfonation of cellulosics. Cellulose 2008; 15: 489 – 496. doi: 10.1007/s10570-007-9193-1.

[33] Suopajarvi T, Liimatainen, Karjalainen M, Upola H, Niinimaki J. Lead adsorption with sulfonated whet pulp nanocelluloses. J Wat Process Eng 2015; 5: 136 – 42. doi: 10.1016/j.jwpe.2014.06.003.

[34] Santana S A A, Vieira A P, da Silva Filho E C, de Melo J C P, Airoldi C. Immobilization of ethylenesulfide on babassu coconut epicarp and mesocarp for divalent cation sorption. J Hazard Mater 2010; 174: 714 – 9. doi: :10.1016/j.jhazmat.2009.09.109.

[35] Sousa K S, da Silva Filho E C, Airoldi C. Ethylenesulfide as a useful agent for incorporation into the biopolymer chitosan in a solvent-free reaction for use in cation removal. Carbohydr Res 2009; 344: 1716 – 23. doi: 10.1016/j.carres.2009.05.028.

Chapter 5

Use of Cellulosic Materials as Dye Adsorbents — A Prospective Study

Fabrícia C. Silva, Luciano C.B. Lima, Roosevelt D.S. Bezerra, Josy A. Osajima and Edson C. Silva Filho

Additional information is available at the end of the chapter

http://dx.doi.org/10.5772/61343

Abstract

Cellulose is the most abundant biopolymer of nature, and it is widely used in the synthesis of new materials as well as in the adsorption of dye. This study reports a literature review (articles) and technology review (patents) about publications and product invention, which contain information on the use of cellulose on the adsorption of dyes in the period 2004–2014. For this work, research database and keywords were used to find articles and patents related to the subject under review. Specific words were used to find articles and patents related to the subject under review. After a demanding research, 1 patent and 23 articles that contain the words "cellulose," "dye," and "adsorption or sorption" in their titles were assessed, and annual evolution studies were performed for publications and countries that publish more.

Keywords: Cellulose, adsorption, dye, review, prospection

1. Introduction

One of the negative consequences of human development is the large amount of waste discharged into water bodies, with varied compositions that may contain heavy metals, dyes, and/or other undesirable chemical compounds [1, 2].

The problem becomes more serious when it comes to textile industrial wastewater, as they are colorful, produced on a large scale, and do not always receive due treatment before being discharged into the environment, carrying various impurities [2, 3].

The aquatic environment coming into contact with these effluents has modified physical and chemical properties due to their color because such dyes become visible at concentrations up to 1 mg L^{-1}, resulting in serious aesthetic problems until their toxicity, leading to hazards to human and animal health [2].

Contamination by dyes prevents the development of fish farming and recreation as well as influence the ecological equilibrium affecting the proliferation of aquatic plants by reducing photosynthetic activity caused by the difficulty passage of sunlight. Thus, the removal of these dyes becomes important environmentally [4, 5].

The removal of these dyes are not always easy because of its complex molecular structures that make them stable to light and heat and resistant to biodegradation [6].

Available wastewater treatment processes mainly involve ion exchange [7], chemical precipitation [8], chemical degradation [9], reverse osmosis [10], oxidation/chemical reduction [11], and adsorption [12].

Adsorption particularly stands out because of its high efficiency combined with low cost. It removes chemical pollutants, returning the natural transparency of the medium, and often enables reusing the reuse of the adsorbent. It is a process that is commonly reversible, where the regeneration of the adsorbent becomes possible, further cheapening the practice of this process, beside the use of natural waste in the trash [12, 13].

The adsorption using low-cost adsorbents is recognized as an effective and economical method for the decontamination of water. Because of this, many studies have been conducted employing the adsorption process using these adsorbents; among them, one can mention corn cob [14], sugarcane bagasse [15], banana peel [16], sawdust [17], wheat straw, [18], orange peel [19], water hyacinth roots [20], peanut hull [21], eucalyptus leaf [22], rice husk [23], corn kernel [24], coconut husk [25], apple pomace [26], and other cellulosic waste used in natural or modified surface [27].

Cellulose is a biopolymer considered as an almost inexhaustible source of raw materials, and it can be cited, as an example, as a promising natural material that has been extensively explored by researchers in the adsorption area. It is even more attractive when it makes modifications in its structure in order to improve their existing properties or adding new potentialities to this material [12, 28–30].

Modifying the surface of the cellulose can varies some of its properties, such as its hydrophilic or hydrophobic character, elasticity, resistance to microbiological attacks, and thermal and mechanical resistance, may also increase its pollutant adsorption capacity in aqueous and/or nonaqueous solutions [31].

The elucidation of the polymeric structure of cellulose was made by the pioneering work of Hermann Staudinger [32]. Through acetylation and deacetylation, he recognized that the structures did not consist merely of an aggregate of D-glucose units. Rather, glycosidic units were discovered by being linked covalently to one other to form long molecular chains.

Figure 1 shows the molecular structure of cellulose as a polymeric carbohydrate generated by β-D-glucopyranose repeating, which is covalently linked by acetal functions between the equatorial OH group of 4-carbon atom (C4) and the 1-carbon atom (C1). Hence, this resulted

http://dx.doi.org/10.5772/61343

in β-1,4-glucan denomination, which is in principle the way in which the cellulose is biogenetically formed. We have the cellulose as a straight-chain extended polymer with a large number of hydroxyl groups: three per anhydroglucose unit (AGU), which is the central unit, see Figure 1, a thermodynamically preferred conformation 4C1 bond between the 4-carbon and the 1-carbon [33].

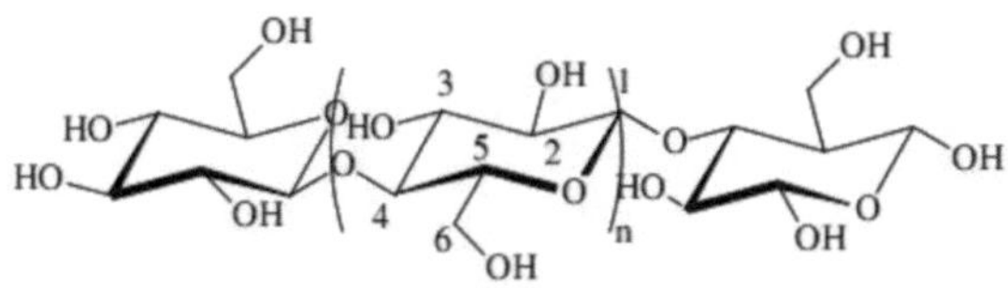

Figure 1. Cellulose molecular structure showing the numbering of the carbon atoms and functional groups per monomer of the polymer.

This chapter aims to present a literature and technology review about the use of cellulosic materials in the adsorption of dyes, published and/or deposited during the period 2004 to 2014.

The prospection performed in this work was based on an extensive electronic search in patent databases and articles databases.

For prospecting patent, a search in the free databases available, that is, the European Patent Office (EPO), the National Institute of Industrial Property (INPI), and the United States Patent and Trademark Office (USPTO), was carried out. The keywords *cellulose, cellulose and dye, cellulose and adsorption, cellulose adsorption, cellulose and adsorption and dye*, and *cellulose and sorption and dye* were used to specify the search in the EPO and USPTO bases, and the related keywords in Portuguese were used for INPI, including "title."

For the prospection of the articles, a survey was conducted to determine the most published database (Web of Science, Science Direct, Scopus, and Scielo) during the period 2004 to 2014. The same keywords used in the search of patents were used, including "title" and exclusively regular articles publications. Review articles, books, and book chapters were left out of the research. The research was conducted in April 2015. The articles found in database with largest amount of publication were subsequently analyzed.

2. Current state of the art in use of cellulosic materials as dye adsorbent

The investigation of patent by databases showed that the EPO was the database that received more product invention associated with the subject studied during the period investigated compared USPTO and INPI, as shown in Table 1.

Table 1 shows that when the more general term "cellulose" was used, the largest number of invention products in all databases was found. As new words are added and crossed, the

Keywords	EPO	USPTO	INPI
cellulose	13,355	823	1
cellulose and *dye*	46	5	0
cellulose and *adsorption*	19	0	0
cellulose and sorption	1	0	0
cellulose and *adsorption* and *dye*	1	0	0
cellulose and sorption and dye	0	0	0
TOTAL	13,422	828	1

Table 1. Total patents registered on the databases EPO, USPTO, and INPI during the period 2004 to 2014.

amount of invention products considerably decreased because the patents containing the word cellulose in their titles have many different purposes.

The patent only found in the databases after refinement was the German patent DE102008026403 (A1)—Dyeing cellulose fibres, e.g., cotton fabric, with reactive vinylsulfone dyes, involves adding dye-bath components in a special sequence to optimize the relation between physical adsorption and chemical reactivity, containing in its title the words "cellulose", "adsorption," and "dye." This patent brings a dyeing method of cellulose fibers with reactive dyes, intending to optimize the relationship between physical adsorption and chemical reactivity in order to improve dye fixation on cellulose fiber [34].

In the prospection of articles, it was observed that the amount of published articles was higher compared to the number of patents, as shown in Table 2.

Articles published between 2004 - 2014						
Keywords Databases	Cellulose	Cellulose and dye	Cellulose and adsorption	Cellulose and sorption	Cellulose and adsorption and dye	Cellulose and sorption and dye
Web of Science	18,428	165	330	97	15	5
Science Direct	4,126	39	120	25	6	3
Scopus	13,985	142	356	99	19	4
Scielo	154	0	7	1	0	0

Table 2. Number of articles published on Web of Science, Science Direct, Scopus, and Scielo databases from 2004 to 2014.

Table 2 shows that the database with the largest number of articles published in the area is Scopus, which showed higher publication when compared with other databases analyzed.

Observing the results obtained in this research using the database Scielo, articles related to the theme in the time period analyzed for their journals are not included. However, 8 articles related to the use of cellulosic materials for adsorption studies have been found. These articles were aimed at removing other species that were not dye. This fact was confirmed when the words "cellulose" and "dye" were used, which were not found articles.

Thus, the research regarding the adsorption of the dye using cellulosic materials as adsorbents was evaluated by analyzing the 23 articles found in this database containing the words "cellulose, adsorption, and dye" and "cellulose, sorption, and dye" in their titles in the last 10-year period from 2004 to 2014.

Figure 2 shows the study of the annual evolution of publications.

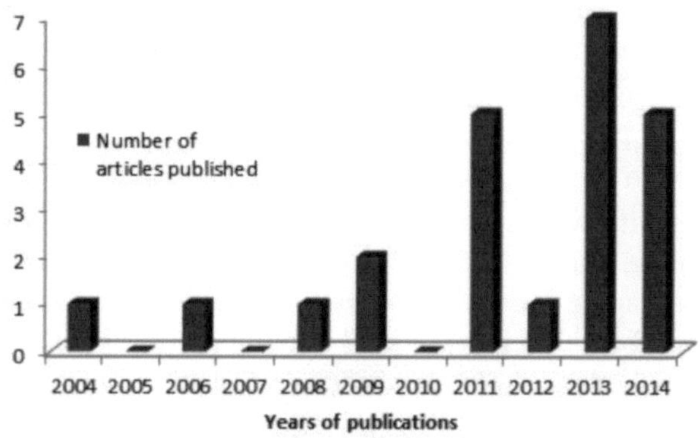

Figure 2. Annual evolution of articles published.

This result proves that the adsorption of dyes using cellulosic material is a subject that is being currently studied, and the year 2013 showed the highest amount of publication about this topic, presenting a number of 7 from the 23 reviews.

Among the countries that published, China was ranked first, followed by the United Kingdom, India, and Iran, showing that the world's major economies have shown interest in this study (Figure 3).

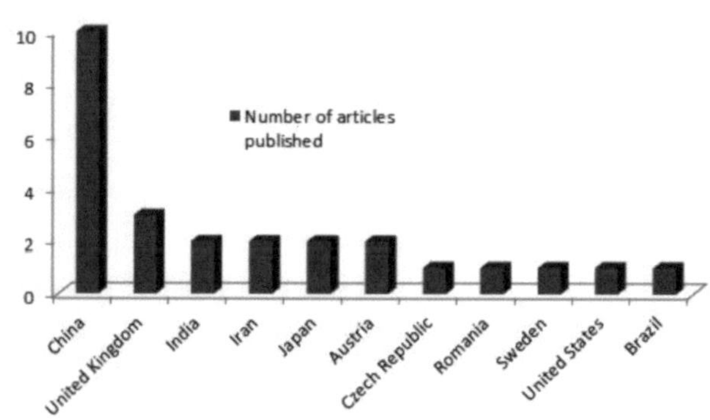

Figure 3. Countries that more published articles.

Adsorption studies often have the aim to evaluate the maximum removal capacity for a given compound (in this case dye) by adsorbent material proposed (in this case, the cellulosic materials). Moreover, there is a wide range of possibility in the variation of experimental parameters and conditions that directly influence in such property, or there are researchers who are concerned with elucidating the various existing mechanisms in this process, making use of theoretical models established or even proposing new theoretical models their investigations, often making use of computational tools.

Thus, it is observed that the publications about this theme can be divided into two major groups: the experimental conditions and parameters for the maximum removal capacity for a given dye and the theoretical studies to elucidate this process, explaining experimental data existing or proposed theoretical models that can be used in the elucidation of this process in general.

Analyzing the articles relating to this theme found in the Scopus database, note that of the 23 articles found (of which 19 articles were found with the keywords "cellulose," "adsorption," and "dye" and 4 items with words keys "cellulose," "sorption," and "dye"), 7 works bring theoretical studies to understand the adsorption process of given dyes on cellulosic materials or do not make use of parameter variation for study of maximum adsorption capacity for the material proposed, thus using different methodologies to other articles.

One of these studies is about reference [35], which brings the validation of proposed new nonlinear mathematical model of adsorption isotherm using a correlation between the classical data dye adsorption CI Direct Blue 1 in bleached cotton and the data obtained by the nonlinear model proposed, showing that the new model is presented with excellent results setting and brings the possibility of calculation of new parameters on the adsorption process.

The articles in references [36,37] made use of computational methods exploring the quantum mechanical method DFT/BPW91/6-31+G(d) using Gaussian 03 software and isotherm theoretical models for a better understanding of the interaction between adsorbent and adsorbate. These studies evaluated the effect of alkaline treatment using NaOH solutions of different concentrations of cellulose II fibers in the adsorption capacity for the Reactive Red 120 dye in the work of reference [36] and the dye CI Reactive Orange 84 and CI Reactive Red 120 [37].

In references [38,39], the authors published about the adsorption process of xanthene dyes on cellulose granules using a different methodology from other publications evaluated here, which made use of a method by column chromatography. Since the first work of Tabara et al. [38], they showed that the maximum adsorption capacity for the dyes erythrosine, rose bengal, and phloxine were 3.75, 3.42, and 4.74 mg g^{-1}, respectively. In their second study [39], they were concerned to elucidate the adsorption mechanism of such systems.

In article of reference [40], the authors have elucidated the structural factors that control the adsorption of acid dyes at the surface of cellulose; for this, they evaluated the adsorption of 15 different acid dyes on cellulose matrix, correlating with their structures and geometric values as the thermodynamic enthalpy and entropy of binding obtained in each process.

A theoretical and experimental study of the adsorption of a direct trisazo dye in mercerized cotton fiber, which use theoretical model by linear regression and nonlinear for the elucidation of such a process, was investigated in reference [41].

The remaining 15 articles have different methodologies, which aim at obtaining the maximum adsorption capacity for one or more dyes cellulose materials by investigating the influence of parameters such as pH, contact time, temperature, and others. Table 3 shows the main parameters obtained by this work, briefly specifying the proposed material as well as some of the values found for them. The main and most used parameters are better discussed, taking into regard the conclusions observed in these publications, in the following topics.

Adsorbent	Dye	pH	Time	Temperature	Removal Capacity	Reference
carboxymethyl cellulose-acrylic acid	Methyl orange	< 4.0	~40 min	35 ºC	84.2%	[42]
	Disperse blue	7.0		45 ºC	79.6%	
	Malachite green chloride	"/> 9.0		40 ºC	99.9%	
partially hydrolyzed polyacrylamide/cellulose nanocrystal	Methylene blue	5.0	240 min	-	326.08 mg g^{1}	[43]
Cellulose Agricultural Wastes	Violet B	11	90 min	-	96 mg g^{-1}	[44]
microcrystalline cellulose gel/nano TiO_2 composites	Methyl orange	-	30 min	-	94%	[45]
IPN Hydrogels based Carboxymethyl Cellulose	Basic fuchsin	7	-	25 °C	2.249 mg g^{-1}	[46]
	Methyl violet				1.723 mg g^{-1}	
carboxymethyl cellulose/organic montmorillonite nanocomposites	Congo red	-	6 h	60 °C	171.37 mg g^{-1}	[47]
Bacterial Cellulose	Direct blue 71	3	120 min	30 °C	25.15 mg g^{-1}	[48]
Plant Cellulose					22.00 mg g^{-1}	
Cellulose modified with quaternary ammonium groups	Reactive red 228	3	6 h	20 °C	190.00 mg g^{-1}	[49]
Cellulose modified with aminoethanethiol	Reactive red	2	100 min	55 °C	78.00 mg g^{-1}	[12]
		9	160 min		26.00 mg g^{-1}	
Surface quaternized cellulose	Congo red	-	-	-	664.00 mg g^{-1}	[50]
	Acid green 25				683.00 mg g^{-1}	
Cellulose-based multicarboxyl	Malachite green	6	120 min	30 °C	458.72 mg g^{-1}	[51]

Adsorbent	Dye	pH	Time	Temperature	Removal Capacity	Reference
adsorbent	Basic fuchsine	6.5	210 min		458.76 mg g^{-1}	
Berinjal plant root powder (celulose)	Malachite green	6	35 mim	70 °C	100.29 mg g^{-1}	[52]
	Methyl orange			40 °C	78.43 mg g^{-1}	
Nano-cellulose hybrid	Yellow B-4RFN	7	240 mim	293 K	16.61 mg g-1	[29]
	Blue B-RN				14.40 mg g^{-1}	
Magnetic cellulose beads	Methyl orange	-	180 min	-	0.045 mmol g^{-1}	[53]
	Methylene blue				0.030 mmol g^{-1}	
Modified cellulose with triazine derivatives	Reactive blue BF-RN	7	-	65 °C	20.00 mg g^{-1}	[30]
Unmodified cellulose					9.00 mg g^{-1}	

Table 3. Adsorptive parameters used in articles according to prospection.

2.1. Modification of cellulose

The presence of free hydroxyl on the surfaces of solids enables the application of a variety of reaction methods, which are explored to obtain new materials exhibited, due to the modification of their surfaces, new properties, or intensifying the existing ones, [54] not only in the reaction developing but also in applicability.

With regard specifically to cellulose, which is a renewable material and low cost, the reactivity is due to the hydroxyl groups, which are in an ideal position, with the primary hydroxyl at carbon 6 (C6), much more reactive than secondary hydroxyl (C2 and C3). Although the reactivity of the hydroxyl depends heavily of the reagents and the reaction conditions, the hydroxyl group present at the 3-carbon is even less reactive. Thus, the cellulose hydroxyl groups have the following reactivity order: C6 >> C2 > C3, this order being explained by the influence of the formation of inter- and intramolecular hydrogen bonds, as well as the degree of crystallinity, by favoring the water absorption in amorphous regions in comparison with the crystalline this material [55]. The main changes occurring in cellulose are due to halogenation, oxidation, etherification, and esterification.

Among the 23 articles found in the Scopus database, 12 present a proposal for the modification of cellulose aimed at intensification the properties of this biopolymer, a proof of importance of the modification of this promising material, to be involved in most of articles published in this area.

An example is the paper in reference [30], in which the authors modified cellulose incorporating cationic and anionic groups from the reaction with compounds of the triazine derivatives. With the proposed modification, the authors were able to increase the adsorption capacity for the reactive blue dye BF-RN, which was 9.00 mg g^{-1} for a value of 20 mg g^{-1} in an unmodified cellulose, justifying the proposed modification.

2.2. pH

The study of pH shows that the adsorption depends on the initial pH of the solution. The same process increases the maximum adsorption capacity because it distributes ions on the surface of the adsorbent by modifying their surface charge, and the dye molecules can be ionized, causing the interaction more efficient between adsorbent and adsorbate [56]. This study is very important (12 of 23 articles found on the prospection assessed the influence of this parameter in its work) and should always be performed in the adsorption tests because whether the pH will influence the adsorption process depends on the adsorbent and/or dye used.

Table 3 shows the reference [12] that evaluated the influence of pH on the reactive red dye adsorption on the surface of the modified cellulose with aminoethanethiol, and it presents the maximum adsorption capacity at pH 2 and pH 9. The authors explained that the maximum adsorption is favored by electrostatic interactions at pH 2 and by the formation of hydrogen bonds and/or covalent interactions with the adsorbent at pH 9.

Another study that verified the influence of this parameter is reference [49], which also studied the adsorption of reactive red dye. This work was conducted to modify the cellulose with incorporation of quaternary ammonium groups, with the aim of adding positive charges on their surface and thus improve the adsorption capacity.

It was observed that this material showed maximum adsorption capacity at pH 3. Considering that the dye molecules used in their studies have negative charges, the increase for dye removed at lower pH values than 6 is explained by favoring attraction electrostatic between dye and this adsorbent medium, cell-R-N^+ $(C_2H_5)_3$ SO_3^-, but at pH values higher than 6, a decrease in the removal of dye different from the work described in reference [12] was noted. Due to the possibility of covalent interactions, they described good results at pH 9.

2.3. Contact time

Twelve of the 23 articles found in the prospection evaluated the influence of this parameter in its work, showing the importance of this parameter in the study of the adsorption process. By studying the influence of the adsorption contact time, it is possible to establish the adsorption equilibrium, which is reached when the adsorption capacity of the material remains constant over time. This study provides further data to establish the removal rate of the solute solution and how the adsorption takes place, making use of the kinetic studies that assess the theoretical model which best fit the experimental data.

The theoretical models for the kinetic study further reported in the literature are the pseudo-first order [57], the pseudo-second order [58], the intraparticle diffusion [59], and the Elovich equation [60].

In most of the articles investigated in the literature, it was observed that experimental data usually adjust better to the pseudo-second order kinetic model because the adsorption process is dependent on both the concentration of the adsorbate with the strength of the adsorbent [48, 52, 12, 51, 61, 62, 63].

An example of this is the work described in reference [29], in which the influence of contact time in the reactive adsorption of yellow dyes 4RFN B-B and reactive blue-NB in its hybrid material of cellulose showed that data fitted to the pseudo-second-order model. The authors of reference [51] also used this model to explain the adsorption of the dye malachite green and basic fuchsin since the experimental data showed a better linear correlation with the parameters of the pseudo-second-order model when compared to pseudo-first-order model.

The authors of reference [46] applied the mathematical models already mentioned and also used the kinetic model Bangham [64], but with a different objective from the other authors cited, as they have not sought the model with which the experimental data better adjusted but a comparison between linear and nonlinear form for each model. The results set is the best nonlinear models, based on the values of the statistical parameters.

2.4. Temperature

Studies on the temperature dependence of the dye adsorption process allow obtaining the experimental data enthalpy and entropy related to these processes. Among the evaluated articles, 12 related the study of the influence of temperature on the adsorption process.

The authors of reference [40] made use of this variable to understand the factors that control the adsorption process evaluating 15 different acid dyes cellulose adsorption. The authors observed the enthalpy of adsorption controls this process for most of dyes and came to the conclusion that the enthalpy of adsorption is controlled by the interaction of dye with the surface of material, i.e., a high enthalpy of adsorption will be observed the higher the approximation (permitted by the structure) of the aromatic nucleus of the dye to the material surface.

Given the dependence of the temperature in a range of 30°C to 70°C, the removal of malachite green dye methyl orange using brinjal plant root powder was evaluated in the paper [52]. For the first dye, the authors have observed an increase in the removal capacity by increasing the temperature, while for the second dye, the opposite trend was observed from 40°C. To explain these trends, the authors make use of the type of interaction between dyes and adsorbent.

The removal of the malachite green is explained by the interaction of the type of Lewis acid–base, thus increasing the temperature and the interaction between the charges of the basic sites of the material and the dye acids. However, in the adsorption of the anionic dye methyl orange, there are only interactions of the type of hydrogen bonds that can be broken easily at elevated temperatures according to the authors.

2.5. Dosage

In this study, 3 of the 23 articles found evaluated the dosage of the adsorbent. Despite the small number of studies among evaluated articles, the dosage of the adsorbent is another very important parameter because once the degree of the adsorption of a material under study is known, it is possible to use the optimal dosage of this adsorbent to determine, for example, the cost of the adsorbent per unit volume.

However, increasing the mass of adsorbent increases the amount of adsorption sites available for the adsorbent–adsorbate interactions on a qualitative analysis, resulting in the increased percentage of dye removal solution [65].

The paper of reference [46] confirmed this fact evaluating cellulose hydrogel dosage in the removal of basic fuchsin dye, finding a good dose of 1.0 g L^{-1}; however, when the dose increased to 2.0 g L^{-1}, the removal percentage continued to increase while the q_e value (mg g^{-1}) decreased.

In several reports, this factor is explained by the availability of sites for adsorption in the adsorbent because it is found to present limited fixed dye concentration and lower dosage, and therefore, each adsorbent achieves its maximum capacity. By contrast, at high dosages, the articles of adsorbent compete to bind the same dye molecules, thus providing a quantitative measure of the amount of adsorbed dye per unit weight of adsorbent decreasing [46, 62, 63].

2.6. Desorption

The possibility of recovery and reuse of biosorbent after going through a desorption process is another relevant factor. Five articles presented results in this parameter; two of them are described below.

The authors of reference [51] evaluated the life cycle of a new material obtained by the modification of cellulose with glycidyl methacrylate and diethylenetriaminepentaacetic acid, observing that, in their dynamic tests adsorption/desorption using a saturated solution of sodium bicarbonate as eluant, the adsorption capacity did not change significantly, showing that the material can be reused in at least four cycles, maintaining the adsorption rate above 85% and 90% for green and basic fuchsin dye malachite, respectively.

Another work that performed desorption testing is the paper of reference [53], in which we evaluated the life cycle of magnetic cellulose beads using NaCl solutions (2 mol L^{-1}) and NaOH (0.05 mol L^{-1}) and found that after three cycles of adsorption, the desorption was maintained at 95% for the dyes methylene blue and methyl orange, on this fact found that the beads can be reused several times.

The paper [52] also conducted a study of desorption, however, for the malachite green dye and methyl orange on the surface of brinjal plant root powder, and by just using water at 80°C, a recovery of 95% for methyl orange and 10% for malachite green was obtained.

The adsorption process involves several attractive interaction forces such as van der Waals forces, hydrogen bonding, covalent, and ionic bonds. Depending on the dye used and the sites available for adsorption on the material, one or more forces will act in the adsorption process, which influences the possibility to reuse the material.

3. Conclusion

Through the presented data, it can be stated that the use of cellulosic materials for the adsorption of dyes it is a promising area because of the number of publications found.

The data presented in the patent deposit prospecting shows the database with largest number of deposit among databases searched was EPO, with 12,815 patent deposits in studied range. Even though only a single patent occurs from the refined search referring to the aim of this study, it results in methods for improving fixation of the dye on cellulose fiber.

In the search for articles, a larger number of searches with limiting terms compared to prospecting patents were observed, in which China published more in this area. Other important aspects refer to the year in which most studies that have been done in this area, which was observed that the year 2013, had the highest number of publications showing currently of this theme.

The 23 items found in the Scopus database, using the limiting terms chosen, elucidate the adsorption of one or more dyes on the surface of cellulosic materials. By analyzing these articles, it can be proved that the factors influencing this process are pH, temperature, dosage, contact time, kind of material and modification. These parameters are of most importance in adsorption studies as well as in directly influencing the maximum adsorption capacity for a given material. Together they make an important tool in the elucidation of the possible interactions between adsorbent and adsorbate, helping to explain the potentiality of the material under study.

Author details

Fabrícia C. Silva[1], Luciano C.B. Lima[1], Roosevelt D.S. Bezerra[2], Josy A. Osajima[1] and Edson C. Silva Filho[1*]

*Address all correspondence to: edsonfilho@ufpi.edu.br

1 LIMAV, Chemistry Department of Piauí Federal University, Teresina, Piauí, Brazil

2 Federal Institute of Education, Science and Technology of Piaui, Campus Teresina Central, Teresina, Brazil

References

[1] Cardoso NF, Lima EC, Pinto IS, Amavisca CV, Royer B, Pinto RB, Alencar SA, Pereira SFP. Application of cupuassu shell as biosorbent for the removal of textile dyes from aqueous solution. Journal of Environmental Management. 2011;92:1237–1247. DOI:10.1016/j.jenvman.2010.12.010.

[2] Kunz A, Mansilla H, Duran N. A degradation and toxicity study three textile reactive dyes by ozone. Environmental Technology. 2002;23: 911–918. DOI: 10.1080/09593332308618358.

[3] Attia AA, Rashwan WE, Khedr SA. Capacity of activated carbon in the removal acid dyes subsequent to its thermal treatment. Dyes Pigmentes. 2006;69:128–136. DOI: 10.1016/j.dyepig.2004.07.009.

[4] Zhou Y, Fu S, Liu H, Yang S, Zhan H. Removal of methylene blue dyes from wastewater using cellulose-based superadsorbenthydrogels. Polymer Engineering and Science. 2011;51:2417–2424. DOI: 10.1002/pen.22020.

[5] Srinivasan A, Viraraghavan T. Decolorization of dye wastewaters by biosorbents: a review. Journal of Environmental Management. 2010;91:1915–1929. DOI: 10.1016/j.jenvman.2010.05.003.

[6] Sanghi R, Verma P. Decolorisation of aqueous dye solutions by low-cost adsorbents: a review. Coloration Technology. 2013;129:85–108. DOI: 10.1111/cote.12019.

[7] Şuteu D, Nacu A, Cristian G. Removal of methyl violet triphenylmethanic dye using ion exchange celluloses. Cellulose Chemistry and Technology. 2001;35:451–457.

[8] Ali M, Sarkar A, Pandey MD, Pandey S. Efficient precipitation of dyes from dilute aqueous solutions of ionic liquids. Analytical Sciences. 2006;22:1051–1053. DOI: 10.2116/analsci.22.1051.

[9] Sharma RK, Gulati S, Pandey A, Adholeya A. Novel, efficient and recyclable silica based organic-inorganic hybrid nickel catalyst for degradation of dye pollutants in a newly designed chemical reactor. Applied Catalysis B: Environmental. 2012;125:247–258. DOI: 10.1016/j.apcatb.2012.05.046.

[10] Abid MF, Zablouk MA, Abid-Alameer AM. Experimental study of dye removal from industrial wastewater by membrane technologies of reverse osmosis and nanofiltration. Journal of Environmental Health Science and Engineering.2012;9. DOI: 10.1186/1735-2746-9-17.

[11] Harriman A, Shoute LCT, Neta P. Radiation chemistry of cyanine dyes: oxidation and reduction of merocyanine 540. Journal of Physical Chemistry. 1991;95:2415–2420. DOI: 10.1021/j100159a055.

[12] Silva LS, Lima LCB, Silva FC, Matos JME, Santos MRMC, Santos Júnior LS, Sousa KS, Silva Filho EC. Dye anionic sorption in aqueous solution onto a cellulose surface chemically modified with aminoethanethiol. Chemical Engineering Journal. 2013;218:89–98. DOI: 10.1016/j.cej.2012.11.118.

[13] Ngah WSW, Teong LC, Hanafiah MAKM. Adsorption of dyes and heavy metal ions by chitosan composites: a review. Carbohydrate Polymers. 2001;83:1446–1456. DOI: 10.1016/j.carbpol.2010.11.004.

[14] Faisal I. Removal of dyes from polluted water by adsorption on maize cob. Iraqi. Journal of Chemical and Petroleum Engineering. 2010;11:55–57.

[15] Zhang Z, O'Hara LM, Kent GA, Doherty WOS. Comparative study on adsorption of two cationic dyes by milled sugarcane bagasse. Industrial Crops and Products. 2013;42:41–49. DOI: 10.1016/j.indcrop.2012.05.008.

[16] Mane RS, Bhusari VN. Removal of colour (dyes) from textile effluent by adsorption using orange and banana peel. International Journal of Engineering Research. 2012;2:1997–2004.

[17] Garg VK, Gupta R, Yadav AB, Kumar R. Dye removal from aqueous solution by adsorption on treated sawdust. Bioresource Technology. 2003;89:121–124. DOI: 10.1016/S0960-8524(03)00058-0.

[18] Bulut Y, Aydin H. A kinetics and thermodynamics study of methylene blue adsorption on wheat shells. Desalination. 2006;194: 259–267. DOI: 10.1016/j.desal.2005.10.032.

[19] Nemr AE, Abdelwahab O, El-Sikaily A, Khaled A. Removal of direct blue-86 from aqueous solution by new activated carbon developed from orange peel. Journal of Hazardous Materials. 2009;161:102–110. DOI: 10.1016/j.jhazmat.2008.03.060.

[20] Low KS, Lee CK, Tan KK. Biosorption of basic dyes by water hyacinth roots. Bioresource Technology. 1995;52:79–83. DOI: 10.1016/0960-8524(95)00007-2.

[21] Gong R, Ding Y, Li M, Yang C, Liu H, Sun Y. Utilization of powdered peanut hull as biosorbent for removal of anionic dyes from aqueous solution. Dyes Pigments. 2005;64:187–192. DOI: 10.1016/j.dyepig.2004.05.005.

[22] Robinson T, Mcmullan G, Marchant R, Nigam P. Remediation of dyes in textile effluent: a critical review on current treatment technologies with a proposed alternative. Bioresource Technology. 2001;77:247–255. DOI: 10.1016/S0960-8524(00)00080-8.

[23] Shih MC. Kinetics of the batch adsorption of methylene blue from aqueous solutions onto rice husk: effect of acid-modified process and dye concentration. Desalination and Water Treatment. 2012;37:200–214. DOI: 10.1080/19443994.2012.661273.

[24] Balathanigaimani M, Shim W, Park K, Lee J, Moon H. Effects of structural and surface energetic heterogeneity properties of novel corn grain-based activated carbons on dye adsorption. Microporous and Mesoporous Materials. 2009;118:232–238. DOI: 10.1016/j.micromeso.2008.08.028.

[25] Kavitha D, Namasivayam C. Experimental and kinetic studies on methylene blue adsorption by coir pith carbon. Bioresource Technology. 2007;98:14–21. DOI: 10.1016/j.biortech.2005.12.008.

[26] Robinson T, Chandran Nigam P. Removal of dyes from a synthetic textile dye effluent by biosorption on apple pomace and wheat straw. Water Research. 2002;36:2824–2810. DOI: 10.1016/S0043-1354(01)00521-8.

[27] Silva Filho EC, Lima LCB, Silva FC, Sousa KS, Fonseca MG, Santana SAA. Immobilization of ethylene sulfide in aminated cellulose for removal of the divalent cations. Carbohydrate Polymers. 2013;92:1203–1210. DOI: 10.1016/j.carbpol.2012.10.031.

[28] Lima IS, Airoldi C. A thermodynamic investigation on chitosan—divalent cation interactions. Thermochimica Acta. 2004;421:133–139. DOI: 10.1016/j.tca.2004.03.012.

[29] Xie K, Zhao W, He X. Adsorption properties of nano-cellulose hybrid containing polyhedral oligomeric silsesquioxane and removal of reactive dyes from aqueous solution. Carbohydrate Polymers. 2011;83:1516–1520. DOI: 10.1016/j.carbpol.2010.09.064.

[30] Xie K, Hou A, Chen Y. The mathematical model of dye diffusion and adsorption on modified cellulose with triazine derivatives containing cationic and anionic groups. Journal of Physics: Conference Series. 2008;96. DOI: 10.1088/1742-6596/96/1/012127. DOI: 10.1088/1742-6596/96/1/012127.

[31] Kobayashi S, Sakamoto J, Kimura S. In vitro synthesis of cellulose and related polysaccharides. Progress in Polymer Science. 2001;26:1525–1560. DOI: 10.1016/S0079-6700(01)00026-0.

[32] STAUDINGER H. Über polymerization. Berichte der deutschenchemischen Gesellschaft. 1920;53:1073–1085. DOI: 10.1002/cber.19200530627.

[33] Klemm D, Heublein B, Fink H-P, Bohn A. Cellulose: fascinating biopolymer and sustainable raw material. Angewandte Chemie International Edition. 2005;44:3358–3393. DOI: 10.1002/anie.200460587.

[34] Manfred H. Dyeing cellulose fibres, e.g. cotton fabric, with reactive vinylsulfone dyes, involves adding dye-bath components in a special sequence to optimise the relation between physical adsorption and chemical reactivity. 2009. DE102008026403.

[35] Xu C, Tang W, Du J.A nonlinear isotherm model for sorption of anionic dyes on cellulose fibers: a case study. Carbohydrate Polymers. 2014;102: 808–812. DOI: 10.1016/j.carbpol.2013.10.086.

[36] Široký J, Blackburna RS, Bechtold T, Taylor J, White P. Alkali treatment of cellulose II fibres and effect on dye sorption. Carbohydrate Polymers. 2011;84:299–307. DOI: 10.1016/j.carbpol.2010.11.037.

[37] Goswami P, Blackburn RS, Taylor J, White P. Sorption of dyes on cellulose II: effect of alkali treatmentoffibre and dye structure. Cellulose. 2011;18:1063–1072. DOI 10.1007/s10570-011-9540-0.

[38] Tabara A, Yamane C, Abe M, Seguchi M. Adsorption of xanthene food additive dyes to cellulose granules. Cellulose. 2011;18:45–55. DOI 10.1007/s10570-010-9462-2.

[39] Tabara A, Yamane C, Seguchi M. Adsorption mechanism for xanthene dyes to cellulose granules. Bioscience, Biotechnology, and Biochemistry. 2012;76:870–874. DOI: 10.1271/bbb.110634.

[40] Bird J, Brough N, Dixon S, Batchelor SN. Understanding adsorption phenomena: investigation of the dye-cellulose interaction. Journal of Physical Chemistry B. 2006;110:19557–19561. DOI: 10.1021/jp063730n.

[41] Simu G, Funar-Timofei S, Hora S, Kurunczi L. Experimental and theoretical study of the adsorption of a trisazo direct dye derived from 4,4-diaminobenzanilide on a cellulose substrate. Molecular Crystals and Liquid Crystals. 2004;416:353–360. DOI: 10.1080/15421400490478236. DOI: 10.1080/15421400490478236.

[42] Zhang G, Yi L, Deng H, Sun P. Dyes adsorption using a synthetic carboxymethyl cellulose-acrylic acid adsorbent. Journal of Environmental Sciences 2014;26:1203–1211. DOI: 10.1016/S1001-0742(13)60513-6.

[43] Zhou C, Wu Q, Lei T, Negulescu II. Adsorption kinetic and equilibrium studies for methylene blue dye by partially hydrolyzed polyacrylamide/cellulose nanocrystal nanocomposite hydrogels. Chemical Engineering Journal. 2014;251:17–24. DOI: 10.1016/j.cej.2014.04.034.

[44] Hashemian S, Shayegan J. A Comparative study of cellulose agricultural wastes (almond shell, pistachio shell, walnut shell, tea waste and orange peel) for adsorption of violet B dye from aqueous solutions. Oriental Journal of Chemistry. 2014;30:2091–2098. DOI: 10.13005/ojc/300478.

[45] Yang F, Song X, Yan L, Wei W. Photocatalysis degradation of azo dye using nano-TiO_2-coated porous cellulose gel:enhancement by adsorption and its self-clean characteristic. Micro & Nano Letters. 2014;9:193–197 DOI: 10.1049/mnl.2013.0691.

[46] Bhattacharyya, R, Ray SK. Kinetic and Equilibrium modeling for adsorption of textile dyes in aqueous solutions by carboxymethylcellulose/poly(acrylamide-co-hydroxyethylmethacrylate) semi-interpenetrating network hydrogel. Polymer Engineering and Science. 2013;53:2439–2453. DOI: 10.1002/pen.23501.

[47] Wang M.-M, Wang L. Synthesis and characterization of carboxymethyl cellulose/organic montmorillonite nanocomposites and its adsorption behavior for Congo red dye. Water Science and Engineering. 2013;6:272–282. DOI: 10.3882/j.issn.1674-2370.2013.03.004.

[48] Ashjarana A, Hajilarib M. Adsorption anionic dye on bacterial and plant cellulose: thermodynamic and kinetic study. Research Journal of Pharmaceutical, Biological and Chemical Sciences. 2013;4:985–994.

[49] Wang L, Li J. Adsorption of C.I. Reactive Red 228 dye from aqueous solution by modified cellulose from flax shive: kinetics, equilibrium, and thermodynamics. Industrial Crops and Products. 2013;42:153–158. DOI: 10.1016/j.indcrop.2012.05.031.

[50] Pei A, Butchosa N, Berglundab LA, Zhou Q. Surface quaternized cellulose nanofibrils with high water absorbency and adsorption capacity for anionic dyes. Soft Matter. 2013;9:2047–2055. DOI: 10.1039/C2SM27344F.

[51] Zhou Y, Zhang M, Hu X, Wang X, Niu J, Ma T. Adsorption of cationic dyes on a cellulose-based multicarboxyladsorbent. Journal of Chemical and Engineering Data. 2013;58:413–421. DOI: 10.1021/je301140c.

[52] Buvaneswaria N, Kannanb C. Plant toxic and non-toxic nature of organic dyes through adsorption mechanism on cellulose surface. Journal of Hazardous Materials. 2011;189:294–300. DOI: 10.1016/j.jhazmat.2011.02.036.

[53] Luo X, Zhang L. High effective adsorption of organic dyes on magnetic cellulose beads entrapping activated carbon. Journal of Hazardous Materials. 2009;171:340–347. DOI: 10.1016/j.jhazmat.2009.06.009.

[54] Jal PK, Patel S, Mishra BK. Chemical modification of silica surface by immobilization of functional groups for extractive concentration of metal ions. Talanta. 2004;62:1005–1028. DOI: 10.1016/j.talanta.2003.10.028.

[55] Martin AI, Sánchez-Chaves M, Arranz F. Synthesis, characterization and controlled release behavior of adducts from chloroacetylated cellulose and α-naphthylacetic acid. Reactive and Functional Polymers. 1999;39:179–187. DOI: 10.1016/S1381-5148(97)00180-6.

[56] Kumar R, Barakat MA. Decolourization of hazardous brilliant green from aqueous solution using binary oxidized cactus fruit peel. Chemical Engineering Journal. 2013;226:377–383. DOI: 10.1016/j.cej.2013.04.063.

[57] Lagergren S. Zurtheorie der sogenannten adsorption gelosterstoffe. Kungliga Svenska Vetenkapsakademiens Handlingar. 1898;24:1–39.

[58] Ho YS, Mckay G. Pseudo-second order model for sorption processes. Process Biochemistry. 1999;34:451–465. DOI: 10.1016/S0032-9592(98)00112-5.

[59] Weber Jr WJ, Morris JC. Kinetics of adsorption on carbon from solution. Journal of Sanitary Engineering Division. 1963;89:31–59.

[60] Aharoni C, Tompkins FC. Kinetics of adsorption and desorption and the Elovich equation. Advances in Catalysis and Related Subjects. 1970;21:1–49.

[61] Rehman MSU, Munir M, Ashfaq M, Rashid N, Nazar MF, Danish M, Han J. Adsorption of brilliant green dye from aqueous solution onto red clay. Chemical Engineering Journal. 2013;228:54–62. DOI: 10.1016/j.cej.2013.04.094.

[62] Argun ME, Güclü D, Karatas M. Adsorption of reactive blue 114 dye by using a new adsorbent: pomelo peel. Journal of Industrial and Engineering Chemistry. 2013;20:1079–1084.DOi:10.1016/j.jiec.2013.06.045.

[63] Tavlieva MP, Genieva SD, Georgieva VG, Vlaev LT Kinetic study of brilliant green adsorption from aqueous solution ontowhite rice husk ash. Journal of Colloid and Interface Science. 2013;409:112–122. DOI: 10.1016/j.jcis.2013.07.052.

[64] Liu Y, Liu Y-J. Biosorption isotherms, kinetics and thermodynamics. Separation and Purification Technology. 2008;61:229–242. DOI: 10.1016/j.seppur.2007.10.002.

[65] Ofomaja AE, Ho YS. Equilibrium sorption of anionic dye from aqueous solution by palm kernel fibre as sorbent. Dyes and Pigments. 2007;74:60–66. DOI: 10.1016/j.dyepig.2006.01.014.

Chapter 6

Microbial Cellulose — Biosynthesis Mechanisms and Medical Applications

Wilton R. Lustri, Hélida Gomes de Oliveira Barud, Hernane da Silva Barud, Maristela F. S. Peres, Junkal Gutierrez, Agnieszka Tercjak, Osmir Batista de Oliveira Junior and Sidney José Lima Ribeiro

Additional information is available at the end of the chapter

http://dx.doi.org/10.5772/61797

Abstract

Currently some principles of sustainability, eco-efficiency and green chemistry are guiding the development of a new generation of materials as an alternative to conventional polymers based on petroleum. Then, in the field of biodegradable polymers one of the most promising investigations is focused on the use of microbial cellulose (MC), biocellulose or bacterial cellulose. MC has received substantial interest since it is synthesized from the bacterium *Gluconacetobacter genus* from a variety of carbon sources such as glucose, fructose, galactose, etc. MC is an interesting emerging biomaterial, with no toxicity, and since its discovery has shown tremendous potential in various fields, because the structural aspect of MC is far superior to those of plant cellulose. Thus, the main focus of the chapter review involves detailed aspects about the biosynthesis and recent advances on microbial production, including mechanism for the biochemistry of the cellulose synthesis, new sources for culture medium, main aspects about static and air-reactor productions and genetic modifications. We also revised microbial cellulose devices for biomedical applications: artificial skin, artificial blood vessels and microvessels, wound dressing of second- or third-degree burn ulcers, scaffolds for tissue engineering, drug delivery systems, dental implants, among others.

Keywords: Microbial cellulose, cellulose synthesis, medical applications

1. Introduction

Microbial cellulose (MC) presents the same chemical formula as plant cellulose, however with the fibers in nanometer dimensions; confer different properties to MC [1]. The MC is a type of exopolysaccharides composed of glucose monomers bound by glycosidic β (1-4) linkages, with the chemical formula $(C_6H_{10}O_5)_n$, as can be seen in Figure 1. [2,3]. This biopolymer is produced extracellularly into nanofibers by several genera of bacteria, such as *Gluconacetobacter,* (formerly *Acetobacter*), *Agrobacterium, Aerobacter, Achromobacter, Azotobacter, Rhizobium, Sarcina,* and *Salmonella* [4].

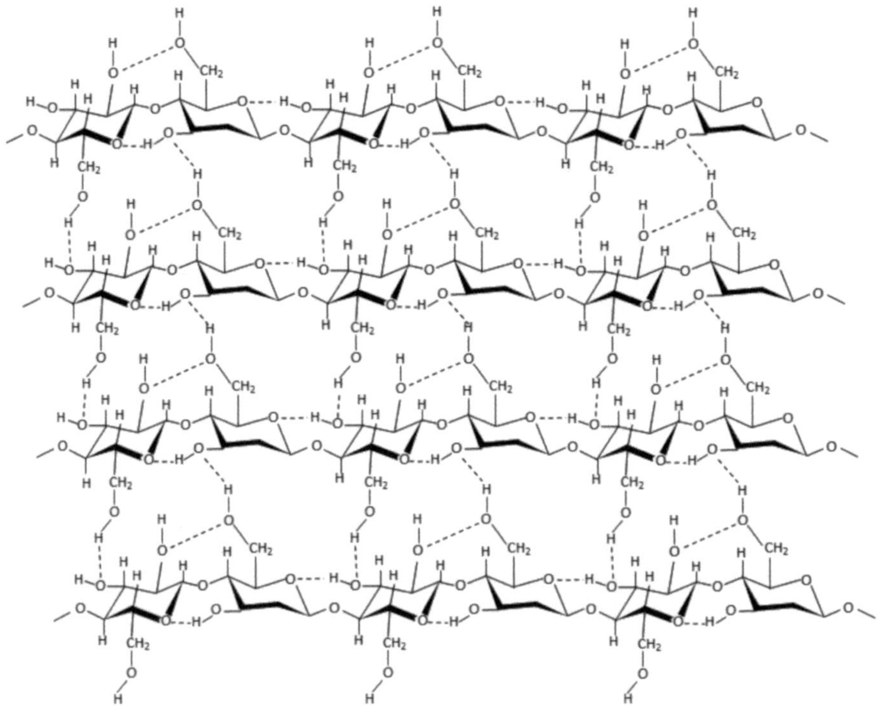

Figure 1. Chemical structure of microbial cellulose.

Historical data show that the MC has been used for a long time in the manufacture of a traditional food in the Philippines, known as the coconut cream [5]. Currently, the MC still remains widely utilized as food in various parts of the world however on the other hand aroused great academic and industrial interest due to its unique properties and diverse opportunity of applications. MC produced by *Acetobacter xylinum* was first reported in 1886 by Brown [6], produced in the presence of oxygen, using glucose as a carbon source.

As proposed by Yamada and colleagues (1997) [7] and subsequently validated by International Journal of Systematic Bacteriology, *Acetobacter xylinum* was reclassified and scientifically cataloged as *Gluconacetobacter xylinus,* due to the characteristics phylogeny based on analysis of partial sequences of 16S ribosomal RNA.

This bacterial species belonging to the family Acetobacteriaceae, being morphologically classified as a bacillus Gran-negative, strictly aerobic, no pathogenic which may be found singly arranged, in pairs or in small sets of chain formation of colonies shiny and smooth in

Mannitol Agar. Bacteria belonging to this family are able to oxidize fully various carbon sources such as glucose, fructose, galactose, sucrose, mannitol, glycerol, inositol, among others [3,8] and alcohols such as ethanol [7] and is capable of extracellularly producing pulp at temperatures between 25 and 30 °C and pH 3 to 7. The bacterial cellulose may provide arrangements in parallel via hydrogen bonds and to form a tridimensional network. The morphology of the membrane depends directly on the environment and the interface culture medium / air where MC form a thick film, which can be easily manipulated according to the size of the vial used for cultivation [1,8].

Researchers have sought new bacterial strains capable to produce biopolymers with potential industrial application [9]. Although many organisms are capable to produce cellulose, *Gluconacetobacter xylinus* bacteria are the only known species able to produce cellulose on the industrial scale [10].

In addition *G. xylinus*, other micro-organisms are considered able to produce cellulose, among others they are *Escherichia coli*, *Salmonella spp.* [11] and *Pseudomonas spp.* [12,13]. The cellulose synthesis genes (*MCsA, MCsB, MCsZ and MCsC*) of these species were similar to those in *G. xylinus* [14]. Although these species are also capable to produce the bacterial cellulose, the fact that many of them are potentially pathogenic limits the commercial use of these biopolymers [12,13].

In nature, microorganisms which produce cellulose are usually found in symbiosis with other microorganisms. In the fermentation of Kombucha, for example, *Zygosaccharomyces* yeasts are used as microorganisms symbionts [15]. The Kombucha, also known as tea fungi and *Haipao* [16], is produced in China for over 2000 years, with a widely varied yeast population [17,18], whose function is to convert sucrose to organic acids, carbon dioxide and ethanol, the latter being used for the cellulose-producing bacteria for the production of acetaldehyde and acetic acid [19,20].

2. Biochemical and molecular mechanisms of bacterial cellulose biosynthesis by *Gluconacetobacter xylinus*

Although several species of microorganisms are capable to produce cellulose, *G. xylinus* is currently considered a model organism for the study these biopolymers [21]. MC biosynthesis consists of a complex process which involves first the polymerization of glucose residues in β1-4-glucan chain [21,22,23], followed by the extracellular secretion of the chains ending the linear arrangement and crystallization of glucan chains through hydrogen bridges and Van der Waals forces hierarchically arranged in strips [22], resulting in formation of a tough three dimensional structure called microfibrils. MC generated by this bacterial species has special characteristics of unidirectional polarity and variable thickness. The crystallization mechanism of the microfibrils in *G. xylinum* can give rise to two cellulose forms, if the microfibrils is oriented parallel arrangement is synthesized cellulose I, while if the arrangement is antiparallel microfibrils is obtained cellulose II [21]. In *G. xylinus*, MC synthesis depends on the cycle of pentoses and of the Krebs cycle [1,21,22] that perform respectively the oxidation function carbohydrates and oxidation of organic acids. A particularity *G. xylinus* is the inability to metabolize glucose anaerobically due to lack of phosphofructokinase-1, an enzyme responsible for catalyzing the reaction of phosphorylation of fructose-6-phosphate, fructose-1,6-bisphosphate, which prevents glycolysis. Thus, MC synthesis by *G. xylinus* results of a metabolic pool hexose phosphate which is produced directly by phosphorylating exogenous hexose or indirectly by the pentose phosphate pathway and gluconeogenesis. The hexose phosphate conversion of cellulose is direct and does not depend on the intermediate divisions carbon skeleton [1,22]. The conversion of glucose, transported from the external environment into the cytoplasm, is catalyzed by four bacterial enzymes, the glucokinase, which is the enzyme responsible for the phosphorylation of the carbon 6 of glucose, yielding glucose-6-phosphate, the phosphoglucomutase, which catalyzes the reaction isomerization of glucose-6-phosphate to glucose-1-phosphate, the UDPG-pyrophosphorylase (also known as glucose-1-phosphate uridylyltransferase), responsible for synthesis of UDP-glucose (UDPG), and cellulose synthase (CS), responsible for the polymerization of cellulose from UDP-glucose. As previously mentioned, the cellulose synthesis can also occur from endogenous sources, for gluconeogenesis. In *G. xylinus*, the synthesis from endogenous sources begins with oxaloacetate, into pyruvate by action of the enzyme pyruvate carboxylase. The transformation of the pyruvate in fosfoenolpiruvado, is produced by action of the enzyme phosphoenolpyruvate carboxykinase [1,22,24]. MC synthesis reaction is costly to the cell, consuming about 10% of the ATP generated in bacterial metabolism. Thus, the energy used for the synthesis of CB comes from aerobic metabolism. There are different proposals for the substrate used by the CS. [1,21]. The enzymatic complex of synthesis of cellulose, termed as terminal complex (TC) [25,26], constitute a kind of membrane protein species that, in *Gluconacetobacter spp.*, corresponds to cellulose synthase complex [27].

One proposed hypothesis is that the UDP-glucose binds to lipids of the plasma membrane [1, 21]. Another one considers that the soluble precursor interacts directly with the CS [28]. The CS is a protein complex consisting of three (AxCcSAB, AxCcSC and AxCcSD) or four (AxCcSA, AxCcSB, AxCcSC and AxCcSD) protein subunits encoded by genes exist in an operon chromosomal called *MCs*.

The two conserved Asp residues (D) invariably are found in loops at the Carboxyl-terminal (C-terminal) ends of predicted strands, a position frequently observed for catalytic residues [21, 39]. The hydrophobic clusters in domain B are more difficult to interpret in terms of secondary structure. AxCcSA and AxCcSAB have a motif consisting of domain, a single conserved residues Asp (D-D-D), presumably important for catalysis, identified along with the conserved sequence motif Gln (Q) Arg-Trp (R-W) in glucotransferases [1, 29]. Through a functional analysis of CS, it appears that the A subunit of this complex with 83kDa, shows catalytic activity. The B subunit of 90kDa, increases the rate of cellulose synthesis by joining a positive allosteric regulator, cyclic diguanosine monophosphate (c-di-GMP). The C subunit (138kDa) and D (17kDa) appear to structural activity. It has been hypothesized that C subunit related to pore formation and extrusion of the cellulose D subunit appears related to the process decrystalization since mutant strains of the gene which encodes the D subunit production are still able to produce cellulose II [23].

http://dx.doi.org/10.5772/61797

In *G. xylinus* Bis-(3'-5')-cyclic dimeric guanosine monophosphate (c-di-GMP) has been identified as an activator of cellulose biosynthesis [30], considered a second global messenger in bacteria [31]. The free c-di-GMP in the cell is considered to allosterically activate the cellulose synthase BcsA. However, 90% of the cellular c-di-GMP is reversibly bound by the c-di-GMP binding protein BcsB, a membrane protein that is structurally associated with the cellulose synthase [32,33]. It is believed that the spatial proximity is necessary to direct c-di-GMP released from BcsB towards the cellulose synthase. The equilibrium between bound and free c-di-GMP is modulated by the intracellular potassium concentration [34]. The level of free c-di-GMP is regulated by the opposing action of two enzymes, diguanylate cyclase (DGC) that cycles two molecules of GTP under the release of two molecules of PPi, and phosphodiesterase A (PDEA) that degrades c-di-GMP to the inactive GTP under the release of two molecules of PPi, and phosphodiesterase A (PDEA) that degrades c-di-GMP to the inactive 5′-pGpG. *G. xylinus* has three distinct operons each containing PDEA/DGC pair, which contribute at different levels to the c-di-GMP turnover [35, 36] indicating that cellulose biosynthesis underlies various control mechanisms in *G. xylinus*. Figure 2 represents a model of the metabolic pathway for the biosynthesis of cellulose by *G. xylinus*.

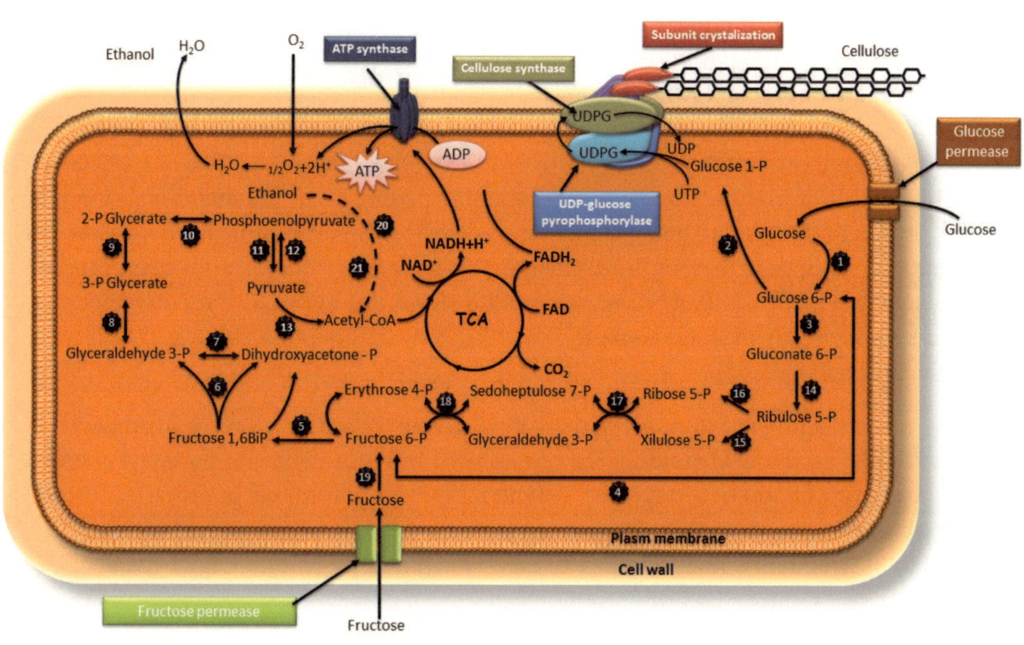

Figure 2. Hypothetical model of the pathway for the biosynthesis of cellulose by G. xylinus from exogenous sources - glucokinase-ATP (1); Phosphoglucomutase (2), glucose-6-phosphate dehydrogenase (3); Phosphoglucoisomerase (4); Fructokinase ATP (5), Aldolase (6); Triosephosphate isomerase (7); Glyceraldehyde 3-phosphate dehydrogenase (8); Phosphoglycerate mutase (9), enolase (10); Pyruvate kinase (11); Pyruvate biphosphate kinase (12), pyruvate dehydrogenase(13); 6-phosphogluconate dehydrogenase (14); Phosphorribulose epimeraase (15); Phosphorribulose isomerase (16); Transaketolase (17); Transaldolase (18); Fructokinase (19); Aldehyde dehydrogenase(20); Alcohol dehydrogenase(21).

3. Cellulose Synthase (CS)

Two bacterial cellulose synthase operons (*bcs*) [23] were identified in the analyzed genome, but only one (*bcs*I) is structurally complete. This operon is composed by seven genes encoding for enzymes endo-1,4-beta-glucanases, a homolog CMCax is a putative beta-glucosidase endoglucanase, CPC is a putative homologues four subunits CS: BCSA, BBRC, BCSC and BCSD and beta-glucosidase. Although the three endoglucanases have been identified [37] its exact function in MC biosynthesis is not well understood. Genomic Analysis showed that *bcs*D gene is conserved as part of the operon and has exactly the same length in all six strains. Interestingly, this gene encodes for CS subunit D, whose function is still speculative, although their crystal structure has been recently resolved.

In cells actively producing cellulose, approximately 50 cellulose-synthesizing multienzyme complexes are organized in a single row along the longitudinal axis of the bacterial rod whereby each complex secretes approx. 12 to 25 glucan chains which assemble into larger microfibrils at the site of synthesis. This so-called linear terminal complex can be visualized by electron microscopy using freeze fracture as 35 Å pores in the outer membrane or as pits when the outer leaflet is fractured away [1,33].

In *G. xylinus* specie the cellulose synthase complex (BCS) is a multicomponent protein complex encoded in an operon containing at least three genes, *bcs*A, *bcs*B, and *bcs*C, which encodes a transmembrane complex over the cytoplasmic and outer membrane whereby the cellulose synthase (BcsA) and the c-di-GMP binding protein (BcsB) are considered to be localized in the cytoplasmic membrane as shown in Figure 3. [32,33,39]. Cellulose synthesis and transport across the inner bacterial membrane is mediated by a complex of the membrane-integrated catalytic BcsA subunit (green) and the membrane-anchored, periplasmic BcsB domain (blue) and membrane-associated regions BcsB transmembrane anchor (blue). The glycosyltransferase domain is shown brown [40]. BcsC is predicted to form a β-barrel in the outer membrane, preceded by a large periplasmic domain containing tetratricopeptide repeats likely involved in complex assembly [41,42].

The gene corresponding to cellulose synthase BcsA is constituted by a long string that presents between 723-880 amino acid residues, as represented in Figure 3, being the most conserved gene of the operon *MCs* between species, although the amino-terminus portions (N-terminus) and carboxyl-terminus portions (C-terminus) is not so conserved, since the homology is not restricted to the frequently analyzed D, D, D35Q (R, Q) and RW motif, which spans domains A and B [39].

The BcsB protein [Figure 3], related to indirect interaction with c-di-GMP [33] is less well conserved among the species. However, direct comparisons of the MCsB proteins with CelB from *A. tumefaciens* and *R. leguminosarum* bv. *trifolii* revealed significant homology (~40% similarity) over the entire length of the proteins with several invariable residues. An alanine/proline rich domain is located at the N-terminus region of all proteins except *A. aeolicus*. One transmembrane domain located at the C-terminus portions has been predicted by various algorithms for all BcsB proteins [39].

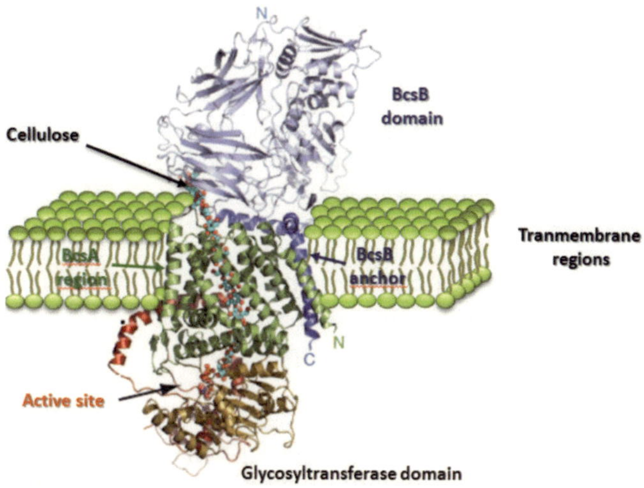

Figure 3. Schematic representation of cellulose synthase.

The proposed model considers that after the transfer and addition of glycosyl terminal residue, the glucose molecule rotates around acetyl glucan binding to align the channel as shown Figure 4. It is believed that allosteric interactions guide the direction of rotation, causing rotation feature 180^0 connecting β-1,4 glucan-glucan between the individual glucose units and intramolecular hydrogen bond between oxygen atoms of the hydroxyl groups of the neighboring unit [40,43]. This phenomenon may be sufficient to allow the polymer to move into the channel [Figure 4]. Alternatively, for the translocation of the elongated glucan occurs, replacement is required to UDP-glucose by UDP. Past the channel, induced glucan chain in the BcsA - twist BcssB interface, interaction with BcsB of CBDs (periplasmic carbohydrate binding domains), or aggregation with other glucans may additionally contribute to a unidirectional motion of the polymer [40].

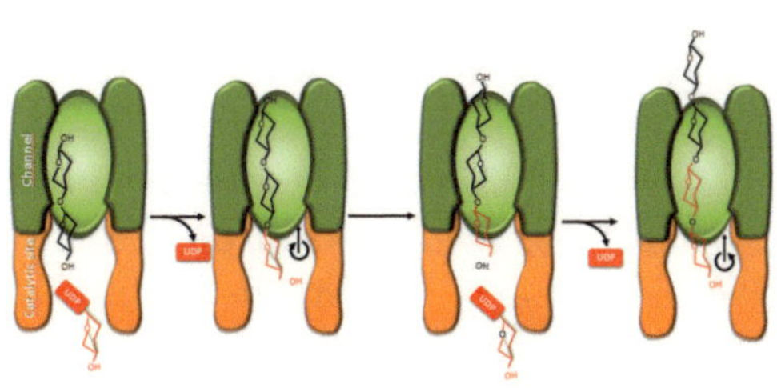

Figure 4. Proposed model for cellulose synthesis and translocation (adapted). After glycosyl transfer, the newly added Glc could rotate around the acetyl linkage into the plane of the polymer. The rotation direction would be determined by steric interactions and formation of the β-1,4 glucan characteristic intramolecular O_3- H••O_5 hydrogen bond. The glucan might translocate into the channel during this relaxation. This process would be repeated with a second UDP-Glc but the rotation direction after glycosyl transfer would be in the opposite direction owing to steric constraints. Alternatively, the glucan might not translocate into the channel until UDP is replaced by UDP-Glc. Trp 383 and Cys 318 mark the entrance to the transmembrane channel [40].

4. Cultivation conditions for production of bacterial cellulose

The MC production depends of the appropriate cultivation conditions, which include the composition of the culture medium (synthetic and natural media), temperature, pH and methods agitated or static cultivation. The choice of condition cultivation or another depend on the purpose, once these conditions have significant influence on the properties of structure, physical and mechanical MC.

The current methods of MC production are static culture [44], submerged fermentation through aerated or agitated cultivation [45], and the airlift bioreactor [46]. Large scale, semi-continuous and continuous fermentation are dominant to meet commercial demand. In all cases, the main objective is to achieve maximum production of MC with optimum form and suitable properties for the application for which it is intended. After all, a wider application of this versatile biopolymer depends on the practical considerations such as the scale-up capability and production costs. *G. xylinus* has two main operative amphibolic pathways: the pentose phosphate cycle for the oxidation of carbohydrates and the Krebs cycle for the oxidation of organic acids and related compounds [12,21,47,48]. Consequently, several studies have been reported that the composition of the culture medium and the fermentation conditions significantly affect the order structure of cellulose [49,50,51].

Static cultivation is a relatively simple and widely used method of cellulose production. The medium is placed into shallow tray or bottles, inoculated, and cultivated for several days until the cellulose nearly fills the tray. *G. xylinus* produces a gelatinous MC membrane, which has a denser surface on the side exposed to air, i.e. capable of generating cellulose as an extracellular product on static media (at the air-medium interface) at temperatures between 25 and 30 °C and pH from 4 to 7 [52]. The traditional static culture represents an expensive way of MC production that may hinder its industrial application since the productivity is low and long cultivation time is required. Consequently, authors have proposed new culture system as strategy to increase the MC productivity to a suitable for commercial applications in simple fed-batch [53], in bioreactor for a semi-continuous production [54], in a modified airlift-type bubble column bioreactor [55].

Nutrients required for the growth of these microorganisms are carbon and nitrogen sources, phosphorus, Sulphur, potassium and magnesium salts [56]. Sometimes a complex medium supplying amino acids and vitamins is also used to enhance the cell growth and production [57]. Between the years 1940 to 1960, researchers at the Hebrew University in Jerusalem intensively investigated the biochemistry of simplified production and quantification of cellulose cellulose for produced by *G. xylinus*. These media, named as HS are widely used nowadays [44]. Typical carbon sources in the production of microbial cellulose include glucose, fructose, sucrose, mannitol, however among others including arabinose, arabitol, citric acid, ethanol, ethylene glycol, diethylene glycol, galactose, glucono lactone, glycerol, inositol, lactose, malic acid, maltose, mannose, methanol, rhamnose, ribose, sorbose, starch, succharide, succinic acid, trehalose, and xylose have been also investigated [58,59,60] to maximize bacterial cellulose production by various *Gluconacetobacter* strains [8]. The best yield was obtained in fed-batch fermentation, 15.3 g/L in 50 hours of cultivation using glucose as carbon source [61] and other examples of different culture conditions can be found in Table 1. The results shown in Table 1 indicate that, glucose seems to be the best carbon source. It demonstrates the validity of the results that various carbon substrates could be converted to monomer glucose by *Gluconacetobacter*, followed by polymerization to MC [60]. Among the factors affecting cost, carbon sources play a major role in fermentation [47,62]. Conversion of 60–80% of the utilized carbon source into crude polymer is commonly found in high yielding polysaccharide fermentations. In order to compensate its low sugar conversion yield and to reduce the feedstock cost of MC production, in recent years, MC has been produced by fermenting the hydrolysates of agricultural wastes such as hemicelluloses [63], konjac powder [62], rice bark [64] and waste cotton fabrics [63,65]. An advantage of using agricultural or industrial residual streams as feedstock is the low or no value of the raw material. Several successful efforts have been made to use certain industrial food wastes as growth medium for the MC producer organisms, which is not only a cheap way but also works as a basin for environmental cleaning [51]. Thin stillage (TS) is a wastewater from rice wine distillery rich in carbon sources and organic acids. [66] discovered that TS, when employ to replace distilled water for preparing Hestrin and Schramm medium (the traditional MC production medium), can enhance the MC production 2.5-fold to a concentration of 10.38 g/L with a sugar-MC conversion yield of 57 % (0.57 g MC/g reducing sugar) after 7 days of static cultivation. In 2012, Ha et al. [48] further improved the MC production, 15.28 g/L of MC was obtained after 15 days of cultivation. Yeast extract and peptone are the most commonly used nitrogen sources in MC production as they provide nitrogen and growth factors for *Gluconacetobacter* strains. Many researchers are trying to find efficient substitutes due to their high cost. Even if various nitrogen sources were added to the HS medium, peptone is found to be the most effective nutrient. However, corn steep liquor (CSL) which produced the second highest production is always chosen as a substitution for the economic viewpoints. Buffering capacity is also important for MC production. Insoluble MC often attaches to pH probe and leads to inaccurate reading [4]. Noro et al. [67] pointed out the buffering capacity of CSL, which could maintain the pH within the optimal range during the production of MC. Jung et al. [68] doubled MC production (from 1.53 to 3.12 g/L) using molasses as carbon source and corn steep liquor as nitrogen source when compared with the results obtained from complex medium. This strategy could not only reduce burden on environment but also achieve the goal of large scale production with low cost. The optimal pH for MC production may vary with carbon source. *G. xylinus* accumulates gluconic acid at low pH, and a preferred environment for both biomass and MC production can be achieved by shifting pH from 4.0 to 5.5 during cellulose production phase in fed-batch cultures [61]. Under static batch cultivation, the pH of the culture medium decreases due to the respiratory metabolism of *G. xylinus*, which involves the ethanol oxidation to acetic acid and the glucose conversion into gluconic acid. This fact makes it very important to control the pH within the optimum range for cell growth and cellulose production [48,69]. *Gluconacetobacter* strains require oxygen as an essential substrate, consequently volumetric oxygen transfer coefficient (kLa) is a key limiting factor in the aerobic fermentation for producing MC. Song et al. [55] investigated the optimum aeration rate for a 50-L spherical type bubble column bioreactor, and it was determined to be 1.0 vvm (30 L/min).

Attempts to enhance MC production by adding different additives in the fermentation medium have been made. The possible mechanisms of these various additives to enhance MC were also proposed such as reduction of the shear force by increasing the viscosity of medium [70]. Different chemical compounds including alcohols [72], glycerol [60,68], organic acids [68], polysaccharides [72] thin stillage from rice wine distillery [66] and thin stillage from beer culture broth [47,53] have been used as additives to the fermentation medium with the aim of increasing MC production.

Microorganism	Carbon source	Cultivation mode	Time culture (days)	MC production ($g\ l^{-1}$)	Reference
G. xylinus (BRC 5)	Glucose	Fed-batch	2	15.3	Hwang *et al.* (1999) [61]
G. xylinus (BPR 2001)	Fructose	Agitated	3	14.1	Bae *et al.* (2004) [70]
G. xylinus (KJ1)	Saccharified food wastes	airlift-type bioreactor	3	5.6	Song et al. (2009)[55]
A. xylinum (ATCC 700178)	CSL-Fru	Agitated	5	13.0	Cheng *et al.* (2011) [72]
Gluconacetobacter sp (F6)	Glucose	Static	6	4.5	Jahan *et al.* (2012) [73]
G. xylinus (MCRC 12334)	TS-Glu	Static	7	10.38	Wu *et al.* (2012)[66]
G. xylinus and Trichoderma ressei	Glucose	Static	14	6.23	Cavka *et al.* (2013) [74]
G. xylinus (PTCC, 1734)	Syrup	Static	14	43.5	Moosavi-Nasab and Yousefi (2011) [75]
G. xylinus (ATCC 23769)	Glucose	Static	15	15.28	Ha and Park (2012) [76]

Table 1. Bacterial cellulose production under different culture conditions

5. Microbial cellulose for biomedical applications

As previously described, intense research has focused on the use of natural biopolymers in a variety of biomedical materials and devices, including wound dressings, medical implants, drug delivery, vascular grafts, and scaffolds for tissue engineering [78]. Consequently, continual efforts from many researchers, led to novel systems that closely mimic the complex and hierarchical structures inherent to the native tissue are sure to emerge.

In the last decade, several nanocellulose-based materials have been created for a diversity of biomedical applications. Some review articles have highlighted the potential applications of cellulose materials [72,79,80,81,82].

MC represents an interesting emerging nanomaterial, with no toxicity, and since its discovery has shown tremendous potential as an effective biopolymer which offers a wide range of applications, especially the biomedical ones, including the use as biomaterial for artificial skin, artificial blood vessels and microvessels, wound dressing of second- or third-degree burn ulcers and dental implants. Other studies with endothelial, smooth muscle cells and chondrocytes have shown that these cells present good adhesion to bacterial cellulose. [83]

5.1. Pristine MC based biomaterials

One of the main direct applications of MC membranes in biomedical field is related to wound dressing. Fontana et al. [84] were the pioneers in describing the use of bacterial cellulose to replace burned skin. Since then, literature shows a great number of papers related to wound dressing. Cellulose dressings are recommended as a temporary covering for the treatment of wounds, including pressure sores, skin tears, venous stasis, ischemic and diabetic wounds, second-degree burns, skin graft donor sites, traumatic abrasions and lacerations, and biopsy sites by the manufacturers [85].

MC based wound dressings are commonly available on the market nowadays, for example: BioFill®, Bioprocess®, XCell® and Gengiflex® (for periodontal diseases reconstruction [86]. The biomembrane BioFill® was one of the first commercial product that fulfills the main prerequisites of an ideal wound dressing, including: low cost, good adherence to the wound, water vapor permeability, elasticity, transparency, durability, it constitutes a physical barrier for bacteria, is hemostatic, it presents easy handling and application with minimum exchanges. BioFill® effectiveness has been proven in more than 300 cases in accelerating the healing process, pain relief, etc. [86,87,88,89].

Despite the analgesic mechanism of action of these dressings has not been fully elucidated, some authors suggest that the healing mechanism involves the capture of ions by means of cellulose hydrogen bonds, or the nano MC 3-D network mimics the skin surface creating optimal conditions for healing or regeneration [87,88,89].

It is important to point out that MC wound dressing clearly shortened the time to heal or wound closure over standard care when applied to non-healing lower extremity ulcers, as observed by many researchers [88,89,90]. As can be seen in Figure 5, novel applications of wet MC as wound dressing in the treatment partial thickness burns were applied by Czaja et al. [88,89] presenting excellent results suggesting that MC as a wound dressing promotes a favorable moist environment for a fast wound cleansing, and consequently for rapid healing.

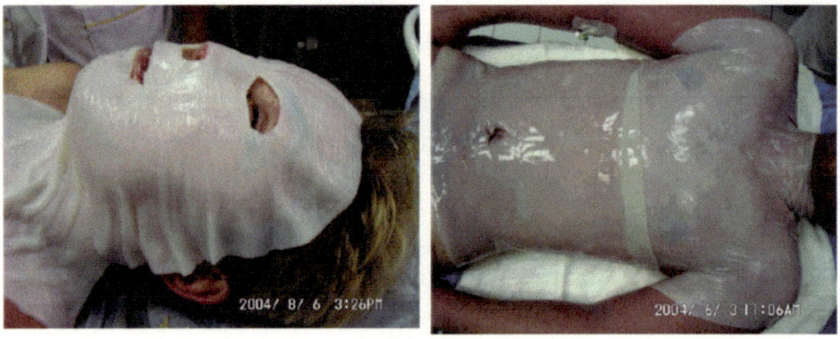

Figure 5. Bacterial cellulose dressing applied on wounded torso and face. (Reprinted with permission from Czaja, W. K. et al. (2007). The future prospects of microbial cellulose in biomedical applications, Biomacromolecules, Vol.8, No.1, pp. 4. Copyright (2007) American Chemical Society).

Other interesting biomedical applications for MC films have emerged. A Brazilian research group had designed and patented a device to manufacture MC contact lens for therapy in cases of regeneration of cornea [91].

In terms of vascular applications, a German group created BASYC® (Bacterial Cellulose Synthetised), consisting of a tubular biomaterial for applications in microsurgery of arteries and veias [92] and [93] applied vascular stents in animals arteries. Bodin et al. [94] obtained MC tubes by modifying the fermentation process of *Acetobacter xylinum* on top of silicon tubes, as shown in Figure 6.

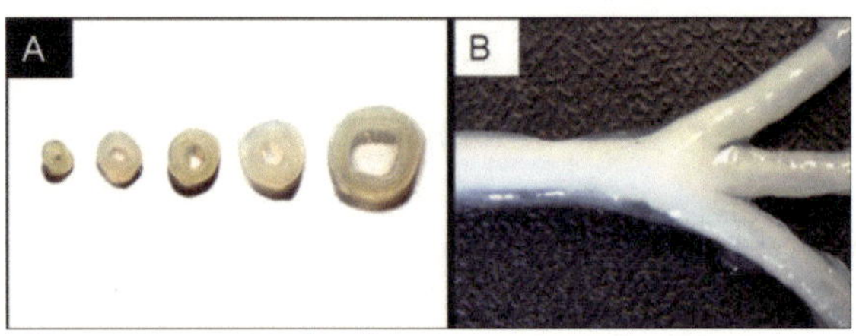

Figure 6. MC tubes presenting different sizes and shapes for applications: a) Mc tubes showing different inner diameters: 1.5mm, 2.4mm, 3.0mm, 4.0 mm and 6.0 mm. b) Branched MC tube fermented on a branched silicone tube. (Reprinted with permission from Bodin et al., 2007 Biotechnology and Bioengineering, Vol. 97, No. 2, June 1, 2007 [94])

Lately, Nimeskern et al. [95] designed and fabricated an ear-shaped pristine MC prototype material applying a Magnetic Resonance Imaging (MRI) scanning technique. This study was extremely important to confirm that MC is a promising tissue engineering material with appropriate mechanical properties for ear cartilage replacement. Thereby, it may be used to create patient-specific ear shapes.

5.2. Microbial cellulose nanocomposites for biomedical applications

Beyond the direct uses, MC can be widely and effectively utilized as either functional reinforcements or excellent matrices due to excellent mechanical properties and biocompatibility which allows it to be engineered in various forms from nano to macro scale. Thus, MC based nanocomposites can be manipulated to improve their properties and/or functionalities becoming one of the reasons that makes MC so exceptional material for biomedical applications.

In relation to biomaterial applications for wound dressing and skin tissue repair several MC based nanobiocomposites were fabricated. Here are some examples: membranes of MC/collagen [96], MC/gelatin [97], MC/ aloe vera films [98], MC/alginate for temporary dressing material [99].

Further, MC composite with kaolin was proved as short-term and long term wound healing materials [100].

Freeze-drying techniques allowed the preparation MC/poly(ethylene glycol) (PEG) composites of by immersing wet MC pellicle in PEG aqueous solution [101]. This same technique was applied by some of us [102] to obtain MC/ silk fibroin (SF) sponge scaffolds. *In vitro* tests proved non-cytotoxic or genotoxic character of these nanobiocomposites. SEM images revealed a greater number of fibroblast cells (L929 cell line) attached at the MC/SF:50% scaffold surface if compare with the surface of pure MC, suggesting that the presence of fibroin improved cell attachment as is possible to see in Figure 7. This could be related to the SF amino acid sequence that act as cell receptors facilitating cell adhesion and growth. Consequently, MC/SF:50% scaffolds configured an excellent option in bioengineering depicting its potential for tissue regeneration and cultivation of cells on nanobiocomposites.

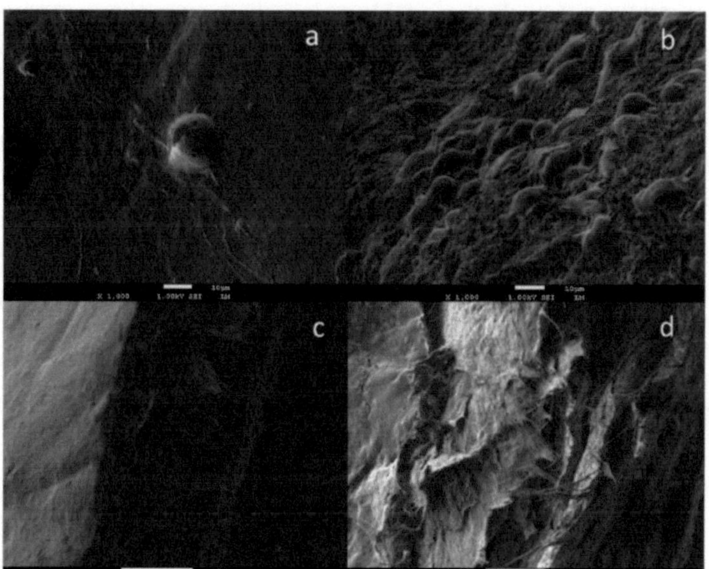

Figure 7. To test the hypothesis that the addition of silk fibroin to cellulose scaffolds increases cell adhesion (48 h), L-929 cells were seeded in MC and MC/SF scaffolds. SEM images of the cells attached to MC (a) and MC/SF (b) scaffolds surface; cross-section SEM images of MC (c) and MC/SF (d) evidenced that the cells did not migrate into the scaffolds. (Reprinted with permission from Oliveira Barud et al., 2015, Carbohydrate Polymers, Vol 128, April, 2015 [102]).

Generally, a scaffold provides a foundation for cell attachment, and several materials have been tested as scaffolds to support growth of cells. The need for bio-mimicking scaffolds has led to the exploration of MC as a scaffold material. There is an increased interest in developing adipose tissue as an *in vitro* model for adipose biology and metabolic disease, and to this end, 2D and 3D porous scaffolds of bacterial nanocellulose and alginate were prepared recently [103].

Bäckdahl et al. [104] also developed MC scaffolds with controlled microporosity by placing paraffin wax and starch particles during culture and removing these particles once the cultivation process was finished. The MC scaffolds were then seeded with smooth muscle cells for investigating the potential tissue engineered blood vessel application.

A variety of surface functionalization through biosynthetic or chemical modification was also investigated. Various approaches to the preparation of functional MC-based nanocomposites by incorporating different guest substrates including small molecules, inorganic nanoparticles or nanowires, and polymers on the surfaces of MC nanofibers are exemplified which can improve the functionality of MC nanomaterials and expand its potential application in the biomedical fields.

Nanocomposites were obtained by the association of nanoparticles presenting antimicrobial activities, including silver nanoparticles [105,106,107,108,109,110]. Additionally, Barud et al. [111] also prepared MC/propol membranes that presented good antimicrobial activities to be used as wound dressing material.

In recent years, several controlled release systems based on nanocellulose material for various pharmaceutical applications have been also investigated to delivery Tetracycline [112], benzalkonium chloride [113], topical release of lidocaine [114] and release of proteins with serum albumin [115].

With respect to bone regeneration in defects of rat tibiae, MC-hydroxyapatite (MCHA) nanobiocomposite were prepared to evaluate the biological properties and performance of the material [116]. The MCHA membranes were effective for bone regeneration and accelerated new bone formation. In addition, reabsorption of the membranes was slow, suggesting that this composite takes time to be completely reabsorbed.

6. Conclusion

Microbial cellulose is a natural renewable polymer synthesized from the bacterium *Gluconacetobacter xylinus* that is the only known species capable to produce cellulose on an industrial scale. In an appropriate culture medium the bacteria secretes about 50-80 cellulose microfibrils from 3.0 to 3.5 mm thick, free of lignin and hemicellulose, which aggregate themselves to form strips arranged in a 3-D hierarchical network. Besides that MC configures one of the most promising investigations in the field of biodegradable polymers. Due to this uniform structure and morphology MC is endowed with unique characteristics such as high purity, high crystallinity and remarkable mechanical properties, good chemical stability, high water holding capacity featuring it as a completely biocompatible polymer. Despite its high water content, MC shows a good mechanical performance and it can be produced in almost any shape due to its high moldability during formation. MC is an interesting emerging biomaterial, with no toxicity, and since its discovery has shown tremendous potential as an effective biopolymer in various fields, because the structural aspect of MC is far superior to those of plant cellulose. Thus, this chapter reviewed involved detailed aspects about the biosynthesis and recent advances on microbial production, including mechanism for the biochemistry of the cellulose synthesis, new sources for culture medium, main aspects about static and air-reactor productions and genetic modifications. We also revised and presented a great number of different MC based materials that were designed for biomedical applications (dressings, scaffolds, drug delivery systems), among others. Additionally, we hope that this book chapter may aggregate high quality information and may be a benchmark to intensify greater interest of the scientific community in microbial cellulose and related devices and also inspire the development of new materials in this field.

Author details

Wilton R. Lustri[2 #], Hélida Gomes de Oliveira Barud[2 #], Hernane da Silva Barud[1,3], Maristela F. S. Peres[3], Junkal Gutierrez[4], Agnieszka Tercjak[4], Osmir Batista de Oliveira Junior[2] and Sidney José Lima Ribeiro[3]

*Address all correspondence to: hernane.barud@gmail.com

1 University Center of Araraquara, UNIARA, Brazil

2 School of Dentistry/Unesp, São Paulo State University – Unesp, Araraquara-SP, Brazil

3 Institute of Chemistry- São Paulo State University, Araraquara, SP, Brazil

4 Group `Materials + Technologies´, Department of Chemical and Environmental Engineering, University of the Basque Country, UPV/EHU, Donostia-San Sebastián, Spain

[#]Lustri, WR and Oliveira Barud, HG present the same contributions in this book chapter manuscript.

References

[1] Donini ÍAN, De Salvi DTB, Fukumoto FK, Lustri WR, Barud HS, Marchetto R, Messaddeq Y, Ribeiro SJL. Biossíntese e recentes avanços na produção de celulose bacteriana. Eclet. Quím. 2010; 35:165-173. DOI: 10.1590/S0100-46702010000400021

[2] Phisalaphong M, Jatupaiboon N. Biosynthesis and characterization of bacteria cellulose–chitosan film. Carbohydr. Polym. 2008;78:482-488. DOI: 10.1016/j.carbpol.2008.04.004

[3] Sheykhnazari S, Tabarsa T, Ashori A, Shakeri A, Golalipour M. Bacterial synthesized cellulose nanofibers. Effects of growth times and culture mediums on the structural characteristics. Carbohydr. Polym. 2011;86:1187-1191. DOI: 10.1016/j.carbpol.2011.06.011.

[4] Shoda M. Sugano Y. Recent advances in bacterial cellulose production. Biotechnol. Bioprocess Eng. 2005;10:1-8. DOI: 10.1007/BF02931175

[5] Lapuz MM, Gallardo EG, Palo MA. The nata organism-cultural requirements, characteristics and identity. Philipp. J. Sci. 1967;96:91-108.

[6] Brown AJ. On an acetic ferment which forms cellulose. J. Chem. Soc. Trans. 1886;49:432-439. DOI: 10.1039/CT8864900432

[7] Yamada Y, Hoshino K, Ishikawa T. The phylogeny of acetic acid bacteria based on the partial sequences of 16S ribosomal RNA: The elevation of the subgenus *Gluconoacetobacter* to the generic level. Biosci. Biotechnol. Biochem. 1997;6:1244-1251. DOI: 10.1271/bbb.61.1244

[8] Ruka DR, Simon GP, Dean KM. Altering the growth conditions of *Gluconacetobacter xylinus* to maximize the yield of bacterial cellulose. Carbohydr. Polym. 2012;89:613-622. DOI: 10.1016/j.carbpol.2012.03.059

[9] Ashtaputre AA, Shah AK. Studies on a viscous, gel-forming exopolysaccharide from *Sphingomonas paucimobilis* GS1. Appl. Environ. Microbiol. 1995;61:1159-1162.

[10] Klemm D, Heublein B, Fink HP, Bohn A. Cellulose: fascinating biopolymer and sustainable raw material. Angewandte Chemie. 2005;44:3358-3393. DOI: 10.1002/anie.200460587

[11] Zogaj X, Nimtz M, Rohde M, Bokranz W, Römling U. The multicellular morphotypes of *Salmonella typhimurium* and *Escherichia coli* produce cellulose as the second component of the extracellular matrix. Mol. Microbiol. 2001;39:1452-63. DOI: 10.1046/j.1365-2958.2001.02337.x

[12] Sutherland IW. Microbial polysaccharides from Gram negative bacteria. Int. Dairy. J. 2001;11:663-674. DOI: 10.1016/S0958-6946(01)00112-1

[13] Souza DM, Garcia-Cruz CH. Produção fermentativa de polissacarídeos extracelulares por bactérias. Semina: Ciências Agrárias. 2004;25:331-340. DOI: 10.5433/1679-0359.2004v25n4p331

[14] Solano C, Garcia B, Valle J, Berasain C, Ghigo JM, Gamazo C, Lasa I. Genetic analysis of *Salmonella enteritidis* biofilm formation: critical role of cellulose. Mol. Microbiol. 2002;43:793-808. DOI: DOI: 10.1046/j.1365-2958.2002.02802.x

[15] Marsh AJ, O'Sullivan O, Hill C, Ross RP, Cotter PD. Sequence-based analysis of the bacterial and fungal compositions of multiple kombucha (tea fungus) samples. Food Microbiol. 2014;38:171-178. DOI: 10.1016/j.fm.2013.09.003

[16] Liu CH, Hsu WH, Lee FL, Liao CC. The isolation and identification of microbes from a fermented tea beverage, Haipao, and their interactions during Haipao fermentation. Food Microbiol. 1996;13:407-415. DOI: 10.1006/fmic.1996.0047

[17] Mayser P, Fromme S, Leitzmann C, Grunder K. The yeast spectrum of the 'tea fungus kombucha'. Mocoses. 1995;38:289-295.

[18] Teoh AL, Heard G, Cox J. Yeast ecology of kombucha fermentation. Int. J. Food Microbiol. 2004;95:119-126. DOI: 10.1016/j.ijfoodmicro.2003.12.020

[19] Greenwalt CJ, Ledford RA, Steinkraus KH. Determination and characterization of the antimicrobial activity of the fermented tea kombucha. Lebensm. Wiss. Technol. 1998;31:291-296. DOI: 10.1006/fstl.1997.0354

[20] Dufresne C, Farnworth E. Tea, Kombucha, and health: a review. Food Res. Int. 2000;33:409-421. DOI: 10.1016/S0963-9969(00)00067-3

[21] Chávez-Pacheco JL, Martínez-Yee S, Contreras-Zentella M, Escamilla-Marván E, Celulosa bacteriana en gluconacetobacter xylinum: biosíntesis y aplicaciones Tip Revista Especializada en Ciencias Químico-Biológicas. 2004;7:18-25.

[22] Ross P, Mayer R, Benziman M. Cellulose biosynthesis and function in bacteria. Microbiol. Rev. 1991;55:35-58.

[23] Saxena IM, Kudlicka K, Okuda K, Brown RM Jr. Characterization of genes in the cellulose-synthesizing operon (acs operon) of *Acetobacter xylinum*: implications for cellulose crystallization. J. Bacteriol. 1994;176:5735-5752.

[24] Krystynowicz A, Koziolkiewicz M, Wiktorowska-Jezierska A, Bielecki S. Klemenska E, Masny A, Płucienniczak A. Molecular basis of cellulose biosynthesis disappearance in submerged culture of *Acetobacter xylinum*. Acta Biochim. Pol. 2005;52:691-698.

[25] Brown Jr RM. Cellulose microfibril assembly and orientation: recent developments. J. Cell. Sci. Suppl. 1985;2:13–32.

[26] Okuda K. Structure and phylogeny of cell coverings. J. Plant. Res. 2002;115:283-288. DOI: 10.1007/s10265-002-0034-x

[27] Sunagawa N, Fujiwara T, Yoda T, Kawano S, Satoh Y, Yao M, Tajima K, Dairi T. Cellulose complementing factor (Ccp) is a new member of the cellulose synthase complex (terminal complex) in *Acetobacter xylinum*. J. Biosci. Bioeng. 2013;115:607-612. DOI: 10.1016/j.jbiosc.2012.12.021

[28] Brown Jr RM, Saxena IM. Cellulose biosynthesis: A model for understanding the assembly of biopolymers. Plant Physiol. Biochem. 2000;38:57-67. DOI: 10.1016/S0981-9428(00)00168-6

[29] Kawano S, Yasutake Y, Tajima K, Satoh Y, Yao M, Tanaka I, Munekata M. Crystallization and preliminary crystallographic analysis of the cellulose biosynthesis-related protein CMCax from *Acetobacter xylinum*. Acta Crystallogr. Sect. F Struct. Biol. Crystall. Commun. 2005;61:252-254. DOI: 10.1107/S174430910500206X

[30] Ross P, Weinhouse H, Aloni Y, Michaeli D, Weinberger-Ohana P, Mayer R, Braun S, de Vroom E, van der Marel GA, van Boom JH, Benziman M. Regulation of cellulose synthesis in *Acetobacter xylinum* by cyclic diguanylic acid Nature 1987;325:279-281. DOI: 10.1038/325279a0

[31] Römling U, Gomelsky M, Galperin MY. C-di-GMP: the dawning of a novel bacterial signalling system. Mol. Microbiol. 2005;57:629-639. DOI: 10.1111/j.1365-2958.2005.04697.x

[32] Kimura S, Chen HP, Saxena IM, Brown JrRM, Itoh T. Localization of c-di-GMP-binding protein with the linear terminal complexes of *Acetobacter xylinum*. J. Bacteriol. 2001;183:5668-5674. DOI: 10.1128/JB.183.19.5668-5674.2001

[33] Mayer R, Ross P, Weinhouse H, Amikam D, Volman G, Ohana P, Calhoon RD, Wong HC, Emerick AW, Benziman M. Polypeptide composition of bacterial cyclic diguanylic acid-dependent cellulose synthase and the occurrence of immunologically crossreacting proteins in higher plants, Proc. Natl. Acad. Sci. USA. 1991;88:5472-5476.

[34] Weinhouse H, Sapir S, Amikam D, Shilo Y, Volman G, Ohana P, Benziman M. C-di-GMP-binding protein, a new factor regulating cellulose synthesis in *Acetobacter xylinum*. FEBS Lett. 1997;416:207-211. DOI: 10.1016/S0014-5793(97)01202-7

[35] Tal R, Wong HC, Calhoon R, Gelfand D, Fear AL, Volman G, Mayer R, Ross P, Amikam D, Weinhouse H, Cohen A, Sapir S, Ohana P, Benziman M, Three *cdg* operons control cellular turnover of cyclic di-GMP in *Acetobacter xylinum*: Genetic organization and occurrence of conserved domains in isoenzymes. J. Bacteriol. 1998;180:4416-4425.

[36] Fujiwara T, Komoda K, Sakurai N, Tajima K, Tanaka I, Yao M. The c-di-GMP recognition mechanism of the PilZ domain of bacterial cellulose synthase subunit A. Biochem. Biophys. Res. Commun. 2013;431:802-807. DOI: 10.1016/j.bbrc.2012.12.103

[37] Koo HM, Song SH, Pyun YR, Kim YS. Evidence that a beta-1,4-endoglucanase secreted by *Acetobacter xylinum* plays an essential role for the formation of cellulose fiber. Biosci. Biotechnol. Biochem. 1998;62:2257-2259. DOI: 10.1271/bbb.62.2257

[38] Legnani C, Vilani C, Calil VL, Barud HS, Quirino WG, Achete CA, Ribeiro SJL, Cremona M. Bacterial cellulose membrane as flexible substrate for organic light emitting devices. Thin Solid Films 2008;517:1016-1020. DOI: 10.1016/j.tsf.2008.06.011

[39] Römling U. Molecular biology of cellulose production in bacteria research in microbiology. 2002;153:205-212.

[40] Morgan JLW, Strumillo J, Zimmer J. Crystallographic snapshot of celulose synthesis and membrane translocation. Nature. 2013;493:181-187. DOI: 10.1038/nature11744

[41] Keiski CL, Harwich M, Jain S, Neculai AM, Yip P, Robinson H, Whitney JC, Riley L, Burrows LL, Ohman DE, Howell PL. AlgK is a TPR-containing protein and the periplasmic component of a novel exopolysaccharide secretin. Structure 2010;18:265-273. DOI: 10.1016/j.str.2009.11.015

[42] Omadjela O, Narahari A, Strumillo J, Mélida H, Mazur O, Bulone V, Zimmer J. MCsA and MCsB form the catalytically active core of bacterial cellulose synthase sufficient for in vitro cellulose synthesis. Proc. Natl. Acad. Sci. USA. 2013;110:17856-17861. DOI: 10.1073/pnas.1314063110

[43] Nishiyama Y, Langan P, Chanzy H. Crystal structure and hydrogen-bonding system-in cellulose Ib from synchrotron X-ray and neutron fiber diffraction. J. Am. Chem. Soc. 2002;124:9074-9082. DOI: 10.1021/ja0257319

[44] Hestrin S, Schramm M. Synthesis of cellulose by *Acetobacter xylinum*. II. Preparation of freeze-dried cells capable of polymerizing glucose to cellulose. Biochem. J. 1954;58:345-352.

[45] Toyosaki H, Naritomi T, Seto A, Matsuoka M, Tsuchida T, Yoshinaga F. Screening of bacterial cellulose-producing *Acetobacter* strains suitable for agitated culture. Biosci. Biotechnol. Biochem. 1995;59:1498-1502. DOI: 10.1271/bbb.59.1498

[46] Cheng HP, Wang PM, Chen JW, Wu WT. Cultivation of *Acetobacter xylinum* for bacterial cellulose production in a modified airlift reactor. Biotechnol. Appl. Biochem. 2002;35:125-32. No tiene

[47] Bielecki S, Krystynowicz A, Turkiewicz M, Kalinowska H. Bacterial Cellulose. Biopolymers Online. Wiley-VCH Verlag GmbH & Co. KGaA 2005. DOI: 10.1002/3527600035.bpol5003

[48] Ha JH, Shah N, Ul-Islam M, Khan T, Park JK. Bacterial cellulose production from a single sugar a-linked glucuronic acid-based oligosaccharide. Process. Biochem. 2011;46:1717-1723. DOI: 10.1016/j.procbio.2011.05.024

[49] Watanabe K, Tabuchi M, Ishikawa A, Takemura H, Tsuchida T, Morinaga Y, Yoshinaga F. *Acetobacter xylinum* mutant with high cellulose productivity and an ordered structure. Biosci. Biotechnol. Biochem. 1998;62:1290-1292. DOI:10.1271/bbb.62.1290

[50] Duta FP, França FP, Lopes LMA. Optimization of culture for exopolysaccharide production in *Rhizobium sp.* using the response surface method. Electron. J. Biotechnol. 2006;9: 391-399. DOI: 10.2225

[51] Khan T, Khan S, Park JK. Simple fed-batch cultivation strategy for the enhance production of a single-sugar glucoronic acid-based oligosaccharides by a cellulose-producing *Gluconacetobacter hansenii* strain. Biotechnol. Bioprocess. Eng. 2008;13:240-247. DOI: 10.1007/s12257-007-0212-8

[52] Iguchi M, Yamanaka S, Budhiono A. Bacterial cellulose - a masterpiece of nature's arts. J. Mater. Sci. 2000;35:261-270. DOI: 10.1023/A:1004775229149

[53] Shezad O, Khan S, Khan T, Park JK. Physicochemical and mechanical characterization of bacterial cellulose produced with an excellent productivity in static conditionsusing a simple fed-batch cultivation strategy. Carbohydr. Polym. 2010;82:173-180. DOI: 10.1016/j.carbpol.2010.04.052

[54] Kralisch D, Hessler N, Klemm D, Erdmann R, Schmidt W. White biotechnology for cellulose manufacturing - The HoLiR concept. Biotechnol. Bioeng. 2010;105:740-747. DOI: 10.1002/bit.22579

[55] Song H-J, Li H, Seo J-H, Kim M-J, Kim S-J. Pilot-scale production of bacterial cellulose by a spherical type bubble column bioreactor using saccharified food wastes. Korean J. Chem. Eng. 2010;26:141-146. DOI: 10.1007/s11814-009-0022-0

[56] Cherian BM, Leão AL, Souza SF, Oliveira GM, Costa LMM, Brandão CVS, Narine SS. Bacterial nanocelulose for medical implants. In: Thomas, S., Visakh, P.M., Mathew, A.P. editors. Advances in natural polymers. Berlin Heidelberg: Springer, 2013. DOI: 10.1007/978-3-642-20940-6_10

[57] Chawla PR, Bajaj IB, Survase SA, Singhal RS. Microbial cellulose: Fermentative production and applications. Food Technol. Biotechnol. 2009;47:107-124.

[58] El-Saied H, El-Diwany AI, Basta AH, Atwa NA, El-Ghwas DE. Production and characterization of economical bacterial cellulose. BioResources. 2008;3:1196-1217.

[59] Hutchens SA, Benson RS, Evans BR, O'Neill HM, Rawn CJ Biomimetic synthesis of calcium-deficient hydroxyapatite in a natural hydrogel. Biomaterials. 2006;27:4661-4670. DOI: 10.1016/j.biomaterials.2006.04.032

[60] Mikkelsen D, Flanagan BM, Dykes GA, Gidley MJ. Influence of different carbon sources on bacterial cellulose production by *Gluconacetobacter xylinus* strain ATCC 53524. J. Appl. Microbiol. 2009;107:576-583. DOI: 10.1111/j.1365-2672.2009.04226.x

[61] Hwang JW, Yang YK, Hwang JK, Pyun Y and Kim YS. Effects of pH and dissolved oxygen on cellulose production by *Acetobacter xylinum* BRC5 in agitated culture. J. Biosci. Bioengineer. 1999;88:183-188. DOI: 1016/S1389-1723(99)80199-6

[62] Hong F, Qiu K. An alternative carbon sources from konjac powder for enhancing production of bacterial cellulose in static cultures by a model strain *Acetobacter aceti* subsp. *xylinus* ATCC23770. Carbohydr. Polym. 2008;72:545. DOI: 10.1016/j.carbpol.2007.09.015

[63] Hong B, Xue G, Weng L, Guo X. Pretreatment of moso bamboo with dilute phosphoric acid. BioResources. 2012;7:4902-4913.

[64] Goelzer FDE, Faria-Tischer PCS, Vitorino JC, Sierakowaki MR, Tischer CA. Production and characterization of nanospheres of bacterial cellulose from *Acetobacter xyli-*

num from processed rice bark. Mater. Sci. Enginner.: C. 2009;29:546-551. DOI: 10.1016/j.msec.2008.10.013

[65] Kuo CH, Lin PJ, Lee CK. Enzymatic saccharification of dissolution pretreated wast cellulosic fabrics for bacterial cellulose production by *Gluconacetobacter xylinus*. J. Chem. Technol. Biotechnol. 2010;85: 1346-1352. DOI: 10.1002/jctb.2439

[66] Wu JM, Liu RH. Thin stillage supplementation greatly enhances bacterial cellulose production by *Gluconacetobacter xylinus*. Carbohydr. Polym. 2012;90:116-121. DOI: 10.1016/j.carbpol.2012.05.003

[67] Noro N, Sugano Y, Shoda M. Utilization of the buffering capacity of corn steep liquor in bacterial cellulose production by *Acetobacter xylinum*. Appl. Microbiol. Biotechnol. 2004;64:199-205. DOI: 10.1007/s00253-003-1457-6

[68] Jung HI, Jeong JH, Lee OM, Park GT, Kim KK, Park HC, Lee SM, Kim YG, Son HJ. Influence of glycerol on production and structural-physical properties of cellulose from *Acetobacter sp.* V6 cultured in shake flasks. Bioresour. Technol. 2010;101:3602-3608. DOI: 10.1016/j.biortech.2009.12.111

[69] Kongruang S. Bacterial cellulose production by *Acetobacter xylinum* strains from agricultural waste products. Appl. Biochem. Biotechnol. 2008;148:245-256. DOI: 10.1007/s12010-007-8119-6

[70] Bae S, Sugano Y, Shoda M. Improvement of bacterial cellulose production by addition of agar in a jar fermentor. J. Biosci. Bioeng. 2004;97:33-38. DOI: 10.1016/S1389-1723(04)70162-0

[71] Lu XK, Shen XY. Solubility of bacteria cellulose in zinc chlorid aqueous solution. Carbohydr. Polym. 2011;86:239-244. DOI: 10.1016/j.carbpol.2011.04.042

[72] Shah N, Ha JH, Park JK. Effect of reactor surface on production of bacterial cellulose and water soluble oligosaccharides by *Gluconacetobacter hansenii* PJK, Biotechnol. Bioproc. Engineer. 2010;15:110-118. DOI: 10.1007/s12257-009-3064-6

[73] Cheng KC, Catchmark JM, Demirci A. Effects of CMC addition on bacterial cellulose production in a biofilm reactor and its paper sheets analysis. Biomacromol. 2011;12:730-736. DOI: 10.1021/bm101363t

[74] Jahan F, Kumar V, Rawat G, Saxena RK. Production of microbial cellulose by a bacterium isolated. Appl. Biochem. Biotechnol. 2012;167:1157-1171. DOI: 10.1007/s12010-012-9595-x

[75] Cavka A, Guo X, Tang SJ, Winestrand S, Jönsson LJ and Hong F. Production of bacterial cellulose and enzyme from waste fiber sludge. Biotechnol. Biofuels. 2013;6:25. DOI: 10.1186/1754-6834-6-25.

[76] Moosavi-Nasab M, Yousefi A. Biotechnological production of cellulose by *Gluconacetobacter xylinus* from agricultural waste. Iranian J. Biotechnol. 2011;9:94-101.

[77] Ha JH, Park JK. Improvement of bacterial cellulose production in *Acetobacter xylinum* using byproduct produced by *Gluconacetobacter hansenii*. Korean J. Chem. Eng. 2012;29:563-566. DOI: 10.1007/s11814-011-0224-0

[78] Reis RL, Neves NM, Mano JF, Gomes ME, Marques AP, Azevedo HS. Natural-based polymers for biomedical applications; CRC: Boca Raton, FL, 2008.

[79] Klemm D, Kramer F, Moritz S, Lindstrom T, Ankerfors M, Gray D, Dorris A. Angew. Chem. Int. Ed. Nanocelluloses: A new family of nature-based materials. 2011;50:5438-5466. DOI: 10.1002/anie.201001273

[80] Fu L, Zhang Y, Zhang J, Yang G. Bacterial cellulose for skin repair materials, In:, Fazel R, editor. Biomedical engineering-frontiers and challenges Rijika: Croatia: InTech; 2011. p. 249-274. DOI: 10.5772/1019

[81] Fu L, Zhang J, Yang G. Present status and applications of bacterial cellulose-based materials for skin tissue repair. Carbohydr. Polym. 2013;92,1432-1442. DOI: 10.1016/j.carbpol.2012.10.071

[82] Rajwade JM, Paknikar KM, Kumbhar JV. Applications of bacterial cellulose and its composites in biomedicine. Appl. Microbiol. Biotechnol. 2015;99:2491-2511. DOI: 10.1007/s00253-015-6426-3

[83] Ul-Islam M, Khan T, Park J. Water holding and release properties of bacterial cellulose obtained by in situ and ex situ modification. Carbohydr. Polym. 2012;88:596-603. DOI: 10.1016/j.carbpol.2012.01.006

[84] Fontana J, De Souza A, Fontana C, Torriani I, Moreschi J, Gallotti B, de Souza S, Narcisco GP, Bichara JA, Farah LF. Acetobacter cellulose pellicle as a temporary skin substitute. Appl. Biochem. Biotechnol. 1990;24-25:253-264.

[85] Kowalska-Ludwicka K, Cala J, Grobelski B, Sygut D, Jesionek-Kupnicka D, Kolodziejczyk M, Bielecki S, Pasieka Z. Modified bacterial cellulose tubes for regeneration of damaged peripheral nerves. Arch. Med. Sci. 2013;9:527-534. DOI: 10.5114/aoms.2013.33433

[86] BIO FILL PRODUTOS BIOTECNOLÓGICOS (BRASIL). Farah LFX. Process for the preparation of cellulose film, cellulose film produced thereby artificial skin graft and its use. US 4912049, 30 Sept. 1990, 10 June 1986.

[87] Wouk AFPF, Diniz JM, Círio SM, Santos H, Baltazar EL, Acco A. Membrana biológica Biofill – estudo comparativo com outros agentes promotores da cicatrização da pele em suínos: aspectos clínicos, histopatológicos e morfométricos. Arch. Vet. Scienc. 1998;3:31-37.

[88] Czaja W, Krystynowicz A, Bielecki S, Brown Jr RM. Microbial cellulose-the natural power to heal wounds. Biomater. 2006;27:145-151. DOI: 10.1016/j.biomaterials.2005.07.035

[89] Czaja W, Young DJ, Kawecki M, Brown Jr RM. The future prospects of microbial cellulose in biomedical applications. Biomacromol. 2007;8:1-12. DOI: 10.1021/bm060620d

[90] Portal O, Clark WA, Levinson DJ. Microbial cellulose wound dressing in the treatment of nonhealing lower extremity ulcers. Wounds 2009;21:1-3.

[91] TRIGGER PESQUISA & DESENVOLVIMENTOS BIOTECNOLOGICOS LTDA (Brasil). Y. Messaddeq; S. J. L. Ribeiro; W. Thomazini. Contact lens for therapy in cases of regeneration of cornea, has bacterial cellulose base. BR 200603704-A, 22 Ago. 2006, 08 Abr. 2008.

[92] Klemm D, Schumann D, Udhardt U, Marsch S. Bacterial synthesized cellulose - artificial blood vessels for microsurgery. Progress Polym. Sci. 2001;26:1561-1603. DOI: 10.1016/S0079-6700(01)00021-1

[93] Wippermann J, Schumann D, Klemm D, Kosmehl H, Satehi-Gelani S, Wahlers T. Preliminary results of small arterial substitute performed with a new cylindrical biomaterial composed of bacterial cellulose. Eur J Vasc Endovasc. 2009; 37(5): 592-96. DOI: 10.1016/j.ejvs.2009.01.007

[94] Bodin A, Backdahl H, Fink H, Gustafsson L, Risberg B, Gatenholm P. Influence of cultivation conditions on mechanical and morphological properties of bacterial cellulose tubes. Biotechnol. Bioeng. 2007;97:425-434. DOI: 10.1002/bit.21314

[95] Nimeskern L, Avila HM, Sundberg J, Gatenholm P, Mueller R, Stok KS. Mechanical evaluation of bacterial nanocellulose as an implant material for ear cartilage replacement. J. Mech. Behav. Biomed. Mater. 2013;22:12-21. DOI: 10.1016/j.jmbbm.2013.03.005

[96] Cai ZJ, Yang G. Bacterial cellulose/collagen composite: Characterization and first evaluation of cytocompatibility. J. Appl. Polym. Sci. 2011;1205:2938-2944. DOI: 10.1002/app.33318

[97] Nakayama A, Kakugo A, Gong JP, Osada Y, Takai M, Erata T, Kawano S. High mechanical strength double–network hydrogel with bacterial cellulose. Adv. Funct. Mater. 2004;14:1124-1128. DOI: 10.1002/adfm.200305197

[98] Saibuatong OA, Phisalaphong M. Novo aloe vera–bacterial cellulose composite film from biosynthesis. Carbohydr. Polym. 2010;79,455-460.

[99] Chiaoprakobkij N, Sanchavanakit N, Subbalekha K, Pavasant P, Phisalaphong M. Characterization and biocompatibility of bacterial cellulose/alginate composite sponges with human keratinocytes and gingival fibroblasts. Carbohydr. Polym. 2011;853:548-553. DOI: 10.1016/j.carbpol.2011.03.011

[100] Wanna D, Alam C, Toivola DM, Alam P. Bacterial cellulose-kaolin nanocomposites for application as biomedical wound healing materials. Adv. Nat. Sci. Nanosci. Nanotechnol. 2013;4:045002. DOI: 10.1088/2043-6262/4/4/045002.

[101] Cai Z, Kim J. Bacterial cellulose/poly (ethylene glycol) composite: Characterization and first evaluation of biocompatibility. Cellulose. 2010;17:83-91. DOI: 10.1007/s10570-009-9362-5

[102] Oliveira Barud HG, Barud HS, Cavicchioli M, Amaral TS, Oliveira Junior OB, Santos DM, Oliveira Almeida Petersen AL, Celes F, Borges VM, de Oliveira CI, de Oliveira PF, Furtado RA, Tavares DC, Ribeiro SJL. Preparation and characterization of a bacterial cellulose/silk fibroin sponge scaffold for tissue regeneration. Carbohydr.Polym. 2015;128:41-51. DOI: 10.1016/j.carbpol.2015.04.007

[103] Krontiras P, Gatenholm P, Hagg D. Adipogenic differentiation of stem cells in three-dimensional porous bacterial nanocellulose scaffolds. J. Biomed. Mater. Res. Part B 2015;103B:195-203. DOI: 10.1002/jbm.b.33198

[104] Bäckdahl H., Helenius G., Bodin A., Nannmark U., Johansson B.R., Risberg B., et al. Mechanical properties of bacterial cellulose and interactions with smooth muscle cells. Biomaterials. 2006; 27(9): 2141-9

[105] Barud HS, Barrios C, Regiane T, Marques RFC, Verelst M, Dexpert-Ghys J, Messaddeq Y, Ribeiro SJL. Self-supported silver nanoparticles containing bacterial cellulose membranes. Mater. Sci. Engineer. C. 2007;1:01-02. DOI: 10.1016/j.msec.2007.05.001

[106] Barud HS, Barrios CE, Regiane T, Marques RFC, Verelst M, Dexpert-Ghys J, Messaddeq Y, Ribeiro SJL. Self-supported silver nanoparticles containing bacterial cellulose membranes. Mater. Sci. Engineer. C: Biomimetic Mater., Sensors and Systems. 2008; 28:515-518.

[107] Barud HS, Regiani T, Marques RFC, Lustri WR, Messaddeq Y, Ribeiro SJL. Antimicrobial bacterial cellulose-silver nanoparticles composite membranes. J. Nanomater. 2011;721631:1-8. DOI: 10.1155/2011/721631

[108] Maria LCS, Santos ALC, Oliveira PC, BARUD, Barud HS, Ribeiro SJL, Messaddeq Y. Synthesis and characterization of silver nanoparticles impregnated into bacterial cellulose. Mater. Lett. 2009;63:797-799. DOI: 10.1016/j.matlet.2009.01.007

[109] Maria LCS, Santos ALC, Oliveira PC, Valle ASS, Barud HS, Messaddeq Y, Ribeiro SJL. Preparation and antibacterial activity of silver nanoparticles impregnated in bacterial cellulose. Polímeros. 2010;20:72-77. DOI: 10.1590/S0104-14282010005000001

[110] Maneerung T, Tokura S, Rujiravanit R. Impregnation of silver nanoparticles into bacterial cellulose for antimicrobial wound dressing. Carbohydr. Polym. 2008;72:43-51. DOI: 10.1016/j.carbpol.2007.07. 025

[111] Barud HS, de Araújo Júnior AM, Saska S, Mestieri LB, Campos JADB, De Freitas RM, Ferreira NU, Nascimento AP, Miguel FG, Vaz MMOLL, Barizon EA, Marquele-Oliveira F, Gaspar AMM, Ribeiro S JL, Berretta AA. Antimicrobial brazilian propolis (EPP-AF) containing biocellulose membranes as promising biomaterial for skin

wound healing. evidence-based complementary and alternative medicine. 2013;703024:1-10. DOI: 10.1155/2013/703024

[112] Stoica-Guzun A, Stroescu M, Tache F, Zaharescu T, Grosu E. Effect of electron beam irradiation on bacterial cellulose membranes used as transdermal drug delivery systems. Nucl. Instrum. Methods B 2007;265:434-438. DOI:10.1016/j.nimb.2007.09.036

[113] Wei B, Yang GA, Hong F. Preparation and evaluation of a kind of bacterial cellulose dry films with antibacterial properties. Carbohydr. Polym. 2011;84:533-538. DOI: 10.1016/j.carbpol.2010.12.017

[114] Trovatti E, Silva NHCS, Duarte IF, Rosado CF, Almeida IF, Costa P, Freire CSR, Silvestre AJD, Neto CP. Biocellulose membranes as supports for dermal release of lidocaine. Biomacromol. 2011;12:4162-4168. DOI: 10.1021/bm201303r

[115] Müller A, Ni Z, Hessler N, Wesarg F, Müller FA, Kralisch D, Fischer D. The biopolymer bacterial nanocellulose as drug delivery system: investigation of drug loading and release using the model protein albumin. J Pharm Sci 2013, 102, 579-592. DOI: 10.1002/jps.23385

[116] Saska S, Barud HS, Gaspar AMM, Marchetto R, Ribeiro SJL, Messaddeq Y. Bacterial cellulose-hydroxyapatite nanocomposites for bone regeneration. Int. J. Biomater. 2011;175362:1-8. DOI: 10.1155/2011/175362

Chapter 7

Crystalline Nanocellulose — Preparation, Modification, and Properties

Mikaela Börjesson and Gunnar Westman

Additional information is available at the end of the chapter

http://dx.doi.org/10.5772/61899

Abstract

Cellulose is a linear biopolymer found naturally in plant cells such as wood and cotton. It is the worlds most abundant polymer in nature and possesses properties such as good biocompatibility, low cost, low density, high strength, and good mechanical properties. By mechanical or chemical treatment, the cellulose fibers can be converted into cellulose nanofibers (CNFs) or cellulose nanocrystals (CNCs) that possess outstanding properties compared with the original cellulosic fiber but also when compared with other materials normally used as reinforcements in composite materials such as Kevlar or steel wires. This review will describe the nanocellulose materials preparation techniques and cellulose sources, chemical modification both on the crystalline surface and during hydrolysis and its many properties and its use in biocomposite materials. Nanocellulose in its different forms shows an increasing interest in application areas such as packaging, paper and paperboard, food industry, medical and hygiene products, paints, cosmetics, and optical sensors

Keywords: Nanocellulose, cellulose nanocrystals, hydrolysis, chemical modification, biocomposites

1. Introduction

Cellulose (*Latin:* rich in small cells) is a biopolymer found naturally in, for example, plant cells such as wood and cotton. It is the most abundant polymer in nature and is the main constituent in the cell wall of trees and plants. Cotton have the highest cellulose content of the plants with about 90% cellulose, compared to wood that has about 40–50% cellulose content or bast fibers, such as flax, hemp, or ramie, which have about 70–80% cellulose content [1–2]. Besides wood and plants, cellulose can also be found in various bacterial species, algae, and tunicates, a sea animal that consists of proteins and carbohydrates.

The cellulose polymer is a linear homo-polysaccharide consisting of D-anhydroglucopyranose units (AGU) linked together by β-1,4-glycosidic bonds. Every other AGU is turned 180° with respect to its neighbor and two AGU next to each other form a cellobios unit, the smallest repeating unit in the polymer (Figure 1). The degree of polymerization (DP) is a measure of how many AGUs there is in the polymer and since no polymer is homogenous in length, the molecular weight distribution will have an important influence on the fibers properties. The DP of a cellulose polymer can be as high as 10,000 in wood cellulose and even higher in, for example, native cotton plant fibers. After degradation reactions and purification processes, the DP is reduced to about 300–1700 in wood cellulose [3–4].

The cellulose molecule contains three different kinds of AGU: a reducing end group that contains a free hemiacetal or aldehyde at the C1 position, a non-reducing end group with a free hydroxyl group at the C4 position, and internal glucose rings joined at the C1 and the C4 positions. The internal glucose units are predominant due to the long chain lengths. Each internal AGU has three hydroxyl groups. The hydroxyl group at the C6 position is a primary alcohol, while the hydroxyl groups at the C2 and C3 positions are secondary alcohols. These hydroxyl groups are all possible sites for chemical modification of cellulose where the hydroxyl group at the C6 position is the most reactive [5].

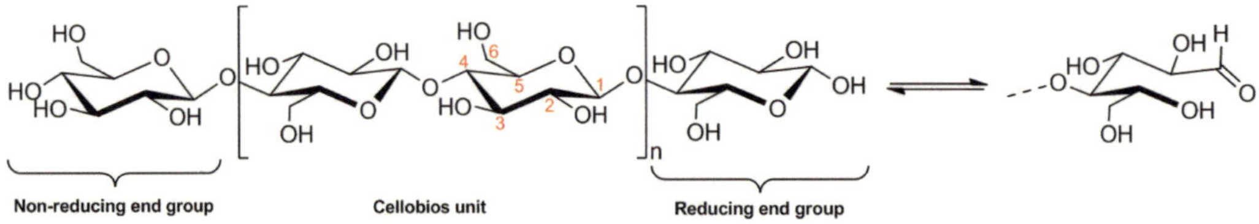

Figure 1. The molecular structure of a cellulose polymer where the cellobios is the smallest repeating unit in the polymer. The reducing end group can be either a free hemiacetal or an aldehyde.

Due to the linear and quite regular structure of cellulose and the many hydroxyl groups in the molecule, cellulose polymers can form ordered crystalline structures held together with hydrogen bonds. These crystalline regions give important mechanical properties to the cellulose fibers. The hydroxyl groups in the cellulose polymer can form hydrogen bonds between different cellulose polymers (intermolecular hydrogen bonds) or within the polymer itself (intramolecular hydrogen bonds). The intramolecular bonds give stiffness to the polymer chain, while the intermolecular bonds allow the linear polymers to form sheet structures. The high crystallinity and the many hydrogen bonds in the cellulose fibers make cellulose insoluble in water and in most conventional organic solvents [6].

The structure of a cellulose fiber can be divided into three different levels: the molecular level, the supramolecular level, and the morphological level. The molecular level was described in the beginning of this section (Figure 1). The supramolecular level is the polymer chains ordered in crystalline and non-crystalline regions due to hydrogen bonds and the morphological level consists of the cellulose fiber and its cell walls (Figure 2).

http://dx.doi.org/10.5772/61899

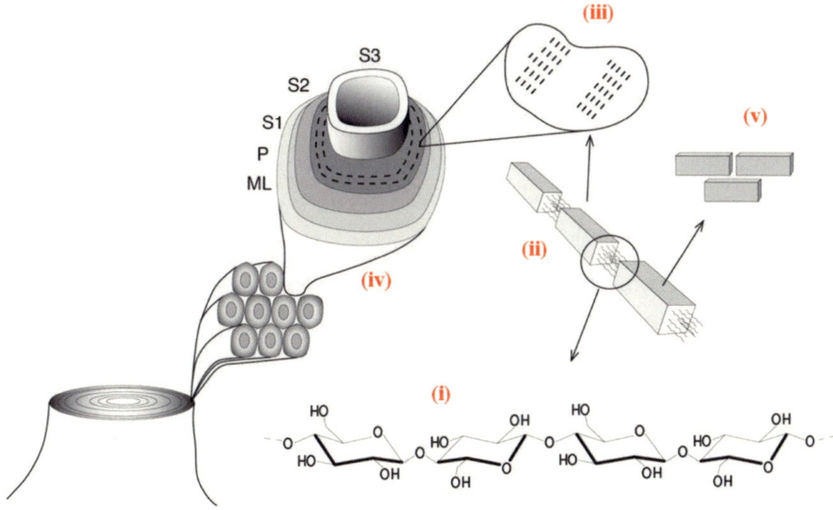

Figure 2. The different levels of the cellulose structure: (i) the molecular structure of a cellulose polymer, (ii) the polymers ordered into microfibrils with crystalline and non-crystalline regions, (iii) several microfibrils assembled together to form a macrofibril, and (iv) the different layers in the cell wall. Numbers (ii) and (v) show cellulose microfibrils (CMFs) and cellulose nanocrystals (CNCs), respectively. *Illustration: Jari Sundqvist.*

Figure 2 shows the different levels of the cellulose structure where the cellulose polymers (i) are aggregated to form microfibrils (ii), which are long bundles of cellulose molecules stabilized by hydrogen bonds. Several microfibrils are assembled together to form a macrofibril (iii), which are oriented in different layers in the cell wall (iv) and the different layers differ in fibril direction, densities, and textures. There are four distinct layers in the cell wall named P (primary), S1, S2, and S3 (secondary) and each layer is mainly composed of a combination of the three polymers: cellulose, hemicellulose, and lignin. The S2 layer in the cell wall of higher plants, for example wood, has the highest quantity of cellulose. Between the cells there is a middle lamella (ML) consisting of pectin, a polysaccharide that binds the cells together providing the fiber stability. Numbers (ii) and (v) in Figure 2 represent nanocellulose, where (ii) shows cellulose microfibrils (CMFs) and (v) shows cellulose nanocrystals (CNCs).

The natural fiber strength and stiffness in cellulose fibers comes from the formation of the microfibrils (ii). Microfibrils have a wide range from 2 to 30 nm depending on cellulose source and a length that can be several micrometers. The fibrils are assembled into long threadlike bundles of cellulose molecules stabilized by hydrogen bonds [7–8].

Cellulose has been known for about 150 years and is a renewable and biodegradable polymer and has for a long time been used as energy source, building material, and clothing. By chemical modification on the cellulose polymers, cellulose derivatives such as cellulose ethers and cellulose ester can be prepared, which have opened up for many novel material and applications for cellulose such as coatings, films, membranes, new building materials, drilling techniques, pharmaceuticals, and food products. Also the regeneration process of cellulose has contributed to novel techniques such as spinning of fibers and the viscose process.

Nanotechnology, in the recent years, has gotten huge interest in many industries and nanotechnology has opened up for many new possibilities, such as in the forest industry and cellulose-based products. Nanotechnology is defined as the understanding and control of matter with at least one dimension measuring from 1 to 100 nm. By mechanical treatments or chemical modifications on cellulose pulp, nanometer-sized cellulose such as cellulose nanofibrils (CNFs) and cellulose nanocrystals (CNCs) can be produced. Nanocellulose has shown extraordinary properties compared with the bulk material but also with other materials such as Kevlar, carbon fibers, or stainless steel [9–10].

Nanocellulose is not yet a fully commercial product but the first factory for production and disposal of CNC was opened in 2012 by CelluForce in Canada that produces about one tonne of CNC per day. There are also some pilot plants for nanocellulose production located all over the world and one of the first pilot plants for nanocellulose was opened by Innventia in Sweden in 2011 and the first pilot plant in the United States opened in 2012 in Madison and is the country's leading producer of nanocellulose materials where they produce both CNCs and TEMPO-based cellulose nanofibrils. The aim of the pilot plant is to aid the commercialization of nanocellulose materials by providing researchers and early adopters in the area with working quantities of nanocellulose.

The most recent news about the production of CNC was announced by MoRe Research on May 2015. A new pilot plant for production of CNC will be started in Örnsköldsvik in north of Sweden on 2016 and will be the first of its kind in Europe [11]. The operations of the pilot plant are based on the Israeli start-up company Melodeas technology. Melodea have developed a unique process that allows utilizing the sludge of the pulp and paper industry as a source for CNC production.

By expanding the knowledge in the area of nanocellulose and finding ways on how to control its properties during processing, new avenues in product development can be opened. Biobased products with outstanding properties can be produced and used as a replacement for fossil fuel-based products. There are some reviews about nanocellulose, most of them about nanocellulose as a reinforcement material in biocomposites or the use of bacterial nanocellulose (BNC) in medical or surgical products [4, 9, 12–15]. This review will deal with the current knowledge of isolation of nanocellulose, mainly derived from wood fibers, and chemical modification of cellulose nanocrystals, as well as the role of nanocellulose in biocomposites.

2. Nanocellulose

Nanocellulose, or variously termed nanocrystals, whiskers, rods, nanofibrils, or nanofibers, is when the cellulose fiber or crystal has at least one dimension within the nanometer size range. In 2011, TAPPI released a roadmap for the development of international standards of nanocellulose [16] where they stated the abbreviations for different nanocelluloses as: cellulose nanocrystals (CNCs), cellulose nanofibrils (CNFs), and cellulose microfibrils (CMFs), which will be the denomination used throughout this review. For cellulose nanocrystals, the noncrystalline regions are hydrolyzed and the remaining crystals will be in nanometer size range in all dimensions (Figure 3). For cellulose nanofibrils CNFs, sometimes also termed cellulose microfibrils (CMFs), some of the interfibrillated hydrogen bonds will break and form fibers with micrometer size in length (non-crystalline regions still present) and nanometer size in width. Bacterial nanocellulose (BNCs), on the other hand, are synthesized by special bacteria and grown as microfibrils in a culture medium. BNC microfibrils can also be hydrolyzed into bacteria nanocrystals by an acid hydrolysis similar to CNC.

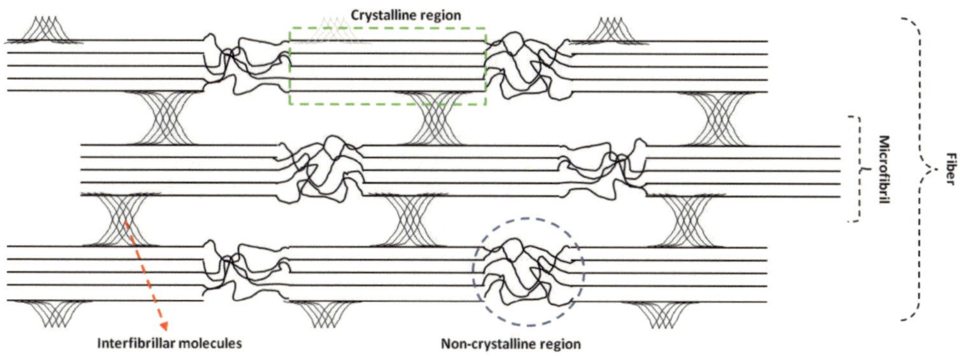

Figure 3. Part of a cellulose fiber where the crystalline and non-crystalline regions are shown. Acid hydrolysis removes the non-crystalline regions and only crystalline parts will remain (CNC). Mechanical treatment of the fibers will remain both the non-crystalline and the crystalline regions but some of the interfibrillar bonds will break, creating fibrils in nanometer size in width and micrometer size in length (CNF).

Nanocellulose has generated a high interest, especially as a filler in biocomposites. Some beneficial attributes of nanocellulose are its sustainability, abundance, mechanical properties such as large surface to volume ratio, high tensile strength and stiffness, high flexibility, and good electrical and thermal properties [17]. Cellulose and nanocellulose have been classified as safe, both to handle and to consume [18–22]. Cellulose and some of its derivatives are approved by the European Food Safety Authority (E-number: E460-E466 and E468-E469) [23] and the U.S. Food and Drug Administration (FDA) for use as additives in food products.

There are, however, some disagreements if nanocellulose should be seen as non-toxic just because its origin, cellulose, is. A recent study by Yanamala et al. found inflammatory responses in the lungs of mice that have been exposed to CNC derived from wood [24]. Since nanomaterials are a quite new area, more studies on health and safety aspects and other analysis methods might be necessary to develop.

Some application areas for nanocellulose are listed below [12–13]:

Paper, paperboard, and packaging: One of the applications for nanocellulose in the paper and paperboard industry is to enhance the fiber-fiber bond strength and have a reinforcement effect on paper materials [25].

Composite materials: Nanocellulose has many beneficial and unique properties and is commonly used as filler or reinforcement in biocomposites (see Section 7 in this review).

Food industry: Nanocellulose can form emulsions and dispersions and is suitable for use in food products as thickeners or stabilizers [22].

Medical and hygiene products: Nanocellulose has good absorption properties and can be used in, for example, tissues, non-woven products, or diapers.

Other applications: Films, painting, cosmetics, automotive, etc.

There are different types of nanocellulose depending on, for example, the cellulose source and treatment. Different cellulose sources give different characteristics and also different aspect ratios (L/d, where L is the length and d is the diameter). The aspect ratio is a measure of length and width of the cellulose crystals or fibers. Almost particular crystals have a low aspect ratio (L/d = 1), while microfibrils can have a very high aspect ratio due to its small diameter (nm) and long fibrils (nm-μm). The higher the aspect ratio, the higher is the reinforcement capacity when incorporated in composite materials. Also, the more surface area of the fillers, the more contact will the filler have with the polymer matrices [17, 26]. Typical aspect ratio for CNC ranges from 1 to 100 [15, 27] and for CNF 15–100 [27].

3. Cellulose Nanocrystals (CNC)

Cellulose nanocrystals (CNCs) generally have a width of about 2–30 nm and could be several hundreds of nanometers in length and are formed during acid hydrolysis of cellulose fibers where a selective degradation of the more accessible, disordered parts takes place. Since the non-crystalline regions (see Figure 3) act as structural defects in the microfibril, it is responsible for the transverse cleavage of the microfibrils into short monocrystals under acidic hydrolysis [28]. In the early stage of the hydrolysis the acid diffuses into the non-crystalline parts of the cellulose fiber and hydrolyzes the glycosidic bonds. After these, more easily accessible glycosidic bonds in the polymer are hydrolyzed and finally hydrolysis occurs at the reducing end group and at the surface of the nanocrystals. The harder it is for the acid to hydrolyze the glycosidic bonds the slower is the reaction [29–30]. The hydrolysis of the reducing end groups and the surface of the nanocrystals will make the nanocrystals charged depending on what acid is used. By using a 64 wt% sulphuric acid solution, 0.5–2% sulfate groups will be attached to the surface of the nanocrystal [31]. Due to the charged sulfate groups, CNC will form stable colloidal dispersion when diluted in water to specific concentrations [32–35].

Cellulose nanocrystals have been hydrolyzed from many different cellulose sources such as hardwood pulp [30], softwood pulp [36–39], microcrystalline cellulose (MCC) [40–41], sisal [42], cotton [39, 43–45], wheat straw [7, 46], rice straw [43, 47], bacterial cellulose [39, 48], algae [40, 49], banana fibers [17], sugar beet [50–53], and tunicin [39, 45, 53–60]. Other biopolymers that have been reported to form nanocrystals during acid hydrolysis are chitin [61–62], potato pulp [63–66], yellow pea [67], and waxy maize [68–69]. The different types of cellulose sources give some different structures of the nanocrystals and the aspect ratio will differ for the different sources. The dimensions and aspect ratio for some different cellulose sources are shown in Table 1.

Cellulose source	Length, L (nm)	Cross section, D (nm)	Axial ratio, L/D	References
Wood	100–300	3–5	30–70	[30, 37]
Cotton	100–400	7–15	10–20	[29–30, 33, 35]
Algae (Valonia)	100 nm to μm	10–20	N/A	[30, 40, 49]
Bacterial cellulose	100 nm to μm	5–10	N/A	[30,48]
Tunicate cellulose	100 nm to μm	10–20	67	[30, 35, 57]
Sugar beet pulp	210	5	40	[35]
Wheat straw	220	5	45	[35]

Table 1. Geometrical characteristics of nanocrystals from different cellulose sources.

3.1. Acidic treatment on cellulose

When cellulose pulp is treated with an acid, the pulp will start to degrade. The degradation will start with the most accessible parts of the fiber, followed by the reducing end groups and the crystal surfaces. Acid concentration, reaction time, and reaction temperature are some of the most important parameters for controlling the acid hydrolysis of wood pulp. A reaction time that is too long will hydrolyze the cellulose crystals completely and a reaction time that is too short will give a high degree of polymerization (DP) due to large undispersed fibers [30]. The reaction temperature and time correlate to each other and a higher reaction temperature, shorten the reaction time. Not only time and temperature affect the properties of nanocellulose, but also the acid concentration and acid to pulp ratio [29]. After hydrolyzation and purification through dialysis, small crystalline rod-like particles will be yielded in an aqueous suspension.

The nanocrystalline cellulose that is formed through the acidic treatment is of colloidal dimensions and forms an aqueous suspension when stabilized. The critical concentration of the colloidal suspension, which is the lowest concentration where the whiskers self-organize, depends on particle size, acidic treatment, preparation conditions, aspect ratio, and ionic strength [29–30, 33, 70–71]. Revol et al. [36] showed that microcrystals from bleached Kraft wood pulp spontaneously ordered in a crystalline phase above the critical concentration. This observation was also observed for cellulose whiskers from cotton and the critical concentration ranging from 2 wt% to 10 wt% depending on the preparation conditions.

The self-ordered nanocrystals form a chiral nematic ordering as seen in Figure 4. This helical, self-ordered structure has catalytic and photonic crystal benefits and researchers have tried to introduce chirality into porous inorganic solids by using CNC as a template to improve properties in other materials for applications such as optical filters or sensors [72]. From many aqueous CNC solutions, a bluish color can be observed that is due to the helical chiral nematic ordering and the length of the pitch gap (p). CNCs are able to absorb visible light and depending on the length of the pitch gap, different wavelengths are absorbed and the reflected light emits different colors. Therefore, different colored CNC film can be seen if the films are sufficiently thin.

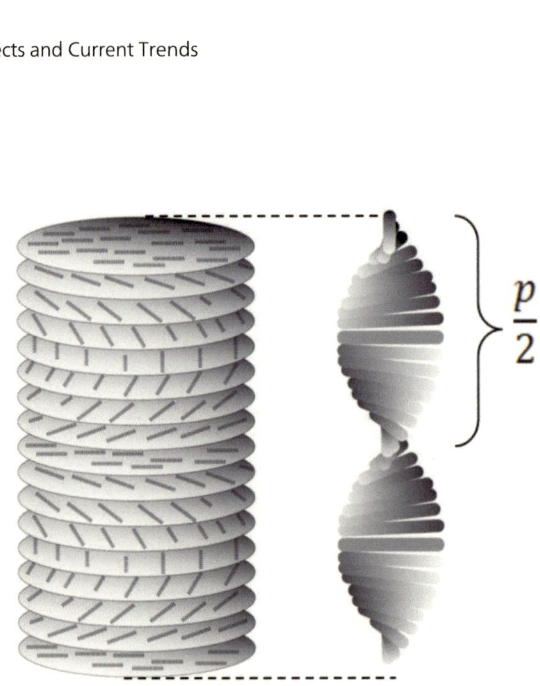

Figure 4. At certain concentrations, the nanocrystals self-organize into a chiral nematic ordering where the length of the pitch gap (p) in the helical structure can absorb different wavelengths and emit wavelengths with different colors. *Illustration: Jari Sundqvist.*

Sulfuric acid (H_2SO_4) is the most common acid for nanocellulose preparation through chemical hydrolysis but it is possible to use other acids as well. Hydrochloric acid (HCl), hydrobromic acid (HBr), and phosphoric acid (H_3PO_4) have been used for CNC preparation but compared to sulfuric acids, hydrochloric acid, and hydrobromic acid, will not have any surface charges and a stable colloidal dispersion is, therefore, harder to form (Figure 5). Phosphoric acid will give charged phosphate groups on the nanocrystal surface (Figure 5c). Regarding the process industry, sulfuric acid is a more suitable choice of acid compared to hydrochloric acid. In 2008, sulfuric acid was the most produced chemical in the U.S., which was almost 10-fold more compared to hydrochloric acid, making sulfuric acid economically beneficial due to its larger quantities [73].

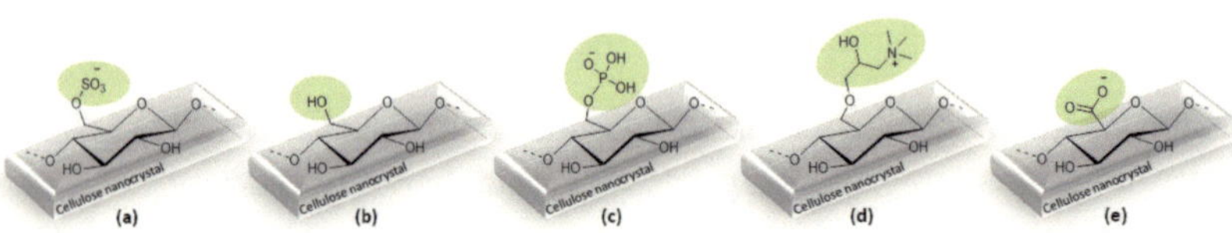

Figure 5. The surface polymer chain of CNC can be modified with different functional groups to give different surface characteristics. The surface modification of CNC depends on the isolation process and further treatment of the nanocrystals. The figure shows some examples of different functional groups attached to the surface cellulose polymers where (a) shows the surface polymer chain after H_2SO_4 hydrolysis, (b) after HCl or HBr hydrolysis, (c) after H_3PO_4 hydrolysis, (d) after H_2SO_4 hydrolysis followed by a surface cationization, and (e) after a HCl/HBr hydrolysis followed by a TEMPO-oxidation.

3.1.1. Isolation of CNC using sulfuric acid

Sulfuric acid hydrolysis of cellulose pulp is a heterogeneous process where the acid diffuses into the pulp fiber and cleaves the glycosidic bonds in the cellulose polymer. Depending on reaction times, the hydrolysis could also occur on the crystalline regions and some of the hydroxyl groups on the crystalline surface will convert into sulfate groups (e.g., conversion of cellulose-OH to cellulose-$OSO_3^-H^+$, Figure 5a). Other side reactions are also possible during acid hydrolysis such as dehydration and oxidation [36, 74]. In the cellulose pulp sample, hemicelluloses or pectin might be present and these polysaccharides will also undergo hydrolysis but at faster rates due to their higher reactivity. As the acid hydrolysis proceeds, the DPs are expected to decrease since the non-crystalline regions in the microfibril will be removed. Due to the loss of non-crystalline regions the crystallinity will increase and also the insolubility against water because the crystalline parts are less accessible (see Figure 6) [74].

The sulfuric acid hydrolyze reaction has been optimized by several researchers and one general way to produce CNC from sulfuric acid is by using a 64 wt% sulfuric acid solution at 45 °C for 45–60 min with constant stirring, followed by quenching the suspension with 10-fold deionized water, concentrate the CNC through centrifugation and dialysis against deionized water until constant neutral pH is achieved. To achieve separate crystals, the suspension has to be sonicated repeatedly [29, 33, 75].

The sulfuric acid hydrolysis reaction conditions on softwood pulp has been evaluated [30, 74]. When using a 64 wt% sulfuric acid solution, an acid to pulp ratio of 8.75 ml/g and treated the pulp at 45 °C at two different reaction times, 25 and 45 min, the longer reaction time showed a less polydisperse length distribution and a higher sulfur content than the shorter reaction time did. When the acid to pulp ratio is increased to 17.5 ml/g and the pulp is treated for 45 min at 45 °C, a smaller length and polydispersity was observed but the effect was not that large compared with the much higher acid to pulp ratio. Also, different sulfuric acid concentrations on softwood Kraft pulp and its effect on DP, crystallinity, crystal size, and yield was observed [74]. Three different sulfuric acid concentrations (16 wt%, 40 wt%, and 64 wt%) with the acid to pulp ratio 8.75 ml/g and three different reaction temperatures (45 °C, 65 °C, and 85 °C) were evaluated at a reaction time of 25 min. Both higher acid concentration and higher reaction temperature resulted in a lower DP, individually. The yield should be lower for CNC due to the loss of the non-crystalline regions and for the pulp samples hydrolyzed with a 64 wt% sulfuric acid, the yield drops at all three temperatures that were tested. Figure 6 shows a diagram where the crystal size, DP, and amount of crystallinity are marked in relation to each other. The highest crystallinity was obtained for the highest concentration of sulfuric acid, which also gives the smallest crystals and lowest DP.

Dong et al. studied the preparation of CNC from cotton fibers where the reaction time and temperature were in focus [29]. A 64 wt%, sulfuric acid was used and the acid to pulp ratio was 8.75 ml/g. The temperatures tested were 26 °C, 45 °C, and 65 °C, and the reaction times were 15 min up to 18 h. For low temperatures (26 °C), the reaction time needed to be very long (18 h) to produce CNC that could form an ordered suspension. At 65 °C, the reaction was hard to control and already after 15 min a color change was noted, indicating side reactions such as dehydration. After 1 h at 65 °C, the whole suspension had turned from white to black. The

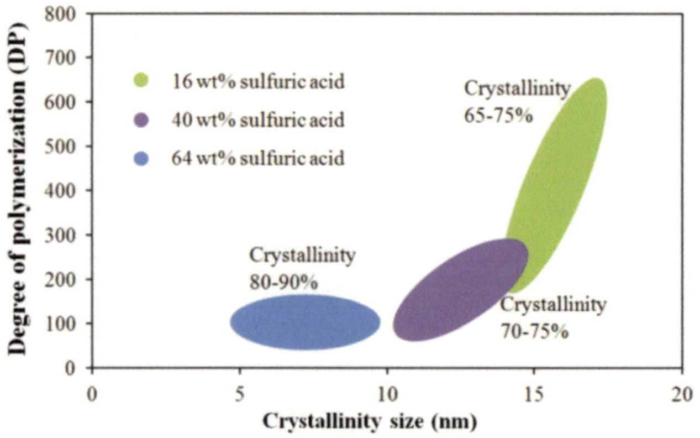

Figure 6. Crystallinity size in relation to the DP. During acid hydrolysis, the DP will decrease and the cellulose crystallinity will increase, due to loss of non-crystalline regions and, as a result, smaller crystal size will be obtained.

most optimal and easy handled reaction parameters for a stable colloidal CNC suspension was performed at 45 °C for 1 h.

Dong and co-workers also studied the relationship between reaction time and crystallinity size of the cellulose particles for hydrolysis with a 64 wt% sulphuric acid at 45 °C for 10–240 min [29]. The reaction was fast from the beginning but slows down in the later stages. The initial rapid decrease was due to the acid that diffused into the non-crystalline regions and hydrolyzed the most accessible glycosidic bonds. The harder it was to hydrolyze the glycosidic bonds the slower was the reaction. After 1 h at 45 °C, the crystallinity size became relatively stable.

3.1.2. *Isolation of CNC using alternative acids*

There are other acids besides sulfuric acid that can be used for the hydrolysis of cellulose fibers into CNC, for example, hydrochloric acid, which when compared to sulphuric acid will not give any charged groups on the cellulose crystal surface (Figure 5b). The lack of charged groups could be suitable for the study of enzymatic degradation, which is hard to study when surface groups can hinder the substrate recognition by the enzymes [37].

The procedure for a hydrochloric acid hydrolysis was described by Araki and co-workers. The hydrolysis was prepared by treating Kraft pulp with 30 ml/g of a 4 N hydrochloric acid at 80 °C for 225 min. The suspension was then centrifuged repeatedly until the sample reached pH 4 and the CNC became non-sedimenting and the supernatant became turbid. The supernatant was collected, purified, and neutralized by dialysis against deionized water and finally sonicated to disperse the suspension [37–38].

Hydrobromic acid and phosphoric acid are two other acids that have been used in the preparation of CNC. Lee and co-workers studied HBr-CNC for their use in PVA composites [76] and the HBr-CNC was prepared by adding 40 ml/g 1.5–2.5 M hydrobromic acid to MCC. The hydrolysis was quenched after 4 h at 100 °C. The HBr-CNC showed a decrease in DP of

the higher acid concentration used and a higher thermal stability both for the CNC and the composite films.

Zhang and co-workers [77] studied the effect of four different acids for CNC hydrolysis on bamboo cellulose where 25 ml/g acid with concentration of 6.5 M was added to the cellulose samples and the hydrolysis occurred for 2 h at 60 °C. The different acids were a sulphuric acid solution, hydrochloric acid solution, phosphoric acid solution, and a mixture of acetic acid and nitric acid solution (ratio 10:1). Sulfuric acid gave the highest crystallinity index followed by phosphoric acid. Hydrochloric acid and the mixed acetic and nitric acid solution gave the lowest crystallinity index due to the higher tendency to promote the breakage of the hydrogen bonds in crystalline regions of cellulose. Both hydrochloric acid and the mixed acetic acid and nitric acid solution have better capability to swell cellulose, thereby facilitating the breakage of intra- and intermolecular hydrogen bonds in the crystalline regions.

Since no surface groups will be attached on the cellulose crystals when prepared from hydrochloric acid or hydrobromic acid, the colloidal dispersion is not as stable for HCl/HBr-CNC as for CNC prepared from sulfuric acid. A comparison between sulfuric acid and hydrochloric acid have been made on waxy maize starch and on softwood Kraft pulp, where it was found that CNC prepared from hydrochloric acid showed a higher risk for agglomerate in aqueous medium. The charged groups on CNC prepared from sulfuric acid limits their ability to flocculate [37, 78]. The CNC prepared from hydrochloric acid showed a thixotropic behavior easier than for CNC from sulfuric acid. This was due to particle aggregation, which is formed in static conditions but is destroyed by shear flow. By increasing the distance between the particles, the ability to aggregate reduces. Colloidal suspensions are often stabilized by electrostatic repulsion or by steric hindrance such as grafted polymers.

Araki et al. described a method to attach sulfate groups to the cellulose crystals prepared from a hydrochloric acid hydrolysis through an esterification reaction with sulfuric acid (post-sulfonation) [38, 79]. The esterification reaction was carried out with a 65 wt% sulfuric acid solution added to CNC prepared from hydrochloric acid and the mixture was reacted in a shaken water bath for 2 h at 60 °C. After reaction, the sample was diluted with a large amount of cold water and washed by repeated centrifugation and decantation cycles, followed by dialysis and sonication. Through this method it was possible to control the surface charge on the cellulose nanocrystals.

Stable aqueous systems of CNC prepared from hydrochloric acid hydrolysis have been achieved by grafting polyethylene glycol (PEG) onto the nanocrystals. The grafting was performed by using an oxidative carboxylation-amidation procedure known as TEMPO [80–81]. The grafting of nanocrystals acts as a steric hindrance in the CNC suspension and inhibits particle aggregation and enhances a stable colloidal CNC solution. An example of charged surface groups from TEMPO-oxidations is shown in Figure 5e.

4. Cellulose Nanofibril (CNF)

Cellulose nanofibrils (CNFs), sometimes termed cellulose microfibrils (CMFs), are unlike CNC produced by mechanical treatment that preserves the non-crystalline parts in the microfibril as well as the length of the fibrils. Microfibrils are long, threadlike bundles of cellulose molecules that are stabilized by hydrogen bonds mainly between the many hydroxyl groups in the cellulose polymer. By using mechanical treatment, forces will peel the fibers and the interfibrillar bonds between the cellulose molecules will break and give nanofibrils with a diameter in nanodimensions and fiber length ranging from nanometer to micrometer. A common mechanical treatment for CNF is refining the pulp, followed by homogenization, which will individualize the nanofibrils and give a stable dispersion when diluted in water. Another technique to produce CNF is through regeneration and electrospinning of a cellulose polymer melt [4].

CNF can be produced from many different types of cellulose sources. A huge area of interest for CNF is in applications such as composite materials where the high aspect ratio of CNF and long flexible microfibrils is suitable for use as reinforcement in a polymer matrix. Bhatnagar and Sain studied the possibility to use cellulose nanofibers from plant cells due to its high abundance and low costs [82]. CNF were prepared from flax fibers, rutabaga, Kraft pulp, and hemp fibers using chemo-mechanical treatments before they were used as a filler in a polyvinyl alcohol (PVA) matrix. The use of plant fibers as reinforcements in composites did improved the mechanical properties in the composite material compared with the pure PVA [82].

4.1. Mechanical treatment of cellulose fibers

The mechanical treatment of pulp fibers consists of refining the pulp, followed by a high-pressure homogenization process to obtain individualized cellulose nanofibrils. In order to purify the cellulose fibers and decrease the amount of lignin, hemicellulose, and pectin, chemical treatments might be necessary before the mechanical treatment. There are mainly two different treatments to extract the fibers before homogenization of the nanofibril solution: using a refiner or by cryochrusching.

In the refining process, a dilute suspension of cellulose (1–2 wt%) is treated with forces in a refiner or blender equipped with bars, against which the fibers are subjected to repeated cyclic stresses which changes the morphology and size of the fibers [4, 52, 64]. The refining process is carried out before the individualization of the cellulose fibers because refining causes external fibrillation by peeling the external cell wall layers (P and S1 layers, see Figure 2) and exposing the S2 layer, which is the layer with the highest quantity of cellulose. Refining also cause internal fibrillation that unlooses the fiber wall, which is suitable in the following homogenization process [4].

Another treatment to extract cellulose fibers is cryochrusching. A liquid nitrogen-frozen pulp is mechanically crushed and the ice crystals, which are formed within the pulp cell wall, leads to release the cellular wall fragments [50]. These smaller fragments are later diluted in water before their homogenization to a CNF suspension in a homogenizer.

The individualization of the nanofibrils takes place in a homogenizer. In the homogenization process, refined and diluted suspensions of cellulose fibers are pumped at high pressure and fed through a spring high-pressure loaded valve assembly. The valve is opened and closed very rapidly, which lead to large pressure drops with shearing and impact forces affecting on the fibers. The number of passes through the valve will vary depending on the starting material. The combination of high pressure and forces on the cellulose fibers gives a high degree of microfibrillation, resulting in a CNF solution [4, 50, 64].

To ease the mechanical treatment of microfibers and to reduce the energy consumption, chemical treatment on the cellulose fibers can be used as a pre-treatment before the refining takes place. The pre-treatment can be enzymatic [83] or by introduction of charged groups to the fiber surface, e.g., through a carboxylation or by TEMPO-oxidation [84]. The pre-treatment will affect the surface of the fibers, which will make it easier to peel the fiber in the refining process and by that reduce the energy consumption..

5. Bacterial Nanocellulose (BNC)

The last kind of nanocellulose that will be discussed is the bacterial nanocellulose (BNC), which is synthesized from special bacteria that build up nanofibers with nanometer size in diameter and up to micrometer size in length. There are several reviews and research reports found in the literature describing BNC and its structure and properties [4, 12–13]. Unlike other cellulose sources, bacterial cellulose is grown to nanofibers by special bacteria, such as *Acetobacter* species, cultivated in a culture medium. The bacteria produce BNC by synthesizing cellulose and building up bundles of microfibrils [85].

Acetobacter is a microorganism present everywhere in nature where sugar fermentation occurs and it is involved in the conversion of ethanol to acetic acid. Acetobacter has for many years been used in the fermentation industry for mass production of acetic acid and is important in several other industries where acetic acid is of importance.

BNC have unique properties like an extremely fine and pure fiber network structure, high degree of polymerization (up to 8,000), good mechanical properties such as high mechanical strength, biocompatibility, and water holding capability [4, 13]. The main application area for bacterial nanocellulose is in medical health and surgical applications such as bandages for wound healing or skin burns or as a substitute for medical materials such as blood vessels. Some other application areas where BNC can be found include the food industry and the paper and packaging industry [86–88]. The use of BNC in composites is also of interest, which have been studied by Grunert and Winter [89], where BNC fibrils first where hydrolyzed to bacterial nanocrystals and then used as a reinforcement in cellulose acetate butyrate (CAB) films.

The procedure that Tokoh et al. used to produce bacterial cellulose from Acetobacter species was to cultivate the bacteria in a culture medium for three days followed by liberation of the bacteria by rinsing the synthesized sheet with distilled water. The cells were gathered by centrifugation and re-suspended in the control medium [85]. The bacterial cellulose is micrometer long threadlike bundles of nanofibrils, which can be hydrolyzed to nanocrystals by e.g., sulphuric acid hydrolysis, as described in section 3.1.1.

Some disadvantages with BNC are the low availability of the bacterial cellulose, the inefficient process in synthesizing bacteria cellulose and the high costs, which makes it hard to make BNC commercially attractive. The traditional process in synthesizing BCN cannot produce the high quantities that would be requested for a commercialization of BNC and further process development needs to be performed for a large scale production of bacteria cellulose [88].

6. Chemical modification on nanocellulose

During sulfuric acid hydrolysis of cellulose, starch or chitin, sulfate groups will cover the surface of the nanocrystals. If using hydrochloric acid instead, the sulfate groups can be attached to the nanocrystal surfaces afterwards by an esterification reaction with sulfuric acid [37–38, 79]. In this way, the amount of charged groups on the cellulose crystal surface can be controlled. By using chemical modification on nanocrystals or nanofibers, the properties of the nanocrystal or nanofiber could be changed and controlled in specific ways.

Figure 5 shows some different surface treatments on CNC where a TEMPO-oxidation or cationization of the CNC surface can give charged side groups which will give electrostatic repulsion between the crystals and prevent aggregation, especially on CNC prepared from hydrochloric acid or hydrobromic acid.

Most of the literature that describes chemical modification of nanoparticles had the aim of improving the interfacial compatibility between the nanocrystals and various polymer matrices. A better compatibility could enhance the mechanical properties of the composite materials and addition of just a few percent of nanocellulose to a polymer matrix could improve the mechanical properties of the significant composite material [27, 35, 62, 90]. Most of the literature about chemical modification on nanocellulose describes a surface modification of the crystals or fibrils. For surface modification of CNC, there are mainly two different procedures, surface compatibilization or co-polymerization. In surface compatibilization the main idea is to attach a small molecular agent to the cellulose polymer, where the molecular agent contains at least one reactive moiety. For co-polymerization the molecular agent needs at least two functional groups, one that can react with the hydroxyl groups on the cellulose polymer and one that can covalently bond to a polymer matrix. The co-polymerization can be performed in different ways such as grafting, radical reactions, or by the use of organometallics [27].

Chemical modification on chitin nanowhiskers has been performed to change the surface properties of the nanocrystals to make it possible to have in organic solvents [62]. The reagents used to modify the chitin nanowhiskers were alkenyl succinic anhydride, phenyl isocyanate, and isopropenyl-α,α'-dimethylbenzyl isocyanate. The chemically modified chitin whiskers showed stable suspensions in toluene and were used in composite films casted from a toluene solution and all of the results showed that the chemically modified whiskers improved the adhesion between the filler and the matrix. The results also showed that the chemical treatment decreased the mechanical performance of the composites. The loss of performance could be due to the destruction of the 3-dimensional network of the chitin whiskers that occurs in the unmodified composites [61–62]. Nanocrystals from waxy maize starch have also been chemically modified with alkenyl succinic anhydride and phenyl isocyanate [90] to study the crystal structure of the nanoparticles after chemical treatment. X-ray diffraction analysis confirmed that the crystalline nanostructure was preserved and contact angle measurements concluded that the surface chemical modification enhancing the non-polar nature of waxy maize starch nanocrystals. A non-polar nanocrystal allows the use of non-polar polymers as a matrix for composite materials.

2,3-epoxypropyltrimethylammonium chloride (EPTMAC) have been used for surface cationization on nanocrystals from cotton cellulose under aqueous alkaline conditions [91]. The surface cationization resulted in an electrostatically stabilized CNC suspension due to the cationic trimethylammonium groups on the crystalline surface (Figure 5d). The morphology and dimensions of the nanocrystals were not affected by the alkaline conditions but the sulfate groups on the nanocrystals surface were hydrolyzed. A decrease in total charge density was observed due to the loss of surface sulfate groups that resulted in an increased tendency for the modified CNC to form thixotropic gels.

Cationic CNC as a reinforcement in fiber networks such as paper or paperboard has been patented [25]. Additions of 1–5 wt% of cationic CNC are expected to increase the strength in the fiber-fiber joints and in that way improve the mechanical performance of the fiber material. Applications for these types of materials can be found in the packaging industry.

6.1. Chemical modification of nanocellulose during acid hydrolysis

There are some literatures available for chemical modification on nanocellulose where the modification takes part on the produced nanocellulose. Most of these modifications are to improve the interface between cellulose and a polymer matrix in biocomposites. There are not that many studies done on chemical modification during acid hydrolysis but some available studies are among other the acidic hydrolysis with sulfuric acid where sulfate groups are attached to the CNC surface. In recent years, studies on the hydrolysis reaction of CNC with phosphoric acid has been performed with the aim of creating phosphoric groups on the surface. During the phosphoric acid hydrolysis, reagents, such as acetic anhydride, have been added to a phosphoric acid hydrolysis to obtain acetylated CNC in a single-step procedure [92].

It would be interesting to study if more traditional organic synthesis on hydroxyl groups in acidic environments could be used for chemical modifications during acid hydrolysis from pulp to CNC. There are at least (in theory) two possible ways to achieve chemically modified nanocrystals during hydrolysis: (1) to find alternative reagents to sulfuric acid that at the same time hydrolyzes the cellulose fiber and modify the nanocrystals in some way, or (2) add a reagent to the sulfuric acid (or other acid that hydrolyze cellulose fibers) that could react with the nanocrystals and is functional in acidic and aqueous conditions.

Attempts of using heteropoly acids (HPA) to catalyze organic reactions in water, was made on cellulose where HPAs were used to hydrolyze cellulose polymers into glucose units [93–94]. HPA is a promising green solid acid catalyst that can replace environmentally harmful liquid acid catalysts. HPA ionic liquids have also been used to catalyze the conversion of sucrose and starch into glucose.

Chemical synthesis that works under acidic conditions and in the presence of water is desirable for CNC modification during hydrolysis. Fischer esterification could be an option where a carboxylic acid reacts with the hydroxyl groups on the nanocellulose. In a Fischer esterification, a carboxylic acid is refluxed with an alcohol in the presence of an acid catalyst. In the case of CNC, a carboxylic acid could react with the hydroxyl groups on cellulose in presence of a sulfuric acid as catalyst.

In the production of acrylamide describe in a patent by McNae [95], one intermediate to acrylamide is acrylamide sulfate, containing 60 wt% sulfuric acid. By increasing the amount of sulphuric acid to a 66 wt% solution and still have the unsaturated acrylate groups left in the solution, unsaturated ester groups could be attached to the cellulose nanocrystal during acid hydrolysis. The authors research group has performed an acid hydrolysis with acrylamide sulfate on microcrystalline cellulose (MCC, Avicel PH-101) with a procedure similar to the one described earlier [29, 33]: the hydrolysis occurred for 2 h at 45 °C with an acid to pulp ratio of 8 g/ml. During the acid hydrolysis, the MCC solution shifted from white to black, which, according to Dong, Revol, and Gray, concluded that side reactions such as dehydration take place when the hydrolysis turns black [29]. After the hydrolysis, the solution was quenched in deionized water before centrifugation, decantation, and dialysis against deionized water.

Figure 7a shows an atomic force microscope (AFM) picture of the acrylamide sulfate hydrolyzed nanocrystals were all the crystals are within the nanometer-sized area in all dimensions. The AFM picture in Figure 7a is similar to CNC produced in the same way but with 64 wt% sulfuric acid (Figure 7b-c). This indicated that the acrylamide sulfate solution could hydrolyze the MCC to nanoparticles, but the process need further optimization since the solution changed color, meaning the hydrolysis went too far. In the AFM picture (Figure 7a), small particles can be seen that further indicates that the hydrolysis went too far. Fourier-Transform Infrared (FTIR) analysis of the acrylamide sulfate hydrolyzed CNC showed new absorption bands at 1,718 cm^{-1} and at 811 cm^{-1} corresponding to unsaturated ester groups and alkene groups, respectively, which correlates well to the expected surface group attachment. More studies on the material needs to be performed for further evaluation.

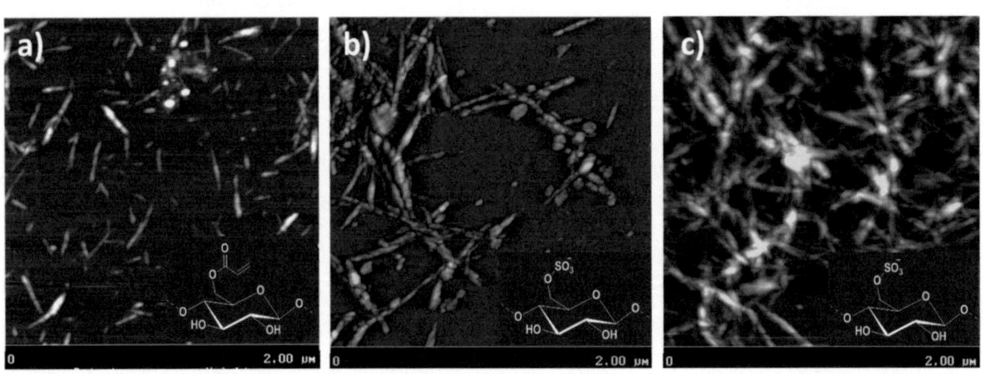

Figure 7. AFM picture of CNC from (a) acrylamide sulfate hydrolyzed MCC and a schematic picture of CNC with an acrylate functional group, (b) AFM picture of CNC that has been fully dried, and (c) dried to a high viscous gel. All the crystal dimensions are within nanometer size.

6.2. Drying of nanocellulose

When preparing nanocellulose from acid hydrolysis, the nanocrystals are in a water suspension. By removing some of the water, a critical concentration will be reached. Above the critical concentration, the CNC will be in a stable colloidal dispersion. The critical concentration for CNC is normally between 2 and 10 wt% [36].

Fully dried CNC prepared, for example, through freeze-drying, will give thin lamellar white flakes that can be re-dispersed in water. Figures 7b and 7c show two different AFM pictures of CNC where different amounts of water have been removed, one fully dried (Figure 7b) and the other one dried to a high viscous gel with a CNC content of 8.9% (Figure 7c). The two samples, which were taken from the same CNC batch, were then re-dispersed in deionized water to the same CNC concentration of 4.15%. The CNC have been prepared from MCC (Avicel PH-101) hydrolyzed in a 64 wt% sulphuric acid at 45 °C for 2 h. Both pictures show crystals in the nanometer size area.

The AFM pictures in Figure 7b–c show no significant differences on the dimensions of the nanocrystals. Both of the two samples show thixotropic behavior, where the fully dried sample showed more tendency to thixotropy compared to the high viscous gel sample. Figure 8 shows the fully dried sample re-dispersed in water to a CNC concentration of 4.15%. In static mode it, behaves as a gel and immediately after agitation, it behaves more like a liquid.

Figure 8. Thixotropic behavior of the fully dried CNC sample re-dispersed in water to a CNC concentration of 4.15%. (a) In static mode and (b) immediately after agitation.

6.2.1. CNC films

CNC films were prepared from the two different CNC samples shown in Figures 7b–c, one sample that had been freeze-dried and one that was dried to a high viscous gel. The two samples were re-dispersed in water to a CNC concentration of ca 1% and sonicated to destroy eventual aggregates. The water was thereafter allowed to evaporate in ambient conditions until two solid air dried CNC films was obtained from the two different samples (Figure 9).

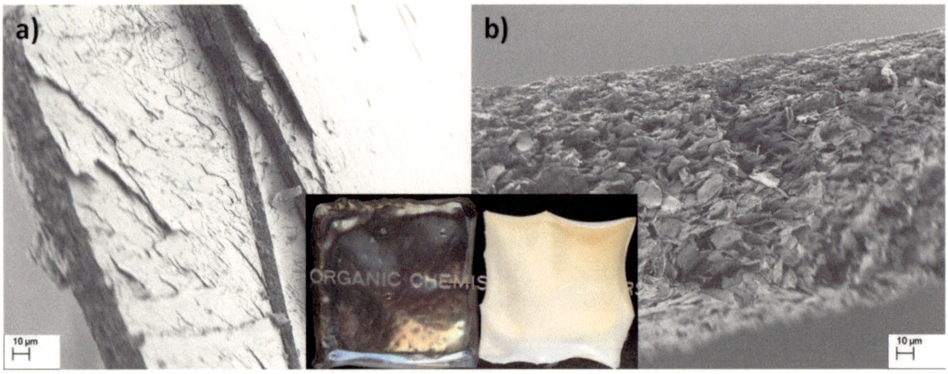

Figure 9. Scanning electron microscope (SEM) pictures of CNC films. (a) CNC from the high viscous gel suspension and (b) CNC from the freeze-dried sample.

The two films shown in Figure 9 show very different appearances, where the film from the high viscous gel sample is transparent and the film from the freeze-dried sample shows no transparency at all. When a never dried CNC suspension evaporates, the CNC concentration will gradually increase and once the critical concentration for a stable colloidal dispersion is exceeded, a self-ordered phase will form. This ordered form will remain as the CNC concentration increase further until all water has evaporated. A solid semitransparent CNC film will be left where the crystals are tightly packed in the same ordered crystalline form. For sufficiently thin films, colors can be seen in the film.

For CNC that has been fully dried or dried far above its critical concentration, particle aggregation occurs, leading to loss of properties or functionality [96]. To overcome this problem, surfactants or surface modification can be used to prevent aggregation during drying [80]. Beck, Bouchard, and Berry studied the difference between freeze-dried CNC films and CNC films from an aqueous CNC solution. The CNC film from the aqueous solution could be re-dispersed again but not the freeze-dried CNC film. Beck and co-workers found out that by adding a sodium cation (Na^+) to the CNC solution before drying, the freeze-dried Na^+-CNC was completely re-dispersed in water to give colloidal CNC suspensions [96].

By studying CNC films with FTIR analysis, Dong and Gray explained the phenomena as intermolecular hydrogen bonding from the cellulose backbone was much stronger in the CNC film without sodium added (the S-H bonding from the sulfate groups in the nanocrystals) than in the CNC film with sodium added (S-Na) [70]. The freeze-dried CNC-film shown in Figure 9b did not have any cations added and this might be the reason why the freeze-dried CNC did not behave like the film from the high viscous gel sample in Figure 9a.

The importance of knowing how to re-disperse fully dried CNC is necessary for many applications of CNC. It is convenient, when handling CNC, to fully dry it to minimize its shipment size, weight, cost, and the inhibition of bacterial and fungal growth. Many applications also require a dried CNC that can be re-dispersed in water or in organic solvents for chemical modification or for nanocomposite manufacture [96].

http://dx.doi.org/10.5772/61899

7. Biobased composites

There are a lot of reviews available regarding nanomaterials as reinforcements in composite materials [4, 9, 12–15, 27, 35, 97–100] and the huge interest of nanomaterials in composite products is due to the many advantages of the biobased nanomaterial. Some of the beneficial properties with nanomaterials from bioresources are listed below [4, 15]:

- Renewable resources from the nature
- Resources available all over the world
- Low costs compared to other nanometer-sized reinforcement materials
- Low energy consumption in manufacturing
- Low density compared to other nanomaterials used as reinforcements, e.g., metal or glass fibers
- High strength, high stiffness, and high E-module
- Good electrical and thermal properties
- Chemical modification can change the properties of the nanocellulose, which make it possible to affect the compatibility between the nanocellulose and different polymer matrices

Composite materials are widely used in many applications today and consists of a polymeric matrix and a filler material as reinforcement. The fillers used are often synthetic such as glass fiber, carbon, or aramid. The addition of a filler to a polymeric material enhances the mechanical and thermal properties of the composite material, compared to the polymeric material itself, because of their high performance and great versatility. Today, there is a strong focus on environmental issues and the synthetic fillers used in composite materials as they cause problems at the end-of-life disposal due to their partial combustibility and the high demand on techniques for recycling of the materials [82]. By replacing the synthetic fillers with natural ones, such as cellulose fibers, starch, or chitin there will be many positive environmental benefits and advantages including lower costs, lower density, good thermal properties, and biodegradability [99].

By using nanomaterials as fillers in composite materials improved stiffness, strength, toughness, barrier properties, and flame retardancy can be achieved compared to the pure polymer material. The addition of only a few percent (1–5 wt%) of nanomaterial is enough for these improvements due to the large surface area of the nanoparticles [98].

Compared to carbon nanotubes, which is the strongest nanofiber produced today, cellulose nanofibers from wood only have 25% of the strength but the costs for cellulose nanofibers are much lower than the costs for carbon nanotubes, which makes the wood-based nanomaterial more attractive for certain applications [4]. The reinforcing ability of the cellulose whiskers comes from its high surface area and good mechanical properties. Table 2 shows the tensile strength and elastic modulus (E-module) of some different reinforcement materials [9].

Material	Tensile strength (GPa)	E-module (GPa)
CNC	7.5–7.7	110–220
Kevlar	3.5	124–130
Steel wire	4.1	210
Carbon fiber	1.5–5.5	150–500
Carbon nanotubes	11–63	270–950

Table 2. Tensile strength and E-module for different reinforcement materials used in composite materials.

The difference between hardwood fibers with diameters in micrometer size and nanometer size (CNF) have been evaluated when used as reinforcement in a polyurethane matrix [101]. Two different filler concentrations were used for both the cellulose fibers (CF, 8.5 and 18.7 wt %) and the CNF (7.5 and 16.5 wt%). Both mechanical and thermal behavior were evaluated for the composite materials and the results shows that both the tensile strength and E-modulus were improved with increased CF and CNF content, compared to a pure polyurethane sample. The CNF composite showed a larger increase in mechanical properties than CF, and are therefore more effective as reinforcement than the micrometer-sized cellulose fibers. The improvement in mechanical properties for CNF compared to CF is due to the smaller fiber dimensions and higher aspect ratio, which gives a better incorporation of the nanofibrils in the polymer matrix.

The cellulosic three-dimensional network of intermolecular hydrogen bonds between the cellulose molecules gives strong interactions between the fillers (fibers or fibrils) or between the fillers and the polymeric matrix, resulting in better composite properties compared to the pure polymer matrix alone. Another important effect of fibers or CNF is the high flexibility of the cellulose polymer that gives a tangling effect with the polymer matrix that contributes to improved mechanical and thermal properties [101–103].

The first reported composites from cellulose whiskers was in 1995 where nanowhiskers from tunica was mixed with a co-polymer of 35 wt% styrene and 65 wt% butyl acrylate [56–57]. Nanowhiskers prepared from tunicate are frequently reported in literature for their use as filler material in biocomposites [54, 56–57, 104].

Azizi Samir et al. compared composite materials where 6 wt% CNF and CNC from sugar beet pulp were used as fillers in a latex composite material [51]. Both CNF and CNC enhanced mechanical and thermal behavior compared to the unfilled polymer matrix. The reinforcing effect for nanocellulose fillers occurs most probably from the cellulose hydrogen bonding network within the polymer matrix. Between the two different nanocellulose fillers, the CNF showed the largest improvements in mechanical and thermal behavior due to its morphology. The more flexible and hairy nanofibrils showed a tangling effect compared to the more particular cellulose crystals. Improved thermal behavior for CNF composites from potato pulp [64] and eucalyptus pulp [105] has also been reported, with similar results as described above.

Table 3 shows some different cellulose sources that have been used as reinforcement fillers in polymer matrices, either as CNF or as CNC. There is a wide variety of the polymer matrixes used, both synthetic polymers and natural polymers, such as starch, polylactic acid (PLA), and cellulose acetate butyrate (CAB).

Source	Filler	Polymer matrix	References
Sugar beet	CNF/CNC	Styrene/butyl acrylate (6 wt% filler)	[51]
Potato pulp	CNF	Starch/glycerol (0-40 wt% filler)	[64]
Flax	CNF/CNC	PVA (10 wt% filler), Waterborne polyurethanes (0-30 wt % filler)	[82, 103]
Rutabaga	CNF	PVA (10 wt% filler)	[82]
Hemp	CNF	PVA (10 wt% filler)	[82]
Wood pulp	CNF/CNC	PVA (10 wt% filler), PLA (5 wt% filler)	[82, 106]
MCC	CNC	PLA (5 wt% filler)	[107]
Cotton	CNC	PVA (0-12 wt% filler)	[108]
Tunicate	CNC	Styrene/butyl acrylate (6 wt% filler), Starch/sorbitol (25 wt% filler) Waterborne epoxy (0.5-5 wt% filler)	[56-57, 104, 109]
Whet straw	CNF	No attempts was made with composites	[110]
Soy hulls	CNF	No attempts was made with composites	[110]
Bacteria	CNC	CAB (0-10 wt% filler)	[89]
Ramie	CNC	Starch/glycerol (0-40 wt% filler)	[102]

*PVA (polyvinyl alcohol), PLA (polylactic acid), CAB (cellulose acetate)

Table 3. Different cellulose sources as reinforcement fillers in polymer matrices.

A problem with using cellulose fibers as reinforcement in polymer composites is the hydrophilic properties of the cellulose fibers when mixed with hydrophobic polymer matrices. To overcome this problem, the cellulose fibers can be modified in different ways, either by coating the cellulose nanoparticles with surfactants or by chemically modifying the cellulose surface with hydrophobic groups. The use of surfactants is the easiest method but a very high amount of surfactants is needed to coat the surface of the fillers, which causes problems in composite applications. Chemical modification of nanocellulose was described earlier (Section 6) and most of the chemical modifications on CNC is to improve the incorporation between CNC and the more hydrophobic polymer matrix.

Researchers have shown that it is possible to use tunicin whiskers in organic solvents without using any modification on the whiskers [60]. The tunicin whiskers were dispersed in dimethylformamide (DMF) without any other additives and the stability of the suspension was found to be as good as in water. The stability is due to the high value of the dielectric constant of DMF and the wettability of the tunicin whiskers. These results open up new ways of using CNC as reinforcements in hydrophobic polymers as matrix.

There is a huge interest in nanocomposites from biobased resources in many different industries today due to the many benefits with nanocellulose compared to other conventional filler materials. Some industries interested in nanocomposites from cellulose-based materials are the automotive, aerospace, medical and health, packaging and forest industries.

8. Summary and outlook

This review has described the properties of nanocellulose and its meaningfulness in the society as a material with extraordinary properties that could be used as reinforcement in paper or composite materials. The addition of few weight percent nanocellulose to a polymer matrix could improve the mechanical properties of the material, compared to pure polymer matrix alone. The use of nanocellulose in the society has also been successful in application areas such as packaging, paper and paperboard, food industry, medical and hygiene products, paints, cosmetics, etc.

The different preparation methods for nanocellulose have been optimized for many years, and good techniques that also work on a larger scale have been developed. For bacterial cellulose, there is an interest to develop a synthesis that could be suitable for larger scale production but the process today is too expensive and takes a long time. The commercialization of nanocellulose is not yet fully developed, but there are manufacturers as well as pilot plants distributed in different places in the world providing nanocellulose for researchers and early adapters to aid in the commercialization of the products.

Different preparation techniques were briefly described and the techniques used depend on what type of nanocellulose is requested and what cellulose source is used. Cellulose nanofibers (CNFs) are prepared from a mechanical treatment while cellulose nanocrystals (CNCs) are hydrolyzed by acids. Bacterial nanocellulose (BNC), on the other hand, is synthesized to very pure and fine microfibrils from bacteria in a cultivated medium. The main focus in the review was on the acid hydrolysis of cellulose pulp to CNC.

The production of CNC hydrolysis could be made from different acids but it is most common to use a sulfuric acid solution due to a better colloidal suspension and high crystallinity. Sulfuric acid will convert some of the hydroxyl groups on the nanocrystal surface into anionic sulfate groups that will aid in stabilizing the dispersion. Phosphoric acid can also be used for CNC production and will, like sulfuric acid, form charged groups on the crystal surface. Hydrolysis from hydrochloric acid or hydrobromic acid will not have any charged surface groups, which makes it harder to form a stable colloidal CNC dispersion. This could, on the other hand, be solved by chemical modification on CNC by esterification reactions or by TEMPO-oxidation. This way, the amount of charged groups on the crystal surface can be controlled.

http://dx.doi.org/10.5772/61899

Chemical modifications on nanocellulose are mainly performed to increase the incorporation of the nanocellulose with a polymer matrix. By improving this incorporation, nanocellulose can be used as reinforcement in biocomposites. It is beneficial to use nanocellulose in composites due to their many outstanding properties and a composite prepared from a nanocellulosic material can exhibit a higher tensile strength than, for example, carbon fibers or aramid fibers such as Kevlar.

There could be an interest in finding new ways for chemical modification on CNC during acidic hydrolysis. Two different paths was suggested in the review, either by finding a substitute to the acids used during hydrolysis that could hydrolyze the cellulose and modify the CNC at the same time, or by adding a reagent to the hydrolyze reaction that will modify the CNC during the acidic hydrolysis. Some demands on these types of reagents are the use in acidic conditions and in the presence of water.

An attempt in using an acrylamide sulfate solution consisting of 66 wt% sulfuric acid was described and the aim of this reaction was to attach unsaturated acrylate groups to the CNC, which could be used as a tool in further reactions. AFM pictures of the acrylamide sulfate-hydrolyzed pulp showed cellulose whiskers in the nanometer size range, which indicated that the solution could hydrolyze cellulose to CNC. FTIR analysis on the nanocellulose showed unsaturated esters and alkenes, which further indicated a successful reaction. However, more studies on the material need to be performed for further evaluation.

After CNC preparation, the nanocellulose is in an aqueous solution and could display a stabilized colloidal behavior above its critical concentration. Knowledge about the degree of drying of CNC is an important issue because the application of CNC often demands dried CNC that could be re-dispersed at the manufacturing sites, either in water or in organic solvents or polymer matrices for composite manufacturing.

The health and safety aspects of nanocellulose were briefly discussed. The cellulose material itself is classified as non-toxic and approved to be used in food products such as in thickening agents. Also for nanocellulose, the health and safety aspects are approved but since nanoparticles are quite new in the health and safety area, other tests might be necessary for further understanding the nanoparticle material.

In conclusion, nanocellulose, in its different forms, shows increasing interest in the industry and society due to its many beneficial properties such as being environmentally friendly, its low cost, and high mechanical performance, there is a bright future for these types of materials. Also, other biobased nanomaterials, such as chitin or starch, will play an important role in the nanoarea future.

Acknowledgements

The authors would like to acknowledge Prof. Herbert Sixta for valuable discussions and critical reading of the manuscript. Jari Sundqvist is acknowledged for his creation of some of the illustrations presented in the review.

Author details

Mikaela Börjesson[1] and Gunnar Westman[1,2*]

*Address all correspondence to: westman@chalmers.se

1 Department of Chemistry and Chemical Engineering, Chalmers University of Technology, Gothenburg, Sweden

2 Wallenberg Wood Science Center, Chalmers University of Technology, Gothenburg, Sweden

References

[1] Han JS, Rowell JS. Chemical Composition of Fibers. In: Rowell RM., Young RA., Rowell J., editors. Paper and Composites from Agro-Based Resources. London: CRC Press; 1996. p. 83–130.

[2] Varshney VK, Naithani S. Chemical Functionalization of Cellulose Derived from Nonconventional Sources. In: Kalia S, Kaith BS, Kaur I, editors. Cellulose Fibers: Bio- and Nano-Polymer Composites. Berlin: Springer; 2011. p. 43–60. DOI: 10.1007/978–3–642–17370–7.

[3] Klemm D, Heublein B, Fink H-P, Bohn A. Cellulose: Fascinating Biopolymer and Sustainable Raw Material. Angewandte Chemie International Edition. 2005;44:3358–3393. DOI: 10.1002/anie.200460587.

[4] Kamel S. Nanotechnology and its applications in lignocellulosic composites, a mini review. eXPRESS Polymer Letters. 2007;1:546–575. DOI: 10.3144/expresspolymlett.2007.78.

[5] Roy D, Semsarilar M, Guthrie JT, Perrier S. Cellulose modification by polymer grafting: A review. Chemical Society Reviews. 2009;38:1825–2148. DOI: 10.1039/B808639G.

[6] Lennholm H, Henriksson G. Cellulose and Carbohydrate Chemistry. In: Ek M, Gellerstedt G, Henriksson G, editors. Ljungberg Textbook Book 1: Wood Chemistry and Wood Biotechnology. Stockholm: Fiber and Polymer Technology, KTH; 2007. p. 73–102.

[7] Helbert W, Cavaille JY, Dufresne A. Thermoplastic Nanocomposites Filled With Wheat Straw Cellulose Whiskers. Part I: Processing and Mechanical Behavior. Polymer Composites. 1996;17:604–611. DOI: 10.1002/pc.10650.

[8] Abe K, Iwamoto S, Yano H. Obtaining Cellulose Nanofibers with a Uniform Width of 15 nm from Wood. Biomacromolecules. 2007;8:3276–3278. DOI: 10.1021/bm700624p.

[9] Moon RJ, Martini A, Nairn J, Simonsen J, Youngblood J. Cellulose nanomaterials review: Structure, properties and nanocomposites. Chem Soc Rev. 2011;40:3941–3994. DOI: 10.1039/c0cs00108b.

[10] Moon RJ, Frihart CR, Wegner T. Nanotechnology Applications in the Forest Products Industry. Forest Products Journal. 2006;56:4–10.

[11] MoRe Research. Europe's first pilot facility for nanocrystalline cellulose [Internet]. 2015. Available from: http://www.more.se/en/news/2015/europes-first-pilot-facility-for-nanocrystalline-cellulose/ [Accessed: 2015–06–05].

[12] Klemm D, Kramer F, Moritz S, Lindström T, Ankerfors M, Gray D, et al. Nanocelluloses: A New Family of Nature-Based Materials. Angew Chem Int Ed. 2011;50:5438–5466. DOI: 10.1002/anie.201001273.

[13] Klemm D, Schumann D, Kramer F, Heßler N, Hornung M, Schmauder H-P, et al. Nanocelluloses as Innovative Polymers in Research and Application. Advance Polymer Science. 2006;205:49–96. DOI: 10.1007/12_097.

[14] Lin N, Huang J, Dufresne A. Preparation, properties and applications of polysaccharide nanocrystals in advanced functional nanomaterials: A review. Nanoscale. 2012;4:3274–3294. DOI: 10.1039/c2nr30260h.

[15] Azizi Samir MAS, Alloin F, Dufresne A. Review of Recent Research into Cellulosic Whiskers, Their Properties and Their Application in Nanocomposite Field. Biomacromolecules. 2005;6:612–626. DOI: 10.1021/bm0493685.

[16] TAPPI's International Nanotechnology Division. Roadmap for the Development of International Standards for Nanocellulose [Report]. October 2011.

[17] Deepa B, Abraham E, Cherian BM, Bismarck A, Blaker JJ, Pothan LA, et al. Structure, morphology and thermal characteristics of banana nano fibers obtained by steam explosion. Bioresource Technology. 2011;102:1988–1997. DOI: 10.1016/j.biortech. 2010.09.030.

[18] Pitkänen M, Honkalampi U, Von Wright A, Sneck A, Hentze H-P, Sievänen J, et al. Nanofibrillar cellulose - In vitro study of cytotoxic and genotoxic properties. Proceedings of the International Conference on Nanotechnology for the Forest Products Industry; 2010; Espoo, Finland.

[19] O'Connor RT, DuPre EF, McCall ER. Infrared Spectrophotometric Procedure for Analysis of Cellulose and Modified Cellulose. Anal Chem. 1957;29:998–1005. DOI: 10.1021/ac60127a003.

[20] Moreira S, Silva NB, Almeida-Lima J, Rocha HAO, Medeiros SRB, Alves C, et al. BC nanofibres: In vitro study of genotoxicity and cell proliferation. Toxicology Letters. 2009;189:235–241. DOI: 10.1016/j.toxlet.2009.06.849.

[21] Vartiainen J, Pöhler T, Sirola K, Pylkkänen L, Alenius H, Hokkinen J, et al. Health and environmental safety aspects of friction grinding and spray drying of microfibrillated cellulose. Cellulose. 2011;18:775–786. DOI: 10.1007/s10570-011–9501–7.

[22] Ström G, Öhgren C, Ankerfors M. Nanocellulose as an additive in foodstuff. Innventia Report No.: 403; June 2013.

[23] Food Standards Agency. Current EU approved additives and their E Numbers [Internet]. 2014. Available from: http://www.food.gov.uk/science/additives/enumberlist [Accessed: 2015–06–05].

[24] Yanamala N, Farcas MT, Hatfield MK, Kisin ER, Kagan VE, Geraci CL, et al. In Vivo Evaluation of the Pulmonary Toxicity of Cellulose Nanocrystals: A Renewable and Sustainable Nanomaterial of the Future. ACS Sustainable Chem Eng. 2014;2:1691–1698. DOI: 10.1021/sc500153k.

[25] Westman G, Hasani M. Modified cellulose fibers. Sweden. Patent No.: EP2428610. 2012–03–14.

[26] Landry V, Alemdar A, Blanchet P. Nanocrystalline Cellulose: Morphological, Physical, and Mechanical Properties. Forest Products Journal 2011;61:104–112. DOI: 10.13073/0015–7473–61.2.104.

[27] Miao C, Hamad WY. Cellulose reinforced polymer composites and nanocomposites: A critical review. Cellulose. 2013;20:2221–2262. DOI: 10.1007/s10570-013–0007–3.

[28] Håkansson H, Ahlgren P. Acid hydrolysis of some industrial pulps: Effect of hydrolysis conditions and raw material. Cellulose. 2005;12:177–183. DOI: 10.1007/s10570-004–1038–6.

[29] Dong XM, Revol J-F, Gray DG. Effect of microcrystallite preparation conditions on the formation of colloid crystals of cellulose. Cellulose. 1998;5:19–32. DOI: 10.1023/A:1009260511939.

[30] Beck-Candanedo S, Roman M, Gray DG. Effect of Reaction Conditions on the Properties and Behavior of Wood Cellulose Nanocrystal Suspensions. Biomacromolecules. 2005;6:1048–1054. DOI: 10.1021/bm049300p.

[31] Liu D, Wang S, Ma Z, Tian D, Gu M, Lin F. Structure - color mechanism of iridescent cellulose nanocrystal films. Royal Society of Chemistry Advances. 2014;4:39322–39331. DOI: 10.1039/c4ra06268j.

[32] Marchessault RH, Morehead FF, Walter NM. Liquid crystal systems from fibrillar polysaccharides. Nature. 1959;184: 632–633. DOI: 10.1038/184632a0.

[33] Dong XM, Kimura T, Revol J-F, Gray DG. Effects of Ionic Strength on the Isotropic-Chiral Nematic Phase Transition of Suspensions of Cellulose Crystallites. Langmuir. 1996;12:2076–2082. DOI: 10.1021/la950133b.

[34] Marchessault RH, Morehead FF, Koch MJ. Some hydrodynamic properties of neutral suspensions of cellulose crystallites as related to size and shape. Journal of Colloid Science. 1961;16:327–344. DOI: 10.1016/0095–8522(61)90033–2.

[35] Dufresne A. Comparing the Mechanical Properties of High Performances Polymer Nanocomposites from Biological Sources. Nanoscience and Nanotechnology. 2006;6:322–330. DOI: 10.1166/jnn.2006.005.

[36] Revol J-F, Bradford H, Giasson J, Marchessault RH, Gray DG. Helicoidal Self-Ordering of Cellulose Microfibrils in Aqueous Suspension. Int J Biol Macromol. 1992;14:170–172. DOI: 10.1016/S0141-8130(05)80008-X.

[37] Araki J, Wada M, Kuga S, Okano T. Flow properties of microcrystalline cellulose suspension prepared by acid treatment of native cellulose. Colloids and Surfaces A: Physicochemical and Engineering Aspects. 1998;142:75–82. DOI: 10.1016/S0927-7757(98)00404-X.

[38] Araki J, Wada M, Kuga S, Okano T. Influence of surface charge on viscosity behavior of cellulose microcrystal suspension. Journal of Wood Science. 1999;45:258–261. DOI: 10.1007/BF01177736.

[39] Orts WJ, Shey J, Imam SH, Glenn GM, Guttman ME, Revol J-F. Application of Cellulose Microfibrils in Polymer Nanocomposites. Journal of Polymers and the Environment. 2005;13:301–306. DOI: 10.1007/s10924-005–5514–3.

[40] Strømme M, Mihranyan A, Ek R. What to do with all these algae? Materials Letters. 2002;57:569–572. DOI: 10.1016/S0167-577X(02)00831–5.

[41] Bondeson D, Mathew A, Oksman K. Optimization of the isolation of nanocrystals from microcrystalline cellulose by acid hydrolysis. Cellulose. 2006;13:171–180. DOI: 10.1007/s10570-006–9061–4.

[42] de Rodriguez NLG, Thielemans W, Dufresne A. Sisal cellulose whiskers reinforced polyvinyl acetate nanocomposites. Cellulose. 2006;13:261–270. DOI: 10.1007/s10570-005–9039–7.

[43] El-Sakhawy M, Hassan ML. Physical and mechanical properties of microcrystalline cellulose prepared from agricultural residues. Carbohydrate Polymers. 2007;67:1–10. DOI: 10.1016/j.carbpol.2006.04.009.

[44] Ebeling T, Paillet M, Borsali R, Diat O, Dufresne A, Cavaillé J-Y, et al. Shear-Induced Orientation Phenomena in Suspensions of Cellulose Microcrystals, Revealed by Small Angle X-ray Scattering. Langmuir. 1999;15:6123–6126. DOI: 10.1021/la990046 +.

[45] Heux L, Chauve G, Bonini C. Nonflocculating and Chiral-Nematic Self-ordering of Cellulose Microcrystals Suspensions in Nonpolar Solvents. Langmuir. 2000;16:8210–8212. DOI: 10.1021/la9913957.

[46] Dufresne A, Cavaille JY, Helbert W. Thermoplastic Nanocomposites Filled With Wheat Straw Cellulose Whiskers. Part II: Effect of Processing and Modeling. Polymer Composites. 1997;18:198–210. DOI: 10.1002/pc.10274.

[47] Lu P, Hsieh Y-L. Preparation and characterization of cellulose nanocrystals from rice straw. Carbohydrate Polymers. 2012;87:564–573. DOI: 10.1016/j.carbpol.2011.08.022.

[48] Roman M, Winter WT. Effect of Sulfate Groups from Sulfuric Acid Hydrolysis on the Thermal Degradation Behavior of Bacterial Cellulose. Biomacromolecules. 2004;5:1671–1677. DOI: 10.1021/bm034519 +.

[49] Imai T, Boisset C, Samejima M, Igarashi K, Sugiyama J. Unidirectional processive action of cellobiohydrolase Cel7A on Valonia cellulose microcrystals. Fed Eur Biochem Soc Lett. 1998;432:113–116. DOI: 10.1016/S0014-5793(98)00845-X.

[50] Dufresne A, Cavaillé J-Y, Vignon MR. Mechanical Behavior of Sheets Prepared from Sugar Beet Cellulose Microfibrils. Journal of Applied Polymer Science. 1997;64:1185–1194. DOI: 10.1002/(SICI)1097–4628(19970509)64:6 < 1185::AID-APP19 > 3.0.CO;2-V.

[51] Azizi Samir MAS, Alloin F, Paillet M, Dufresne A. Tangling Effect in Fibrillated Cellulose Reinforced Nanocomposites. Macromolecules. 2004;37:4313–4316. DOI: 10.1021/ma035939u.

[52] Dinand E, Chanzy H, Vignon MR. Suspensions of cellulose microfibrils from sugar beet pulp. Food Hydrocolloids. 1999;13:275–283. DOI: 10.1016/S0268-005X(98)00084–8.

[53] Heux L, Dinand E, Vignon MR. Structural aspects in ultrathin cellulose microfibrils followed by 13C CP-MAS NMR. Carbohydrate Polymers. 1999;40:115–124. DOI: 10.1016/S0144-8617(99)00051-X.

[54] Hajji P, Cavaille JY, Favier V, Gauthier C, Vigier G. Tensile Behavior of Nanocomposies From Latex and Cellulose Whiskers. Polymer Composites. 1996;17:612–619. DOI: 10.1002/pc.10651.

[55] Favier V, Dendievel R, Canova G, Cavailé JY, Gilormini P. Simulation and modeling of threedimensional percolating structures: Case of a latex matrix reinforced by a network of cellulose fibers. Acta Materiala. 1997;45:1557–1565. DOI: 10.1016/S1359-6454(96)00264–9.

[56] Favier V, Canova GR, Cavaillé JY, Chanzy H, Dufresne A, Gauthie C. Nanocomposite Materials from Latex and Cellulose Whiskers. Polymer for Advanced Technologies. 1995;6:351–355. DOI: 10.1002/pat.1995.220060514.

[57] Favier V, Chanzy H, Cavaillé JY. Polymer Nanocomposites Reinforced by Cellulose Whiskers. Macromolecules. 1995;28:6365–6367. DOI: 10.1021/ma00122a053.

[58] Schroers M, Kokil A, Weder C. Solid Polymer Electrolytes Based on Nanocomposites of Ethylene Oxide–Epichlorohydrin Copolymers and Cellulose Whiskers. Journal of Applied Polymer Science. 2004;93:2883–2888. DOI: 10.1002/app.20870.

[59] Azizi Samir MAS, Alloin F, Gorecki W, Sanchez J-Y, Dufresne A. Nanocomposite Polymer Electrolytes Based on Poly(oxyethylene) and Cellulose Nanocrystals Journal of Physical Chemistry B. 2004;108:10845–10852. DOI: 10.1021/jp0494483.

[60] Azizi Samir MAS, Alloin F, Sanchez J-Y, El Kissi N, Dufresne A. Preparation of Cellulose Whiskers Reinforced Nanocomposites from an Organic Medium Suspension. Macromolecules. 2004;37:1386–1393. DOI: 10.1021/ma030532a.

[61] Nair KG, Dufresne A. Crab Shell Chitin Whisker Reinforced Natural Rubber Nanocomposites. 1. Processing and Swelling Behavior. Biomacromolecules. 2003;4:657–665. DOI: 10.1021/bm020127b.

[62] Nair KG, Dufresne A, Gandini A, Belgacem MN. Crab Shell Chitin Whiskers Reinforced Natural Rubber Nanocomposites. 3. Effect of Chemical Modification of Chitin Whiskers. Biomacromolecules. 2003;4:1835–1842. DOI: 10.1021/bm030058g.

[63] Dufresne A, Cavaillé J-Y. Clustering and Percolation Effects in Microcrystalline Starch-Reinforced Thermoplastic. Journal of Polymer Science: Part B: Polymer Physics. 1998;36:2211–2224. DOI: 10.1002/(SICI)1099–0488(19980915)36:12 < 2211::AID-POLB18 > 3.0.CO;2–2.

[64] Dufresne A, Vignon MR. Improvement of Starch Film Performances Using Cellulose Microfibrils. Macromolecules. 1998;31:2693–2696. DOI: S0024-9297(97)01532–5.

[65] Dufresne A, Dupeyre D, Vignon MR. Cellulose Microfibrils from Potato Tuber Cells: Processing and Characterization of Starch–Cellulose Microfibril Composites. Journal of Applied Polymer Science. 2000;76:2080–2092. DOI: 10.1002/(SICI)1097–4628(20000628)76:14 < 2080::AID-APP12 > 3.0.CO;2-U.

[66] Dufresne A, Cavaillé J-Y, Helbert W. New Nanocomposite Materials: Microcrystalline Starch Reinforced Thermoplastic. Macromolecules. 1996;29:7624–7626. DOI: S0024-9297(96)00273–2.

[67] Dubief D, Samain E, Dufresne A. Polysaccharide Microcrystals Reinforced Amorphous Poly(â-hydroxyoctanoate) Nanocomposite Materials. Macromolecules. 1999;32:5765–5771. DOI: 10.1021/ma990274a.

[68] Angellier H, Choisnard L, Molina-Boisseau S, Ozil P, Dufresne A. Optimization of the Preparation of Aqueous Suspensions of Waxy Maize Starch Nanocrystals Using a Response Surface Methodology. Biomacromolecules. 2004;5:1545–1551. DOI: 10.1021/bm049914u.

[69] Angellier H, Molina-Boisseau S, Lebrun L, Dufresne A. Processing and Structural Properties of Waxy Maize Starch Nanocrystals Reinforced Natural Rubber. Macromolecules. 2005;38:3783–3792. DOI: 10.1021/la047530j.

[70] Dong XM, Gray DG. Effect of Counterions on Ordered Phase Formation in Suspensions of Charged Rodlike Cellulose Crystallites. Langmuir. 1997;13:2404–2409. DOI: S0743-7463(96)00724-X.

[71] Stroobants A, Lekkerkerker HNW, Odijk T. Effect of Electrostatic Interaction on the Liquid Crystal Phase Transition in Solutions of Rodlike Polyelectrolytes. Macromolecules. 1986;19:2232–2238. DOI: 10.1021/ma00162a020.

[72] Lagerwall JP, Schütz C, Salajkova M, Noh J, Park JH, Scalia G, et al. Cellulose nanocrystal-based materials: from liquid crystal self-assembly and glass formation to multifunctional thin films. NPG Asia Materials. 2014;6:1–12. DOI: 10.1038/am.2013.69.

[73] Wittcoff HA, Reuben BG, Plotkin JS. Industrial Organic Chemicals. 3rd ed. John Wiley & Sons, Inc.; 2013. p. 44–45.

[74] Hamad WY, Hu TQ. Structure-Process-Yield Interrelations in Nanocrystalline Cellulose Extraction. The Canadian Journal of Chemical Engineering. 2010;88:392–402. DOI: 10.1002/cjce.20298.

[75] Cranston ED, Gray DG. Morphological and Optical Characterization of Polyelectrolyte Multilayers Incorporating Nanocrystalline Cellulose. Biomacromolecules. 2006;7:2522–2530. DOI: 10.1021/bm0602886.

[76] Lee S-Y, Mohan DJ, Kang I-A, Doh G-H, Lee S, Han SO. Nanocellulose Reinforced PVA Composite Films: Effects of Acid Treatment and Filler Loading. Fibers and Polymers. 2009;10:77–82. DOI: 10.1007/s12221-009–0077-x.

[77] Zhang PP, Tong DS, Lin CX, Yang HM, Zhong ZK, Yu WH, et al. Effects of acid treatments on bamboo cellulose nanocrystals. Asia-Pacific Journal of Chemical Engineering. 2014;9:686–695. DOI: 10.1002/apj.1812.

[78] Angellier H, Putaux J-L, Molina-Boisseau S, Dupeyre D, Dufresne A. Starch nanocrystal fillers in an acrylic polymer matrix. Macromolecular Symposia. 2005;221:95–104. DOI: 10.1002/masy.200550310.

[79] Araki J, Wada M, Kuga S, Okano T. Birefringent Glassy Phase of a Cellulose Microcrystal Suspension. Langmuir. 2000;16:2413–2415. DOI: 10.1021/la9911180.

[80] Araki J, Wada M, Kuga S. Steric Stabilization of a Cellulose Microcrystal Suspension by Poly(ethylene glycol) Grafting. Langmuir. 2001;17:21–27. DOI: 10.1021/la001070m.

[81] Isogai A, Kato Y. Preparation of polyuronic acid from cellulose by TEMPO-mediated oxidation. Cellulose. 1998;5:153–164. DOI: 10.1023/A:1009208603673.

[82] Bhatnagar A, Sain M. Processing of Cellulose Nanofiber-reinforced Composites. Journal of reinforced Plastics and Composites. 2005;24:1259–1268. DOI: 10.1177/0731684405049864.

[83] Pääkkö M, Ankerfors M, Kosonen H, Nykänen A, Ahola S, Österberg M, et al. Enzymatic Hydrolysis Combined with Mechanical Shearing and High-Pressure Homoge-

nization for Nanoscale Cellulose Fibrils and Strong Gels. Biomacromolecules. 2007;8:1934–1941. DOI: 10.1021/bm061215p.

[84] Saito T, Kimura S, Nishiyama Y, Isogai A. Cellulose Nanofibers Prepared by TEMPO-Mediated Oxidation of Native Cellulose. Biomacromolecules. 2007;8:2485–2491. DOI: 10.1021/bm0703970.

[85] Tokoh C, Takabe K, Fujita M, Saiki H. Cellulose synthesized by Acetobacter xylinum in the presence of acetyl glucomannan. Cellulose. 1998;5:249–261. DOI: 10.1023/A:1009211927183.

[86] Geyer U, Heinze T, Stei A, Klemm D, Marsch S, Schumann D, et al. Formation, derivatization and applications of bacterial cellulose. International Journal of Biological Macromolecule. 1994;16:343–347. DOI: 10.1016/0141–8130(94)90067–1.

[87] Shah J, Brown Jr. RM. Towards electronic paper displays made from microbial cellulose. Appl Microbiol Biotechnol. 2005;66:352–355. DOI: 10.1007/s00253-004–1756–6.

[88] Czaja W, Krystynowicz A, Bielecki S, Brown Jr. RM. Microbial cellulose - the natural power to heal wounds. Biomaterials. 2006;27:145–151. DOI: 10.1016/j.biomaterials.2005.07.035.

[89] Grunert M, Winter WT. Nanocomposites of Cellulose Acetate Butyrate Reinforced with Cellulose Nanocrystals. Journal of Polymers and the Environment. 2002;10:27–30. DOI: 10.1023/A:1021065905986.

[90] Angellier H, Molina-Boisseau S, Belgacem MN, Dufresne A. Surface Chemical Modification of Waxy Maize Starch Nanocrystals. Langmuir 2005. 2005;21:2425–2433. DOI: 10.1021/la047530j.

[91] Hasani M, Cranston ED, Westman G, Gray DG. Cationic surface functionalization of cellulose nanocrystals. Soft matter. 2008;4:2238–2244. DOI: 10.1039/b806789a.

[92] Yan M, Li S, Zhang M, Li C, Dong F, Li W. Characterization of Surface Acetylated Nanocrystalline Cellulose by Single-Step Method. Bioresources. 2013;8:6330–6341. DOI: 10.15376/biores.8.4.6330–6341.

[93] Heravi MM, Fard MV, Faghihi Z. Heteropoly acids-catalyzed organic reactions in water: doubly green reactions. Green Chemistry Letters and Reviews. 2013;6:282–300. DOI: 10.1080/17518253.2013.846415.

[94] Lanzafame P, Temi DM, Perathoner S, Spadaro AN, Centi G. Direct conversion of cellulose to glucose and valuable intermediates in mild reaction conditions over solid acid catalysts. Catalysis Today. 2012;179:178–184. DOI:10.1016/j.cattod.2011.07.018.

[95] McNae CJ. Production of Acrylamide. U.K. Patent No.: GB19640005603. 1964.

[96] Beck S, Bouchard J, Berry R. Dispersibility in Water of Dried Nanocrystalline Cellulose. Biomacromolecules. 2012;13:1486–1494. DOI: 10.1021/bm300191k.

[97] Kalia S, Dufresne A, Cherian B, Kaith BS, Avérous L, Njuguna J, et al. Cellulose-Based Bio- and Nanocomposites: A Review. International Journal of Polymer Science. 2011;2011:1–35. DOI:10.1155/2011/837875.

[98] Alexandre M, Dubois P. Polymer-layered silicate nanocomposites: preparation, properties and uses of a new class of materials. Materials Science and Engineering. 2000;28:1–63. DOI:10.1016/S0927-796X(00)00012–7.

[99] Eichhorn SJ, Baillie CA, Zafeiropoulos N, Mwaikambo LY, Ansell MP, Dufresne A, et al. Current international research into cellulosic fibers and composites. Journal of Material Science. 2001;36:2107–2131. DOI: 10.1023/A:1017512029696.

[100] Siqueira G, Bras J, Dufresne A. Cellulosic Bionanocomposites: A Review of Preparation, Properties and Applications. Polymers. 2010;2:728–765. DOI: 10.3390/polym2040728.

[101] Seydibeyoglu MÖ, Oksman K. Novel nanocomposites based on polyurethane and micro fibrillated cellulose. Composites Science and Technology. 2008;68:908–914. DOI:10.1016/j.compscitech.2007.08.008.

[102] Lu Y, Weng L, Cao X. Morphological, thermal and mechanical properties of ramie crystallites—reinforced plasticized starch biocomposites. Carbohydrate Polymers. 2006;63:198–204. DOI:10.1016/j.carbpol.2005.08.027.

[103] Cao X, Dong H, Li CM. New Nanocomposite Materials Reinforced with Flax Cellulose Nanocrystals in Waterborne Polyurethane. Biomacromolecules. 2007;8:899–904. DOI: 10.1021/bm0610368.

[104] Mathew AP, Dufresne A. Morphological Investigation of Nanocomposites from Sorbitol Plasticized Starch and Tunicin Whiskers. Biomacromolecules. 2002;3:609–617. DOI: 10.1021/bm0101769.

[105] Xu X, Liu F, Jiang L, Zhu JY, Haagenson D, Wiesenborn DP. Cellulose Nanocrystals vs. Cellulose Nanofibrils: A Comparative Study on Their Microstructures and Effects as Polymer Reinforcing Agents. ACS Applied Materials & Interfaces. 2013;5:2999–3009. DOI: 10.1021/am302624t.

[106] Bondeson D, Oksman K. Polylactic acid/cellulose whisker nanocomposites modified by polyvinyl alcohol. Composites: Part A. 2007;38:2486–2492. DOI:10.1016/j.compositesa.2007.08.001.

[107] Kvien I, Tanem BS, Oksman K. Characterization of Cellulose Whiskers and Their Nanocomposites by Atomic Force and Electron Microscopy. Biomacromolecules. 2005;6:3160-3165. DOI: 10.1021/bm050479t.

[108] Roohani M, Habibi Y, Belgacem NM, Ebrahim G, Karimi AN, Dufresne A. Cellulose whiskers reinforced polyvinyl alcohol copolymers nanocomposites. European Polymer Journal. 2008;44:2489–2498. DOI: 10.1016/j.eurpolymj.2008.05.024.

[109] Ruiz MM, Cavaillé J-Y, Dufresne A, Graillat C, Gerard J-F. New waterborne epoxy coating based on cellulose nanofillers. Die Makromolekulare Chemie. 2001;169:211–222. DOI: 10.1002/1521-3900(200105)169:1<211::AID-MASY211>3.0.CO;2-H0

[110] Alemdar A, Sain M. Isolation and characterization of nanofibers from agricultural residues – Wheat straw and soy hulls. Bioresource Technology. 2008;99:1664-1671. DOI: 10.1016/j.biortech.2007.04.029.

Chapter 8

Current Trends in the Production of Cellulose Nanoparticles and Nanocomposites for Biomedical Applications

John Rojas, Mauricio Bedoya and Yhors Ciro

Additional information is available at the end of the chapter

http://dx.doi.org/10.5772/61334

Abstract

The goal of this chapter is to review the most recent trends to produce cellulose nanoparticles and nanocomposites with biomedical applications. These particles could be named as bacterial cellulose, cellulose nanofibers, and cellulose nanocrystals. The production of these nanoparticles with diameters below 100 nm is challenging because of the strong agglomeration tendency which occur upon drying aqueous cellulose suspensions or during the compounding process with hydrophobic polymers. Typically, the physical and mechanical properties of these nanoparticles depend on the source of cellulose and the extraction process employed. Cellulose nanoparticles are obtained by mechanical, chemical, or enzymatic process treatments to open the structure of the cellulose source and facilitate accessibility to its microstructure. Usually, a combination of these processes makes the extraction more efficient.

On the other hand, cellulose and polymer nanocomposites are commonly produced by techniques such as solvent evaporation, melt compounding, compression molding, impregnation, and electrospinning. The most salient nanocellulose applications discussed in this chapter deal with the production of bandages, implants, skins replacements for burnings, face masks, artificial blood vessels, cuffs for nerve surgery, drug delivery, cell carriers, and support matrices for enzyme immobilization, and silver nanoparticles as antimicrobial agents in wound dressing.

Keywords: Cellulose nanofibers, cellulose nanocrystals, bacterial cellulose, biomedical composites

1. Introduction

Cellulose is the most abundant, renewable, and sustainable biopolymer on earth. It is present in plants, tunicates, and some bacteria. For instance, it is present in the cell wall of wood fibers along with hemicellulose and lignin. The cellulose fibers are composed of microfibrils that, in turn, are composed of elementary fibrils, which are the basic structural units. These elementary fibrils or nanofibers are about 2–20 nm in diameter and a few micrometers in length. About 30 to 100 cellulose chains aggregate into an elementary fibril. There are regions within each of these elementary fibrils, where the cellulose chains are arranged in highly ordered (crystalline) structures and regions that are disordered (amorphous) [1]. These elementary fibrils are formed during cellulose biosynthesis. Each microfibril is a flexible hair strand composed of cellulose nanocrystals (CNC) linked along with cellulose nanofibers (CNF) [2] (Fig. 1). The terms nanofibrils, nanofibers, and elementary fibrils are usually employed as synonyms. A CNF is a bulky, moderately degraded cellulose with a large surface area with diameter ranging from 20 to 60 nm and length of several micrometers. It presents a weblike structure, and the length/diameter ratio is very high. They are usually extracted by mechanical treatment without using acid hydrolysis. Conversely, when subjected to acid hydrolysis, cellulose microfibrils undergo transverse cleavage along the amorphous regions, and the use of sonication results in CNC also referred as cellulose nanowhiskers, or nanorods. Thus, CNCs are described as the crystalline region of cellulose and exhibits a rodlike shape having a low aspect ratio [2, 3]. Its elastic modulus can be compared to the modulus of crystalline cellulose (up to 140 GPa) due to its high hydrogen bonding capability. Further, cellulose nanoparticles degrade faster than other nanoparticles such as fullerenes and carbon nanotubes, which do not degrade at all [4].

Another type of nanocellulose is bacterial cellulose (BC), which is produced as an extracellular primary metabolite by bacteria belonging to the genera *Acetobacter* (*Gluconacetobacter*), *Agrobacterium, Acanthamoeba, Achromobacter, Zooglea, Aerobacter, Azotobacter, Rhizobium, Sarcina, Salmonella, Escherichia, Pseudomonas,* and *Alcaligenes* [5]. However, the most efficient producer of bacterial cellulose (BC) is *Acetobacter xylinum* (or *Gluconacetobacter xylinus*). BC is secreted as a ribbon-shaped fibril, of less than 100 nm wide, which is composed of nanofibrils of 2–4 nm in diameter. These microfibril bundles have excellent intrinsic properties due to their high crystallinity (84–89%), elastic modulus of ~78 GPa, high water holding capacity, and high degree of polymerization (up to 8,000) [6]. BC, when compressed into sheets, exhibits a highly planar orientation [7]. BC is degraded by few bacterial strains such as *Trichoderma viride* at pH values ranging from 4.5 to 6.0 [8–12]. BC is highly hydrated and no chemical treatments are needed to remove lignin and hemicellulose [13, 14]. For those reasons, BC is preferred for tissue and bone growth. The goal of this chapter is to describe and discuss the state of the art for the production of cellulose nanoparticles such as cellulose nanofibers, cellulose nanocrystals, bacterial cellulose, and their composites intended for biomedical applications.

http://dx.doi.org/10.5772/61334

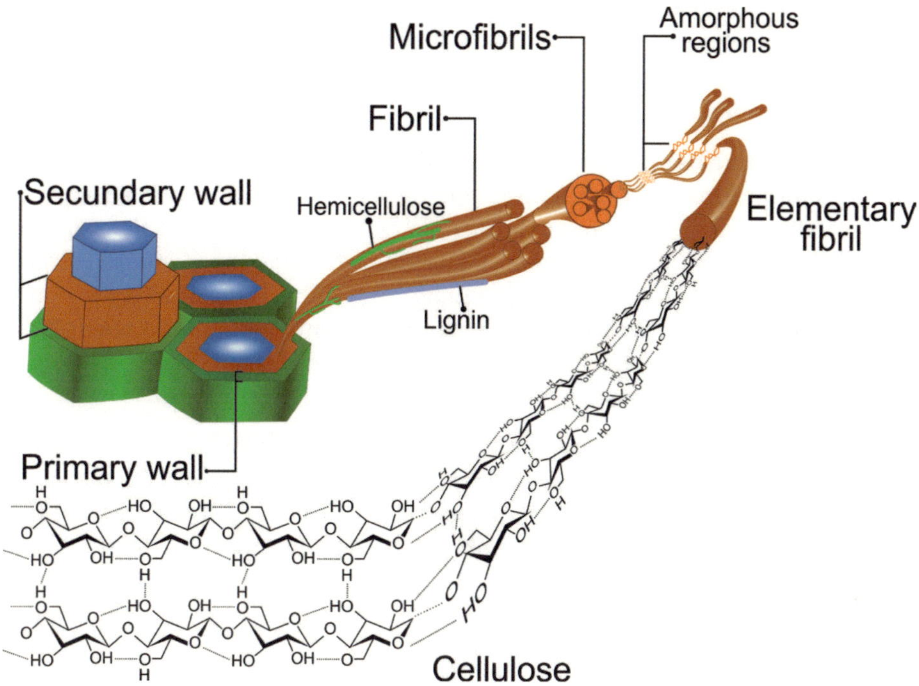

Figure 1. Hierarchical structure of cellulose extracted from plants.

2. Preparation of cellulose nanofibers

CNF is mainly extracted from wood. However, it can be extracted from natural resources such as such as sisal [15], flax, hemp, grass [16], sorghum, barley, sugar cane [17], pineapple leaf fibers [18], banana rachis [19], soy hulls [20], algae [21], bacterial cellulose, kenaf stem, swede root, wheat straw [22], carrots, empty fruit bunches, potato pulp, branch bark of mulberry [23], bagasse, rice straw, chardonnay grape skins [24], stems of cacti, coconut husk [25], bamboo, pea hull fiber, cotton and industrial bioresidues [26]. Pineapple leaf and jute fibers are the best sources for its extraction due to the low cost, abundance, and high cellulose content [60–70%) [27]. The size of CNF depends on the source and exhibits an entangled morphology with an aspect ratio over 250. For instance, CNF obtained from wheat straw, soy hull, and soybean stock have diameters ranging from 10 to 80 nm, from 20 to 120 nm, and from 50 to 100 nm, respectively [28, 29]. Nevertheless, other researchers have obtained CNF from sisal, carrots, beet pulp, and Luffa cylindrical with smaller diameters of 20–65 nm, 3–36 nm, 30–100 nm, and 55 nm, respectively. These nonwooden sources contain less lignin and require less processing steps and energy consumption due to the less tightly bound microfibril in the primary cell wall than wood. There are several extraction methods to obtain CNF [30–32] (Fig. 2). They can be performed by mechanical techniques such as grinding, cryocrushing with liquid nitrogen, high-pressure homogenization, etc. In addition, different chemical alkali and enzymatic hydrolyses can be utilized before mechanical processes in order to promote the accessibility of hydroxyl groups, increase the inner surface, alter crystallinity, break cellulose hydrogen bonds, and therefore boost the reactivity of the fibers.

2.1. Mechanical treatments

The mechanical treatments can isolate CNF from the primary and secondary cell wall without severely degrading cellulose. For instance, microfluidization and high-intensity ultrasonic treatments produce a high shear gradients causing transverse cleavage along the longitudinal axis of the cellulose fibers, and as a result, they tend to damage the microfibril structure by reducing the molar mass and degree of crystallinity. Depending on the mechanical force levels and types of mechanical treatment, interfibrillar hydrogen bonding are broken [2, 32, 33]. However, the mechanical methods exhibit high production costs (tools and materials); they are also less efficient and require greater energy than the chemical methods [34].

For this reason, a chemical pretreatment reduces energy consumption and makes the surface more hydrophobic. Further, the mechanical treatment usually reduces the degree of polymerization (DP) from 1,200 to 1,400 to a DP between 850 and 500. A high cellulose DP is desirable since this is correlated with the nanofiber tensile strength, which can be at least 2 GPa [22, 35]

2.1.1. High-Pressure Homogenization (HPH)

In this process, dilute slurries of cellulose fibers (2–7% w/v) are passed through a spring-loaded valve assembly, at high pressure (8,000 psi), low velocity and exposed to a pressure drop to atmospheric condition while the valve opens and closes in a cyclic motion. This results in high shear and impact forces generated in a minute gap of the valve maintained at a temperature of 70–80°C [30]. As a result, the cell wall is peeled off and the DP is reduced [35, 36]. For instance, the DP is reduced from 2720 to 740 when cotton is used as a source of cellulose. Usually, this method produces fibers with diameters between 20 and 100 nm and lengths of several tens of micrometers. However, this method present some problems such as clogging of the homogenizer, high energy consumption, and mechanical damage of the crystalline microfibril structure [34, 36, 37]. HPH also decreases the crystallinity of nanofibers by increasing the number of passes.

2.1.2. Microfluidization

It is a process by which a fiber suspension is pumped through thin z-shaped chambers under a high pressure (~30,000 psi). The slurry is accelerated and led into the interaction chamber where it passes through geometrically fixed microchannels at very high velocities. Thin Z-shaped chambers with different sizes generate a high shear rate and impact forces against colliding streams. A microfluidizer generates CNFs with several micrometers in length and less than 100 nm in diameter.

2.1.3. Grinding

This is a single process by which the cellulose suspension is passed through an ultrafine grinder where the upper stone is static and the lower stone is rotating at 1400–1500 rpm. As a result, the cell wall structure is broken down by shear forces generating a gel due to the heat generated by friction while evaporating water. However, a mechanical damage of the fiber would occur [38]. This process has been used to extract CNFs from wheat straw and soy hulls [39]. Compression is a modified grinding system by which delignified fibers of cellulosic materials are placed in a bed of stripes placed between the two plates and subjected to a constant load of 10 tons for several seconds. However, in this process fibers in the micrometer rather in the nanometer size range are obtained [40].

2.1.4. Cryocrushing

In this process, swollen cellulosic fibers are immersed in liquid nitrogen. These brittle fibers are subsequently crushed by high shear and impact forces. As a result, ice crystals exert pressure on the cell walls, causing them to rupture. Usually, this method produces CNFs with diameters ranging from 30 to 80 nm [3].

2.1.5. High-intensity ultrasonication

It is a mechanical process in which oscillating power is used to isolate CNFs by hydrodynamic forces of ultrasound [41]. During the process, cavitation leads to a formation of powerful oscillating high intensive waves. These microscopic gas bubbles expand and implode breaking down cellulose fibers. However, a large feed concentration and a large distance from probe to beaker is not advantageous for fibrillation. A typical treatment requires a cylindrical titanium alloy probe tip of 1.5 cm in diameter, high temperatures, 1000 W power, and 20–25 kHz for ~30 min [42].

2.1.6. Steam explosion

It is a thermomechanical process (200–270°C) that exposes cellulose to a high pressure of steam (14–16 bar). As a result, it penetrates the biomass by diffusion for short periods of time (20 s to 20 min), followed by a sudden decompression (explosion) generating shear forces which hydrolyze the glycosidic and hydrogen bonds between the glucose chains [31].

All the above-described mechanical methods demands a high energy consumption (20,000–30,000 kWh/tonne), which prevents their successful commercialization. Therefore, by combining the mechanical treatment with enzymatic or chemical pretreatments, it is possible to decrease the high energy consumption.

2.2. Electrospinning

In this electromechanical method, a cellulose dispersion is extruded and electrospun under the effect of a high electric field. Thus, a charged stream of cellulose dispersion is ejected following a 3D spiral trajectory. Once the solvent evaporates, it leaves behind a randomly oriented nanofibers in the collector [43]. This is a quite simple and cost-effective process. The morphology of the CNFs produced by this technology depends on factors such as the electric field strength, solution feed rate, tip-to-collector distance, etc. [43].

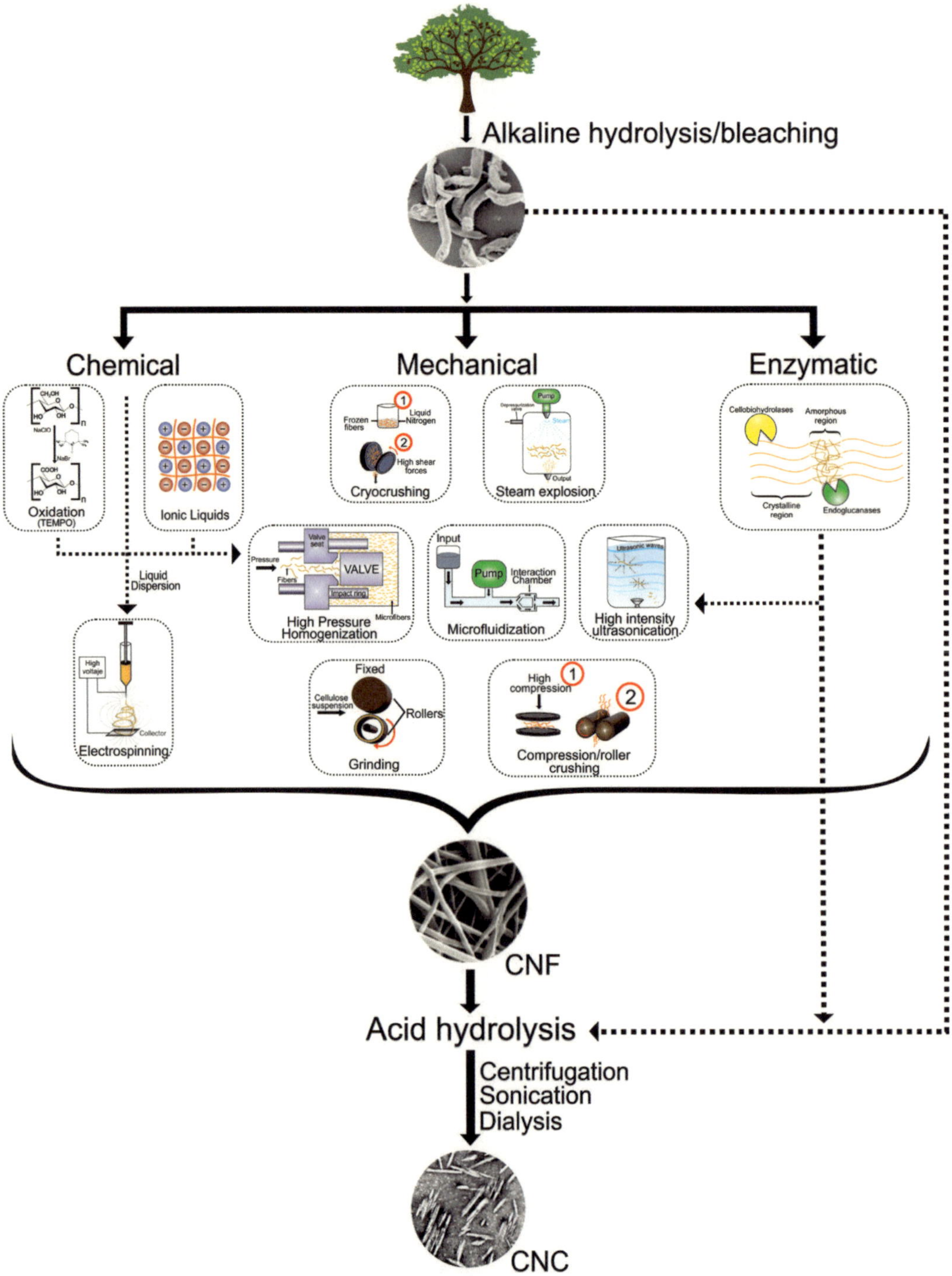

Figure 2. Conventional treatments to obtain cellulose nanoparticles.

2.3. Enzymatic hydrolysis

In this treatment, an enzyme is used to modify and/or degrade the lignin and hemicelluloses, while preventing the cellulose region. These enzymes are produced by cellobiohydrolases, which are A- and B-type cellulases able to attack the crystalline portion of cellulose, and endoglucanases C and D type, which are able to attack the disordered structure (amorphous) of cellulose [12]. Cellobiohydrolases and endoglucanases have strong synergistic effects. Thus, pretreated fibers subjected to the lowest enzyme concentration (0.02%) disintegrate, while molecular weight and fiber length are preserved. Endoglucanases cleave the noncovalent internal bonds, whereas exoglucanases attack the terminal glycosidic bonds [44]. Furthermore, *Trichoderma reesei* and *A. xylinum* produce enzymes which are able to reduce the size of microcrystalline cellulose. Enzymatic methods are highly costly due to the isolation process of the enzymes and the long enzymatic treatment time required for a successful hydrolysis [31].

3. Preparation of cellulose nanocrystals

CNCs are commonly isolated from cellulose fibers by acid hydrolysis. Tunicin is a cellulose extracted from sea animal sources made up of highly crystalline nanofibers and has an helical organization. It has a high modulus, a high aspect ratio, and good compatibility with matrix materials [45]. This cellulose is obtained by cutting into small fragments followed by bleaching. Subsequently, CNC is extracted from bleached samples by acid hydrolysis with 64 v/v% H_2SO_4, for 5 h, at 50°C. Typically, the diameter, the length, and the aspect ratio of CNC are 4–25 nm, 100–500 nm, and 15–50, respectively. Among the many cellulosic sources used for its isolation, cotton constitutes the main source. It exhibits an elongated crystalline rodlike shape and has a limited flexibility since it has no amorphous regions. These CNC have a degree of crystallinity from 55% to 90%. Moreover, the degree of crystallinity, aspect ratio, and morphology depends on the source of cellulosic material and preparation conditions [46].

A typical production process involves acid hydrolysis, washing, centrifugation, dialysis, and sonication to form a suspension followed by drying [47]. The main process for the preparation of CNCs is based on strong acid hydrolysis under strictly controlled conditions of concentration, temperature, agitation, and time. The mineral acid breaks the β-1,4 glycosidic bonds in cellulose. During hydrolysis, the amorphous regions are attacked, leaving the crystalline regions intact [48]. The resulting suspension is washed and centrifuged, and dialysis is performed to remove any free acid molecules. Microbial hydrolysis has also been utilized to produce CNCs. This microbial hydrolysis is eco-friendly and does not require any surface modification [44]. The charge and colloidal behavior of CNCs depends on the acid used for the production. Sulfuric and hydrochloric acids are the most commonly used, but phosphoric and hydrobromic acids have also been used [7].

For instance, hydrolysis with 63.5 v/v% H_2SO_4 for 2 h leads to a 30% yield, a width narrower than 10 nm and length ranging between 200 and 400 nm [3]. Smaller CNCs are obtained by increasing hydrolysis time and acid concentration [15]. Another treatment employs wood pulp boiled with 2.5 N sulfuric acid for 12 h, generating CNCs with lengths between 50 and 60 nm and widths between 5 and 10 nm [49].

The physical characteristics of CNCs depend on the origin of cellulose sources, concentration of acid, types of acid, reaction time, and temperature [3]. If sulfuric acid (50–70% w/v), temperature of 20–70°C, rate of 500 rpm, and time of 0.5–6 h are employed, esterification also occurs forming "cellulose sulfate," resulting in a negatively charged surface on the cellulose crystallites. Longer hydrolysis time and higher temperature generate shorter nanocrystals with higher surface charge, high crystallinity (~80%), and narrower polydispersity [50–53]. However, the very limited commercial availability of CNC is due to the time consuming production process and the low yield produced, especially if the initial amount of amorphous cellulose is very high. The aggregation of CNC occurs with HCl, but sulfuric acid creates charged sulfate esters promoting the dispersion of the CNCs in water preventing aggregation. The combination of both sulfuric and hydrochloric acids during hydrolysis generates spherical nanoparticles with improved thermal stability due to the reduced presence of sulfate groups on their surface. The negative surface charge of CNCs stabilizes the aqueous suspension against flocculation, but this charge also compromises the thermostability of nanocrystals. Therefore, the increase in the sulfate group content decreases the temperature at which thermal degradation takes place [54]. CNCs also exhibit chiral nematic liquid crystalline alignments, which are seen as a flow of birefringence between two crossed polarizing films.

4. Pretreatment processes

The aim of the pretreatment process is to remove ashes, waxes, lignin, hemicellulose, and other noncellulosic compounds, which are crucial to produce pure cellulosic products such as CNFs and CNCs [30–32]. A pretreatment also reduces the energy demand of mechanical processes from 20,000 to 30,000 kWh/tonne to 1000 kWh/tonne. The types of pretreatments applied on different raw materials such as tunicate, algae, and bacteria cellulose have been reported previously [21, 55]. The alkaline delignification and organosolvation with acetic acid, aqueous methanol, or ethanol are also considered as pretreatment processes [56, 57].

4.1. Alkaline hydrolysis

Alkaline treatments are conducted when a more effective lignin, hemicellulose, and pectin solubilization and removal is needed. Alkaline extraction needs to be controlled to avoid cellulose degradation [3]. A typical treatment involves dipping of fibers in a 5% sodium hydroxide solution for ~48 h at 30°C. At pH >12, NaOH reduces super oxide radicals ($-O_2$), wherein lignin and hemicellulose are hydrolyzed [15]. However, if lignin content is high in the cellulosic source, the nanocellulose yield is low [15, 58].

4.2. Bleaching

Pulp can be bleached to improve ageing resistance avoiding yellowing and brittleness. These two defects are mainly related to the presence of lignin. Different compounds are commonly used for bleaching. These include hydrogen peroxide (H_2O_2), chlorine dioxide (ClO_2), ozone (O_3), peracetic acid, and $NaClO_2$. Sulfite pulps are more readily bleached and results in high yields [59].

4.3. Oxidation

The TEMPO-mediated surface oxidation is the most commonly used chemical pretreatment conducted under aqueous and mild conditions. It converts the primary hydroxyl group (C6) to a charged aldehyde or carboxylate functional group, whereas the secondary hydroxyl moieties present in the cellulose molecule remain unaffected [1, 12]. The oxidation of cellulose fibers occurs in the presence of NaClO and catalytic amounts of 2,2,6,6 tetramethyl-1 piperidinyloxy radical (TEMPO) and NaBr as catalyst at a pH between 9 and 11 and room temperature. The higher the amount of NaClO in the reaction medium, the larger is the number of carboxylic groups formed at the surface of the CNFs and the stronger is the decrease in DP [44]. This oxidation creates negative charges on the surface of CNC [31] without changing the original fibrous morphologies [61–66]. The reaction by-product is only sodium chloride. Other N-oxyl compounds, such as the 4-hydroxy TEMPO derivative (less expensive than TEMPO), have been proposed. The residual aldehyde groups causes discoloration. In order to avoid depolymerization or discoloration of the oxidized cellulose, a TEMPO/NaClO/$NaClO_2$ system is employed under neutral or slightly acidic conditions [60]. This treatment also prevents the postaggregation of nanoparticles during the drying step.

A variation of this condition oxidizes wood cellulose rendering CNF with a higher molecular weight and with no aldehyde groups using a TEMPO/NaClO/$NaClO_2$ system at pHs ranging from 5 to 7 [67]. The TEMPO pretreatment eases the separation of the nanofibrils from each other due to the repulsive forces of the ionized carboxylate groups, which overwhelm the hydrogen bonds. The TEMPO oxidation pretreatment is usually followed by a mechanical treatment. Other less commonly used processes include oxidation at 60°C with ammonium persulfate and the sequential periodate and chlorite oxidation [68].

4.4. Ionic liquids

Ionic liquids are organic salts having no corrosive properties, no flammability, a melting point below 100°C, low vapor pressure, and low viscosities [34]. Ionic liquids dissolve cellulose and render a wide range of particle morphologies after precipitation. The ionic liquid breaks intramolecular hydrogen bonds, whereas the cations attack the O atom of the –OH, and anions attack the hydrogen atoms of the -OH group [34]. PF_6^-, BF_4-, $(CF_3CO_2)^-$, $(SbF_6)^-$, $(OTS)^-$, $(ClO_4)^-$, $(GeCl_3)^-$, $(Al_2Cl_7)^-$, and $(AlCl_4)^-$ are the most common anions employed [69, 70].

5. Nanocellulose derivatization

This chemical treatment is performed on the surface of BC, CNC, or CNF for making them more hydrophobic reducing the agglomeration tendency of these materials. The goal of the derivatization is to endow nanocellulose with a hydrophobic character in order to improve its compatibility with nonpolar polymers [60, 71–73].

5.1. Carboxymethylation

This process makes the surface negatively charged, promotes the formation of a stable suspension, and increases the breakup of lignocellulosic fibers [2]. If carboxymethylation is conducted before mechanical treatment, the fibers become more dispersible having a lower degree of crystallinity [72].

5.2. Acetylation

In this reaction, the C6 hydroxyl groups of cellulose are selectively converted to carboxylate groups and only NaClO and NaOH are consumed. The amount of carboxylate groups formed increases with the amount of NaClO and by employing long reaction times [74]. This reaction causes plasticization of lignocellulosic fibers [31] [54]. Further, a reaction of CNF with acetic anhydride at 105°C for 30 min causes a degree of substitution (DS) of 0.43. As a result, the contact angle increases from 33° for nonacetylated nanofibers to 115° for acetylated ones. The acetylated fibers have a lower crystallinity due to degradation of crystalline regions during the reaction.

5.3. Isocyanate

Isocyanates, in particular, octadecyl isocyanate, can generate covalent bonds with hydroxyl groups on the particle surface, rendering a degree of substitution of 0.07 and 0.09 for CNC and NFC, respectively [75].

5.4. Silylation

The reaction between silanol and OH groups of cellulose at high temperature is initiated by water. Surface silylation of CNFs from bleached softwood pulp using chlorodimethyl isopropylsilane renders a degree of surface substitution from 0.6 to 1. Conversely, silylation of CNFs by isopropyl dimethylchlorosilane renders a CNF that forms suspensions with a shear-thinning behavior [54].

6. Physical properties

6.1. Morphology

Depending upon the source of the cellulose and the method of production, a CNF displays similar morphologies but several dimensions. Typically, a CNF and a CNC have typical diameters of 2–100 nm and between 2 and 30 nm, respectively. For instance, CNFs from wheat straw have diameters from 10 to 80 nm [22] as compared to soy hulls (20–120 nm) [20], kenaf bast (2–6 nm), wood (15 nm), bagasse (5–15 nm), rice straw (4–13 nm), soybean stock-based (50–100 nm), cotton (10– 25 nm), and empty fruit bunch (10–30 nm) [31, 76–85].

6.2. Crystallinity

Crystallinity is highly dependent on the lignocellulosic source. For instance, crystallinity of flax, rutabaga, and wood CNFs are 59%, 64%, and 54%, respectively, whereas a crystallinity of 85.9%, 76%, 84.9%, 94%, 80.6%, and 81.7% has been found for CNCs obtained from sisal, rice husk, flax, cotton, corn stover, and commercial MCC, respectively. On the other hand, a DC of 78% and 70% has been obtained for wheat straw [22] and soy hull CNFs, respectively. Further, very low values have been obtained for beet pulp ~30–40% [20]. The degree of crystallinity ranges in the order: pineapple > banana > jute, and this order agrees with the values of cellulose content determined in these samples. Usually, CNCs prepared from H_2SO_4 have lower crystalline values than those prepared from HCl. In addition, the increase in hydrolysis time also increases crystallinity due to the elimination of amorphous regions [15].

6.3. Thermal properties

The thermal degradation of lignocellulosic materials begins with an early decomposition of hemicelluloses, followed by an early stage of pyrolysis of lignin, depolymerization, active flaming combustion, and char oxidation. Further, CNF has a high degradation temperature onset (350°C) and better thermal behavior than hemicellulose, pectin, and lignin. On the contrary, the onset of the thermal degradation of CNC typically occurs at 200–300°C[86]. CNC with lower sulfate content have better thermal stability [53]. On the other hand, banana CNF exhibited three main weight loss regions. The initial weight is mainly due to moisture evaporation followed by thermal depolymerization of hemicellulose and the cleavage of glycosidic linkages of cellulose. The broad peak in the region from 200°C to 500°C is due to residual lignin components. A convection drying of a CNF removes water slowly causing the formation of aggregates. Therefore, the dried CNF presents a lower degree of thermal stability than that of the original fibers. [19].

6.4. Degree of Polymerization (DP) and mechanical properties

DP is strongly correlated with the aspect ratio of the nanofibers. As explained previously, any pretreatment process also reduces the DP. On the other hand, the mechanical properties of cellulose nanoparticles depend on morphology, geometrical dimensions, crystal structure, crystallinity, and the process used to produce CNCs and CNFs. For instance, the DP of softwood is 2249, but the DP of sulfated CNF is 825. Further, the tensile strength of native CNCs ranges from 7.5 to 7.7 GPa. Further, the Young's modulus of CNCs is estimated to lie between 130 and 250 GPa.

6.5. Hornification

Drying of individual cellulose nanoparticles creates irreversible agglomeration affecting their dimension and, therefore, their unique properties [87]. This irreversible agglomeration is known as hornification and is related to the hydrogen bonds formed [81]. If freeze-drying or supercritical dying with CO_2 is used, agglomeration is avoided [87–89].

6.6. Film properties

CNF gels can be diluted and either cast or vacuum filtered followed by drying to form stiff films due to the formation of an interfibrillar hydrogen bonding network [63, 90–93]. For instance, CNF films obtained from sugar beet have values of tensile strength and modulus of 104 MPa and 3.5 GPa, respectively [37, 61, 94].

6.7. Surface area

Cellulose nanoparticles have a high specific surface area (SSA). Typically, cellulose nanoparticles have SSA ranging from 50 to 200 g/m^2. Conversely, the SSA of nanocellulose aerogels ranges higher from 250 to 350 m^2/g and have a very low density (0.02 g/cm^3) and a high porosity of 98% [61, 63, 94].

6.8. Rheological properties

CNF suspensions exhibit a shear-thinning behavior and pseudoplasticity, which in turn depends on the pH medium [95]. Further, sulfate cellulose shows a pH-dependent viscosity profile due to electrostatic interactions. Moreover, a CNF suspension has a decreasing viscosity with increasing shear rates. Further, CNF also has a high elastic modulus due to the entangled network structure [96].

6.9. Water sorption and permeability

CNFs are able to form films with a low moisture diffusivity due to a rigid fiber network [90] and, thus, have excellent barrier properties [90]. The water vapor transfer rate (WVTR) of CNFs films are 20% smaller than those made of macrofibers (from 20% to 30%). CNC films are expected to provide a better barrier to water since CNC films have a more crystalline nature than CNF [90].

6.10. Oxygen barrier

The oxygen transmittance rate of CNF at 0% RH is in the range 17–18 mL/m^2/day. The increase of RH increases the oxygen permeability due to the limited hydrogen bonding and loose network caused by the incoming water molecules. The water and oxygen permeability decreased with increasing film thickness due to the lack of interconnectivity of pores [63].

7. Toxicity

CNCs have low toxicity and low environmental risk according to ecotoxicological tests with several aquatic species (e.g., Daphnia, rainbow trout and fathead minnow). Further, cytotoxicity (intracellular toxic effect) and proinflammatory response are significantly lower than those for MWCNT (multiwalled carbon nanotubes) and CAF (crocidolite asbestos fibers). CNF shows no toxicity and genotoxicity in vitro [97]. The toxicity of BC nanofibers has been successfully evaluated in vitro through cell viability and flow cytometric assays and in vivo using C57/B16 mice surgeries. Further, BC shows no toxicity in human umbilical vein endothelial cell culture, fibroblasts, and chondrocytes. The in vitro evaluation also shows that 95% of the mesenchymal stem cells aggregate to cellulose membrane [98, 99] (Fig. 3).

8. Biomedical applications

8.1. Biological benefits

Nanocellulose has been mainly used as a filler in nanocomposites because of its good mechanical properties due to their biodegradability, renewability, availability, sustainability, lower cost, lower weight, higher mechanical strength, biocompatibility, high hydrophilicity, and high surface area [94, 100, 101–105]. It evades adverse tissue reactions, and unlike proteins, its polysaccharide nature makes it less immunogenic and nonhemolytic. It also promotes cellular interaction and tissue development. It is a slow/nondegrading material in vivo and in vitro, which makes it suitable for use as a scaffold providing a long-term support, sustains high loads, and has a high wear resistance. The biomedical industry includes skin replacements for burnings and wounds; drug releasing system; blood vessel growth; nerves, gum, and dura mater reconstruction; scaffolds for tissue engineering; stent covering; and bone reconstruction. The cellulose nanoparticle surface dictates cellular response by interfering with cellular adhesion, proliferation, migration, and functioning. On the other hand, cells support, hold, synthesize the matrix for the new tissues, and keep the proper growth ambient, whereas the growth factors promote the cell regeneration [97, 101, 106, 107].

In one study, a charged CNC–FITC and newly synthesized CNC–rhodamine B isothiocyanate (RBITC) were synthesized, and in vitro cellular uptake studies showed that the positively charged CNC-RBITC was taken up by human embryonic kidney (HEK) and *Spodoptera frugiperda* (Sf9) cells without any noticeable cytotoxic effect on the two cell lines [108].

8.2. Biosensors and diagnostics

CNFs serve as a suitable platform for immobilization of bioactive molecules (e.g., enzymes, antibodies, etc.), which is useful in biosensors and diagnostics. For example, novel gold bacterial cellulose (Au-BC) nanocomposites have been prepared by a one-step biotemplated method in aqueous suspensions. This material shows excellent biocompatibility, good conductivity, and ultrafine nanofiber network structure, which makes it able to entrap horseradish peroxidase (HRP), maintaining enzyme bioactivity. HRP biosensors allow detection of H_2O_2 with a detection limit lower than 1 μM [109]. CNF films carboxlylated with TEMPO and activated via EDC/NHS coupling have been used to immobilize the antibody (antihuman immunoglobulin G (anti-IgG)) by physical adsorption. This surface can detect positively charged molecules. A TEMPO-activated film has also been used to conjugate Avidin for selectively capturing biotinylated molecules (anti-IgG) [109, 110].

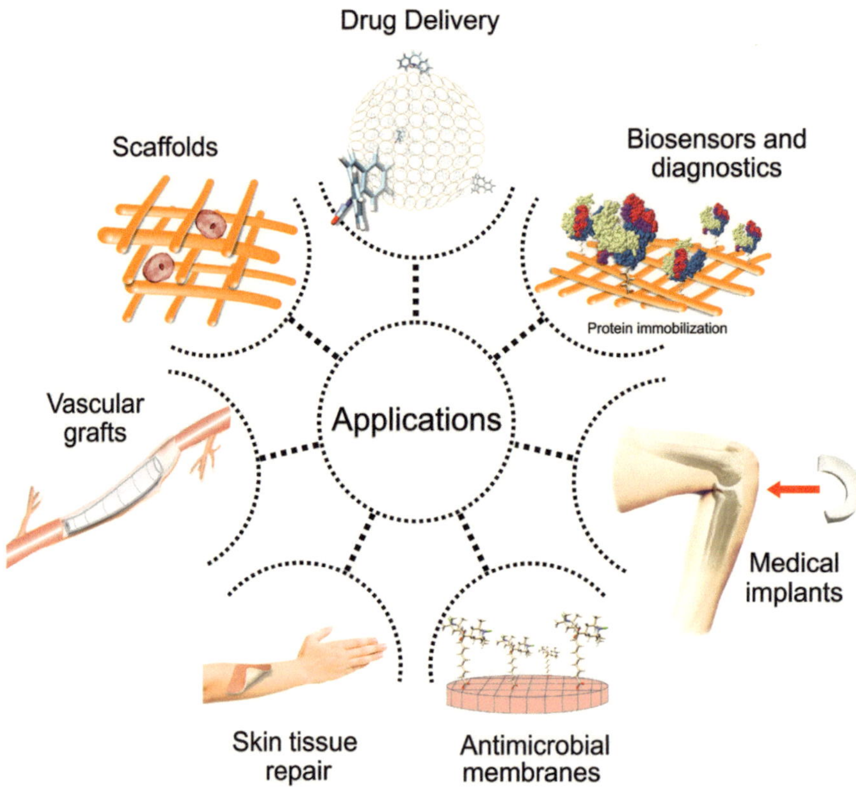

Figure 3. Biomedical applications of cellulose nanoparticles.

Another study showed CNF to prepare support films with carboxyl groups, which are then converted to amine-reactive species. These substrates were then used to bind polyclonal anti-IgG. The CNF surface can also be activated by copolymer grafting. Thus, a peptide with specific affinity to human IgG is conjugated to the grafted polymer having a high selectivity.

Proteins such as collagen, elastin, hyaluronan, and growth factors such as the basic fibroblast growth factor (B-FGF), human epidermal growth factor (H-EGF), and keratinocyte growth factor (KGF) have been immobilized on macroporous BC to improve biocompatibility. The attachment of cells can be improved by utilizing adhesive amino acid sequences, such as Arg-Gly-Asp (RGD) found in several extracellular matrix proteins [111, 112].

Peptides such as the HWRGWV peptide can be immobilized on the TEMPO-activated film to detect human IgG. Thus, the acetylated peptide is covalently immobilized to the spacers on CNF via amide reaction. For instance, chitosan can be physically adsorbed onto the CNF and used as a spacer. The resulting biosensor has a very high specific binding capability for IgG and exhibits excellent resistance for nonspecific protein adsorption [113]. In another study, Edwards et al. created biosensors based on CNC by peptide conjugation for detection of human neutrophil elastase [114].

On the other hand, CNCs have been used as electrochemical sensors to selectively detect DNA hybridization. DNA oligomers were grafted onto TEMPO-oxidized CNC produced from cotton. This DNA-grafted CNC is able to self-assembling into larger aggregates as compared to the unmodified CNC. In another study, silver nanoparticles were obtained onto TEMPO-mediated oxidized CNC by using $NaBH_4$ as a reducing agent. The presence of CNC prevented the aggregation of the nanoparticles [115].

Moreover, the layer by layer (LbL) assembly technique has been used to adsorb collagen onto the surface of CNF. The LbL technique has also been used to prepare a luminescent single-walled carbon nanotube–CNC films enhancing their water dispersibility [116].

Dong et al. synthesized folic acid (FA)-grafted CNCs and explored their folate-receptor-mediated uptake by human and rat brain tumor cells. First, CNCs were labeled with fluorescein isothiocyanate (FITC) for detection in the cells and were then conjugated with FA. In vitro studies showed that the cellular binding of the FITC–CNC–FA by the folate receptor (which is overexpressed by cancer cells) was higher than that of the free FA [117].

8.3. Skin tissue repair

Nanocellulose membranes could serve as an infection barrier, prevent loss of fluids, have a painkiller effect, allow drugs to be easily applied, and also absorb the purulent fluids during all inflammatory stages, expelling them later on in a controlled and painless manner [118].

Properties such as biocompatibility, high superficial area, high water absorption capacity, high elastic modulus, low thermal expansion coefficient, optical transparency, anisotropy, and flexible surface chemistry endow nanocellulose with suitable wound dressing applications. For instance, a membrane has been developed with BC and propolis extract, rendering antimicrobial and anti-inflammatory activities in chronic wounds absorbing purulent exudates. Further, it eases the BC removal from a wound surface after recovery [97].

Traditionally, skin tissue repair materials have been absorbent, permeable materials such as gauze, which can adhere to desiccated wound surfaces inducing trauma upon removal [119, 120]. BC controls wound dressing since it can control wound exudates and can provide a moist environment to a wound resulting in better wound healing. For instance, Biofill® is a commercial product of BC used as a temporary substitute for human skin in cases of second and third degree burns.

Czaja et al. showed that the skin of the patients whose burns were covered with a BC membrane healed faster than the wounds of patients who received conventional wound dressings. This is explained by the faster tissue regeneration, capillary formation and cell proliferation [121].

In a different study, BC wound dressing materials were compared to two different commercial dressings, Vaseline gauze, and Algisite M in a rat model. This study showed that BC-dressed animals had more rapid wound healing within 14 days without any evidence of toxicity [122]. Further, BC, gelatin, and alginate composite membranes showed the successful growth of NIH/3T3-type cells and, hence, proved potential as a skin tissue regeneration template [123].

The addition of chitosan to a tempo-oxidized BC renders a composite with high superior mechanical properties, water holding capacity, and water release rate, and thus, these composites can be used for wound dressing [124].

BC has also been used in a surgery of the lateral wall of the nose preventing nasal bleeding, surgical wound infections, local pain, and clotting [125]. Therefore, it provides a more rapid healing without the formation of crusts and prevent infection without the need for removal as compared to commercial nasal packing which causes a great discomfort upon removal [125].

A nanocellulose membrane has also been implanted into the subcutaneous tissue of diabetic rats for 12 weeks. Rats showed no macroscopic signs of inflammation around the implants, no formation of fibrotic capsule or giant cells, and fibroblasts were fully integrated to the cellulosic membrane and started to synthesize collagen [126].

8.4. Cortical implants

Neural interfaces are able to record neural signals from individual neurons or small groups of neurons in the brain. The most common neural interfaces are made of iridium oxide, silicon, platinum, titanium, glassy carbon, gold, and stainless steel. However, they cause glia encapsulation at the electrode interface, leading to neuron death near the surface of the implanted electrode. For this reason, adaptive cellulose nanoparticles interfaces should be stiff enough to be easily implanted into the brain but subsequently soften under in vivo conditions to closely match the stiffness of the brain tissue.

This nanocellulosic interface relies on stiff collagen fibers dispersed throughout a soft fibrillin matrix. Because of the abundance of surface hydroxyl groups, CNCs strongly interact with each other through hydrogen bonding and/or van der Waals' forces, but exposure to water reduces CNC–CNC interactions because of competitive hydrogen bonding or interfacial interactions with intermolecular van der Waal's forces. For instance, CNCs isolated from tunicate sea creatures have been successfully integrated into a rubbery ethylene oxide-epichlorohydrin copolymer matrix having a lower stiffness than conventional electrodes [127].

Another study reported the development of poly(vinyl acetate) (PVAc) and CNCs composites showing a dual responsive behavior. Upon exposure to physiological conditions, the composites are plasticized and the CNC network loses its stiffness capabilities due to the loss of hydrogen bonding. Once these composites were implanted in the pia mater of the cerebral cortex of a rat, the initial stiffness rapidly decreased matching that of the brain tissue [128, 129]. These adaptive microelectrodes implanted into a rat cortex for 8 weeks increased the cell density at the electrode–tissue interface. After 16 weeks of implantation, there was no neuron death surrounding PVAc/CNC implants as compared to PVAc-coated microelectrodes [130]. Another cortical implant in rats containing CNC/PVA and curcumin showed that after 4 weeks, curcumin promoted higher neuron survival and a more stable blood–brain barrier than the neat PVA controls [131].

8.5. Vascular grafts

BC-based implants have been developed to replace synthetic by-pass implants made of polytetrafluoroethylene, poly(ethylene terephthalate), polyethylene, and polyurethane.

Further, BC tubes have been successfully used to replace carotid arteries in rats, pigs, and sheep without any rejection after 4 weeks [127].

One study showed that the mechanical properties of CNC-PVA composites are similar to that of cardiovascular tissues, such as aorta and heart valve leaflets. For instance, the stress–strain properties for porcine aorta and heart valve tissue matches those of PVA-nanocellulose composites in the circumferential and the axial tissue directions [132].

The nanocellulose implants have been attached in the carotid artery of rats for 1 year, resulting in incorporation nanocellulose forming neointima and ingrowth of active fibroblasts. In another study, the grafts were used to successfully replace the carotid arteries of pigs [133]. Further, cellulose and chitosan composites have been successfully used to produce hollow tubes with a compliance compared to that of human coronary arteries showing potential for coronary artery bypass graft applications [134].

In one study, part of the carotid artery (4–6 mm) of a rat was replaced using BC and after 4 weeks, the BC complex was wrapped up with connective tissue and was infused with small vessels [28, 135]. BC porous surfaces have also been reported to maintain fiber network arrangement viable in endothelial cells for 20 days [136]. Further, the surface modification of BC by nitrogen-containing plasma improved cell adhesion and proliferation of the endothelial and neuroblast cells [137].

8.6. Medical implants

Medical implants must have mechanical characteristics as the tissue it replaces. It must also show nonthrombogenicity, sterilizability, durability, lesser degree of calcification, and good processability. The implant should be biocompatible with the host tissues in terms of chemical, mechanical, surface chemistry, and pharmacological properties.

BC is a good candidate for use as medical implants since it is nondegradable under physiological conditions and provides durable mechanical properties and chemical stability. One study showed that the mechanical properties of the BC gel with collagen meniscal implants were similar in magnitude to the ones of pig menisci [138].

Further, an ear-shaped BC prototype material showed suitable mechanical properties for ear cartilage replacement for a customized patient-specific ear shapes [139].

A nanocellulose membrane has also been implanted in the nasal dorsum of 22 rabbits as an excellent substitute of bone cartilage. After 6 months, residual inflammation was attributed to the surgical procedure itself and not to the cellulosic membrane [140].

The cartilage that covers the trochlear groove in dogs is composed of chondrocytes. Further, BC was utilized successfully in experimental trochleoplasty in dogs, showing advantage in respect to conventional treatment for osteochondral injuries [141]. Moreover, BC membrane was applied in the tissue formation of fibrocartilage ripe, resulting in a good integration of the newly formed tissue.

Other researchers have successfully incorporated BC and polytetrafluoroethylene (PTFE) membranes in rats to correct abdominal wall defects [142]. Further, BC implanted in the peritoneum of dogs formed an integrated net along the conjunctive tissue after 6 months [143].

On the other hand, the biggest challenge in dental applications is the loss of alveolar bone. For this reason, synthetic hydroxyapatite (HAP-91) was implanted in the dental cavities and covered by nanocellulose membranes. Nanocellulose promoted faster bone regeneration and resembled those of the original tissue after 50 days [144].

A bandage product derived from BC (Gengiflex®) restores the osseous defects. It consists of an inner layer of microbial cellulose, which offers rigidity to the membrane, and an outer layer of alkali cellulose. A greater amount of bone formation was present in bone defects protected by the BC membrane, when compared to the control sites [145, 146].

BC membranes were tested as physical barriers used to treat bone defects. In this scenario, two osseous defects (8 mm in diameter) were performed in each hind foot of four adult rabbits. After 3 months, the bone defects showed lamellar bone formation resulting in partial bone deposition [147].

8.7. Cell culture and scaffolds

The asymmetric structure of a scaffold is composed of a fine network of nanofibrils, which is similar to a collagen network, which promotes the adhesion and proliferation of muscle cells [97, 101]. BC has shown significantly higher levels of chondrocyte growth, suggesting the potential application of scaffolds for cartilage tissue engineering [148]. The beneficial properties of CNF are based on its unique nanofibrillar structure, mimicking properties of the extracellular matrix, and thus, a CNF scaffold promotes hepatocyte 3D cell culture without added bioactive components.

Nanocellulose and hydroxyapatite (HA) are both capable of bone replacement because of their properties, including biocompatibility with the human body, bioactivity, osteoconductivity, and noninflammatory properties [74]. Further, nanocellulose has also been soaked into HA to develop a composite scaffold for bone regeneration. Thus, CNC/HA scaffolds were prepared by absorption of HA onto the BC surface to induce nucleation of calcium-deficient HA. The presence of calcium-deficient HA crystals on the BC surface increased cell attachment and alkaline phosphatase activity on bone cells. Further, nanocellulose has been combined with polyacrylamide and gelatin, yielding hydrogels with improved toughness [149].

Moreover, enzymatically modified gelatin (EMG) and nanocellulose composites have been prepared to improve the rehydration properties and, thus, can be used as scaffold for the cornea tissue since the stromal cells are able to grow into the scaffold [150].

In one study, BC and poly(3-hydroxubutyrate-co-4-hydroxubutyrate) (P(3HB-co-4HB) composite scaffolds showed excellent biocompatibility in Chinese hamster lung (CHL) fibroblast cells [151]. Further, the BC and alginate composite (80:20 w/w) dried by supercritical carbon dioxide formed a nanoporous structure, which supports the proliferation of keratinocytes and gingival fibroblasts [152, 153].

http://dx.doi.org/10.5772/61334

In another study, nanocellulose/PEG composite scaffolds were prepared by soaking a nanocellulose hydrogel with a PEG solution forming scaffold with improved thermal stability. Results indicated that the Young's modulus and tensile strength tended to decrease while the elongation at break had a slight increase. Thus, the prepared nanocellulose/PEG composite scaffolds were suitable for cell attachment [154].

The functionalization of the BC surface with recombinant proteins containing a bioactive peptide (IKVAV) and a carbohydrate-binding module (CBM3) has improved their biocompatibility with neuronal and mesenchymal cells [155]. BC cross-linked with heparin is able to prevent the formation of blood clots [156]. A natural peptide called as polylysine (PLL) has been cross-linked to the surface of BC resembling the collagen fibers and composition of natural bone [157].

CNCs have been coated with 2-hydroxyethylmethacrylate and methacrylic functional groups forming hydrogels with excellent mechanical properties comparable to articular cartilage with hydrogel-like properties [158].

One study reported the creation composites based on BC and type I collagen (COL) for potential bone tissue engineering, in which collagen was covalently introduced into the BC network. Further, cell culture with osteogenic cells revealed that collagen did not affect cell adhesion and proliferation or its morphology [159]. Heparin and nanocellulose scaffolds have been prepared with anticoagulant properties for potential use in vascular tissue engineering [160].

Further, grafted zwitterionic carboxybetaine improved CNF membrane blood compatibility. In another study, dialdehyde BC membranes supported the epidermal cell adhesion and proliferation [161].

BC has also been used to treat wounds in diabetic foot ulcers. The mean time for 75% epithelization was achieved in 79 days, and BC shortened the epithelization time as compared to XeroformTM® Petrolatum gauze [162]. Another study showed a complete closure of the facial wound within 44 days with no significant signs of extensive scarring [121]. Further, the release of BC dressing from the wound is a painless operation due to the moisture still present in the cellulose structure [163].

A nanogel made of poly(*N*-isopropyl acrylamide-co-butyl methacrylate), BC, and a surfactant has been produced by emulsion polymerization. This nanogel showed a thermal responsive behavior from a swollen to shrunken gel with increasing temperature. This is explained by the decrease of hydrogen bonding interactions with temperature [164].

8.8. Antimicrobial activity

Nanocellulose intrinsically does not possess any antimicrobial property. Therefore, it needs to be functionalized with antimicrobial agents. For instance, chemical grafting of aminoalkyl groups [165], 2-benzyl-4- chlorophenol [166], and L-cysteine [167], onto the surface of the cellulose backbone has been reported [168]. In one study, a BC film was soaked in a benzalkonium chloride solution, resulting in antimicrobial activity against *Staphylococcus aureus* and *Bacillus subtilis,* which are bacteria generally found on contaminated wounds [169].

The electrospun composite nanofiber membrane containing bis(*N*-chloro-2,2,6,6-tetramethyl-4-piperidinyl) sebacate (Cl-BTMP) showed significant antimicrobial activity against *S. aureus, Escherichia coli,* and *Pseudomonas aeruginosa* attributed to the aggregation of Cl-BTMP. Further, nanocomposites with curcumin rendered antimicrobial activity against *E. coli* and *S. aureus* over a period of 24 h [170].

The immersion of BC in a silver nitrate solution, followed by reduction of absorbed silver ions (Ag+) with sodium borohydride, formed metallic silver nanoparticles with antimicrobial activity against *E. coli* and *S. aureus* [169]. Further, aminoalkyl-grafted bacterial nanocellulose (BC-NH_2) membranes were prepared by hydrolysis of the silane derivative, adsorption of the hydrolyzed species onto BC nanofibrils, followed by a chemical condensation reaction. These BC–NH_2 membranes showed antimicrobial activity against *E. coli* and *S. aureus* after 24 h [167].

In another study, composites of CNF and chitin nanocrystals formed a 3D network with bactericidal activity against *E. coli.* On the other hand, the in situ sol-gel formation of silver and gold particles within CNF endowed it with antibacterial activity against *E. coli, S. aureus,* and *Klebsiella pneumonia* strains. The assembly of the CNFs and silver nanoparticle (AgNP) composites occurs through electrostatic interactions. Another strategy of incorporation of silver is through magnetic interaction where the 3D structure of cellulose provides plenty of sites for heterogeneous nucleation of magnetite rendering a high antimicrobial activity against *E. coli* and *B. subtilis* [2]. Further, nanocomposite films prepared by the addition of cellulose nanocrystals, with silver nanoparticles, in a PLA matrix generate an antibacterial film against *S. aureus* and *E. coli* [171]. Silver ions interfere with the respiratory chain causing a decrease in bacterial viability [172–174]. Moreover, Ag and Au nanoparticles have been synthesized on CNC using cationic surfactants such a CTABr [175, 176].

In one study, BC was first homogenized with a ferric and ferrous salt mixture followed by soaking in dopamine and silver nitrate solution. The resulting magnetic silver/BC nanocomposites had antimicrobial activity against *E. coli* and *B. subtilis* were developed.

8.9. Drug delivery

The abundant surface hydroxyl groups in nanocellulose provide a site for the surface modification by a variety of methods. Surface modification modulates the loading and release of nonionized or hydrophobic drugs that would not normally bind to nanocellulose. For instance, poly(caprolactone) chains might be conjugated onto CNC for drug release [31].

PVA and methyl cellulose form aninterprenetrating polymer network through cross-linking with epichlorohydrin and serve as a carrier to load a drug for controlled release [31]. Further, coating of CNC with a cationic surfactant such as cetyltrimethylammonium bromide (CTABr) has been useful to load significant quantities of anticancer agents for controlled released [177–179]. In one study, nanoparticles of itraconazole were stabilized by the nanostructured cellulose matrix during freeze-drying and storage increasing its dissolution rate and in vivo performance [177].

On the other hand, spray-dried CNFs were produced in order to increase the long-term stability of drugs due to a better ability to pack having a low porosity and forming fast disintegrating tablets [179]. Further, BC membranes loaded with lidocaine rendered lower permeation rates in the skin than traditional drug delivery systems. The greatest advantage of the BC membrane is the combination of its wound healing capacity and the ability to absorb exudates with the release of antimicrobial and anti-inflammatory drugs [180].

In another study, freeze-dried BC and serum albumin composites were investigated as potential drug delivery systems for proteins [110]. In one report, CNC was oxidized with periodic acid to graft a spacer molecule (aminobutyric acid), and then syringyl alcohol was attached. In another investigation, calcium peroxide (CPO) was embedded into highly porous CNCs to produce H_2O_2, whereas catalase was added to convert the generated H_2O_2 to O_2, increasing cell survival up to 5 days [181].

Weng et al. created biodegradables cellulose microspheres loaded with doxorubicin for arterial embolization applications. They showed a burst release profile within 8 h followed by a release plateau over a 24-h period in rabbits [182].

Further, docetaxel-loaded CMC-based nanoparticles have been produced for enhanced cytotoxicity against cancer cells releasing 100% of drug within 3 weeks inhibiting 90% of tumor growth [183]. In another study, CMC gels were produced by polymerization of oligo(ethylene oxide)-methacrylate (OEOMA) in the presence of CNC. These gels have a dual drug release in response to acidic pH and thiol-reducing agents [184].

Zoppe and collaborators applied CNC-based systems as viral inhibitors (alphavirus infectivity) and suggested that CNC can be used for inhibition of HIV [185].

9. Conclusion

The attractive properties of nanocellulosic materials such as biodegradability, biocompatibility, renewability, low density, high strength, good stiffness, low thermal expansion, and high aspect ratio make them suitable for biomedical applications. For this reason, the number of publications and patents related to these applications has skyrocketed in the last 5 years. However, a great effort has still to be made to reduce the high cost involved with the production process of nanocellulose intended for biomedical used.

Author details

John Rojas*, Mauricio Bedoya and Yhors Ciro

*Address all correspondence to: jrojasca@gmail.com

Department of Pharmacy, School of Pharmaceutical Chemistry, University of Antioquia, Medellín, Columbia

References

[1] Li J, Xu QH, Jin LQ. Research development on hydrophobic modification of cellulose nanofibrils. Adv Mater Res [Internet]. 2013;785–786:440–3. Available from: http://www.scientific.net/AMR.785-786.440

[2] Robles NB. Tailoring cellulose nanofibrils for advanced materials. KTH Royal Institute of Technology; 2014. p. 82.

[3] Zheng H. Production of fibrillated cellulose materials—effects of pretreatments and refining strategy on pulp properties. Aalto University; 2014. p. 74.

[4] Deepa B, Abraham E, Cordeiro N, Mozetic M, Mathew Ap OK, Faria M, Thomas S, Pothan LA. Utilization of various lignocellulosic biomass for the production of nanocellulose: a comparative study. Cellulose. 2015;22(2):1075–90.

[5] Lin SP, Loira Calvar I, Catchmark JM, Liu JR, Demirci A, Cheng KC. Biosynthesis, production and applications of bacterial cellulose. Cellulose. 2013;20(5):2191–219.

[6] Lee KY, Buldum G, Mantalaris A, Bismarck A. More than meets the eye in bacterial cellulose: biosynthesis, bioprocessing, and applications in advanced fiber composites. Macromol Biosci. 2014;14(1):10–32.

[7] Lee KY, Tamelin T, Schulter K, Kiliskinen H, Samela J, Bismarck A. High performance cellulose nanocomposites: comparing the reinforcing ability of bacterial cellulose and nanofibrillated cellulose. ACS Appl Mater Interfaces. 2012;4(8):4078–86.

[8] Rajwade JM, Paknikar KM, Kumbhar J V. Applications of bacterial cellulose and its composites in biomedicine. Appl Microbiol Biotechnol [Internet]. 2015;ASAP. Available from: http://link.springer.com/10.1007/s00253-015-6426-3

[9] De Campos A, Correa AC, Cannella D, Teixeria EM, Marconcini JM, Dufresne A, Mattoso LHC, Cassland P, Sanadi AR. Obtaining nanofibers from curauá and sugarcane bagasse fibers using enzymatic hydrolysis followed by sonication. Cellulose. 2013;20(3):1491–500.

[10] Liu LL, Tian YB, Tang CC, Zhang H, Liu X. The preparation of nanocrystalline cellulose by acid and enzymatic method. Food Ferment Ind. 2011;

[11] De campos A, Correa AC, Cannella D, Teixeira EM, Marconcini JM, Dufrense A, Mattoso LHC, Cassland P, Sanadi AR. Obtaining nanofibers from curauá and sugarcane bagasse fibers using enzymatic hydrolysis followed by sonication. Cellulose. 2013;20(3):1491–500.

[12] Anderson SR, Esposito D, Gillete W, Zhu JY, Baxa U, McNeil SE. Enzymatic preparation of nanocrystalline and microcrystalline cellulose. Tappi J. 2014;13(5):35–42.

[13] Lai C, Zhang S, Sheng L, Liao S, Xi T, Zhang Z. TEMPO-mediated oxidation of bacterial cellulose in a bromide-free system. Colloid Polym Sci. 2013;291(12):2985–92.

[14] Lin D, Li R, Lopez-Sanchez P, Li Z. Physical properties of bacterial cellulose aqueous suspensions treated by high pressure homogenizer. Food Hydrocoll. 2015;44:435–42.

[15] Siqueira G, Tapin-Lingua S, Bras J, da Silva Perez D, Dufresne A. Morphological investigation of nanoparticles obtained from combined mechanical shearing, and enzymatic and acid hydrolysis of sisal fibers. Cellulose. 2010;17(6):1147–58.

[16] Amiralian N, Annamalai PK, Memmott P, Taran E, Schmidt S, Martin DJ. Easily deconstructed, high aspect ratio cellulose nanofibres from *Triodia pungens*; an abundant grass of Australia's arid zone. RSC Adv. 2015;5:32124–32.

[17] Lu H, Gui Y, Zheng L, Liu X. Morphological, crystalline, thermal and physicochemical properties of cellulose nanocrystalsobtained from sweet potato residue. Food Res Int. 2013;50(1):121–8.

[18] Dos Santos RM, Flauzino WP, Silvério HA, Ferreira D, Oliveira N, Pasquin D. Cellulose nanocrystals from pineapple leaf, a new approach for the reuse of this agrowaste. Ind Crops Prod. 2013;50:707–14.

[19] Deepa EA, Koshy RR, Pothan LA, Thomas S. Extraction and characterization of cellulose nanofibers from banana plant. Handbook of Polymer Nanocomposites Processing, Performance and Application. 2015. p. 65–80.

[20] Flauzino WP, Alves H, Oliveira N, pasquini D. Extraction and characterization of cellulose nanocrystals from agro-industrial residue–soy hulls. Ind Crops Prod. 2013;42:480–8.

[21] Hua K, Carlsson DO, Ålander E, Lindström T, Strømme M, Mihranyan A, Ferraz N S. Translational study between structure and biological response of nanocellulose from wood and green algae. RSC Adv. 2014;4:2892–903.

[22] Rahimi M, Behrooz R. Effect of cellulose characteristic and hydrolyze conditions on morphology and size ofnanocrystal cellulose extracted from wheat straw. Int J Polym Mater. 2011;60(8):529–41.

[23] Li R, Fei J, Cai Y, Li Y, Feng J, Yao J. Cellulose whiskers extracted from mulberry: a novel biomass production. Carbohydr Polym. 2009;76(1):94–9.

[24] Hsieh YL. Cellulose nanocrystals and self-assembled nanostructures from cotton, rice straw and grape skin: a source perspective. J Mater Sci. 2013;48(22):7837–46.

[25] Xu C, Zhu S, Xing C, Li D, Zhu N, Zhou H. Isolation and properties of cellulose nanofibrils from coconut palm petioles by different mechanical process. PLoS One. 2015;10(4).

[26] Mao J, Osorio-Madrazo A, Laborie MP. Preparation of cellulose I nanowhiskers with a mildly acidic aqueous ionic liquid: reaction efficiency and whiskers attributes. Cellulose. 2013;20(4):1829–40.

[27] Fortunati E, Puglia D, Monti M, Peponi L, Santulli C, Kenny JM, Torre L. Extraction of cellulose nanocrystals from *Phormium tenax* fibres. J Polym Env. 2013;21(2):319–28.

[28] Lin N, Dufrense A. Nanocellulose in biomedicine: current status and future prospect. Eur Polym J. 2014;59:302–25.

[29] Jonoobi M, Mathew AP, Oksman K. Producing low-cost cellulose nanofiber from sludge as new source of raw materials. Ind Crops Prod. 2012;40(1):232–8.

[30] Li Q. Nanocellulose: preparation, characterization, supramolecular modeling, and its life cycle assessment. Virginia Polytechnic Institute & State University; 2012. p. 186.

[31] kalia S, Boufi S, Celli A, Kango S. Nanofibrillated cellulose: surface modification and potential applications. Colloid Polym Sci. 2014;292(1):5–31.

[32] Isogai A. Wood nanocelluloses: fundamentals and applications as new bio-based nanomaterials. J Wood Sci. 2013;59(6):449–59.

[33] Wang Q, Zhao X, Zhu JY. Kinetics of strong acid hydrolysis of a bleached kraft pulp for producing cellulose nanocrystals (CNCs). Ind Eng Chem Res. 2014;53(27):11007–14.

[34] Li J, Wei X, Wang Q, Chen J, Chang G, Kong L, Su J, Liu Y. Homogeneous isolation of nanocellulose from sugarcane bagasse by high pressure homogenization. Carbohydr Polym. 2012;90(4):1609–13.

[35] Chaker A, Mutjé P, Vilar MR, Boufi S. Agriculture crop residues as a source for the production of nanofibrillated cellulose with low energy demand. Cellulose. 2014;21(6):4247–59.

[36] Liu LL, Tian YB, Tang CC, Zhong C, Liu X. Preparation of nanocrystalline cellulose from soybean dregs by acid hydrolysis followed by high-pressure homogenization. Food Sci. 2011;

[37] Wang Y, Wei X, Li J, Wang Q, Wang F, Kong L. Homogeneous isolation of nanocellulose from cotton cellulose by high pressure homogenization. J Mater Chem Eng. 2013;1:49–52.

[38] Hassan ML, Mathew AP, Hassan EA, El-Wakil NA OK. Nanofibers from bagasse and rice straw: process optimization and properties. Wood Sci Technol. 2012;46(1):193–205.

[39] Xie C, Liu ZM, Wu P, Fang GZ, Zhao X. Optimization of preparation technology of alkali pretreated reed pulp nano-cellulose. Chem Ind For Prod. 2013;

[40] Stelte W, Sanadi AR. Preparation and characterization of cellulose nanofibers from two commercial hardwood and softwood pulps. Ind Eng Chem Res [Internet]. 2009;48(24):11211–9. Available from: http://dx.doi.org/10.1021/ie9011672

[41] He W, Jiang X, Sun F, Xu X. Extraction and characterization of cellulose nanofibers from *Phyllostachys nidularia munro* via a combination of acid treatment and ultrasonication. BioResources. 2014;9(4):6876–87.

[42] Chen W, Yu H, Liu Y, Hai Y, Zhang M, Chen P. Isolation and characterization of cellulose nanofibers from four plant cellulose fibers using a chemical-ultrasonic process. Cellulose. 2011;18(2):433–42.

[43] Frey MW. Electrospinning cellulose and cellulose derivatives. Polym Rev. 2008;48(2): 378–91.

[44] Satyamurthy P, Jain P, Balasubramanya RH, Vigneshwaran N. Preparation and characterization of cellulose nanowhiskers from cotton fibres by controlled microbial hydrolysis. Carbohydr Polym. 2011;83(1):122–9.

[45] Sacui LU, Nieuwendaal RC, Burnett DJ, Stranick SJ, Jorfi M, Weder C, Foster EJ, Olsson RT, Gilman JW. Comparison of the properties of cellulose nanocrystals and cellulose nanofibrils isolated from bacteria, tunicate, and wood processed using acid, enzymatic, mechanical, and oxidative methods. ACS Appl Mater Interfaces. 2014;6(9):6127–38.

[46] Sinko R, Qin X, Keten S. Interfacial mechanics of cellulose nanocrystals. MRS Bull [Internet]. 2015;40(04):340–8. Available from: http://www.journals.cambridge.org/abstract_S0883769415000676

[47] Habibi Y, Lucia LA, Rojas OJ. Cellulose nanocrystals: chemistry, self-assembly, and applications. Chem Rev. 2010;110(6):3479–500.

[48] yue Y, Zhou C, French AD, Xia G, Han G, Wang Q, Wu Q. Comparative properties of cellulose nano-crystals from native and mercerized cotton fibers. Cellulose. 2012;29(4):1173–87.

[49] Charreau H, L. Foresti M, Vazquez A. Nanocellulose patents trends: a comprehensive review on patents on cellulose nanocrystals, microfibrillated and bacterial cellulose. Recent Pat Nanotechnol. 7(1):56–80.

[50] Ioelovich M. Optimal conditions for isolation of nanocrystalline cellulose particles. Nanosci Nanotechnol. 2012;2(2):9–13.

[51] Sadeghifar H, Filpponen I, Clarke SP, Brougham DF, Argypoulos DS. Production of cellulose nanocrystals using hydrobromic acid and click reactions on their surface. J Mater Sci. 2011;46(22):7344–55.

[52] Brito BSL, Pereira FV, Putaux JL, Jean B. Preparation, morphology and structure of cellulose nanocrystals from bamboo fibers. Cellulose. 2012;19(5):1527–36.

[53] Zhang J, Elder TJ, Pu Y, Ragauskas AJ. Facile synthesis of spherical cellulose nanoparticles. Carbohydr Polym. 2007;3(25):607–11.

[54] Cheng ZL, Xu QH, Gao Y. Research progress in nano-cellulose modification. Adv Mater Res. 2012;627:859–63.

[55] Hai LV, Son HN, Seo YB. Physical and bio-composite properties of nanocrystalline cellulose from wood, cotton linters, cattail, and red algae. Cellulose. 2015;22(3):1789–98.

[56] Yang D, Peng XW, Zhong LX, Cao XF, Chen W, Sun RC. Effects of pretreatments on crystalline properties and morphology of cellulose nanocrystals. Cellulose. 2013;20(5):2427–37.

[57] Yingkamhaeng N, Sukyai P. The potential of mango peel utilization for cellulose extraction by hydrothermal pretreatment. Annual Meeting of the Thai Society for Biotechnology and International Conference. 2014.

[58] Penttilä PA, Várnai A, Fernández M, Kontro I, Liljeström V, Lindner P, Siika-aho M, Viikari L, Serimaa R. Small-angle scattering study of structural changes in the microfibril network of nanocellulose during enzymatic hydrolysis. Cellulose. 2013;20(3): 1031–40.

[59] Kargarzadeh H, Ahmad I, Abdullah I, Dufrense A, Zainudin SY, Sheltami RM. Effects of hydrolysis conditions on the morphology, crystallinity, and thermal stability of cellulose nanocrystals extracted from kenaf bast fibers. Cellulose. 2012;19(3):855–66.

[60] Lam E, Male KB, CHong JH, Leung ACW, Luong JHT. Applications of functionalized and nanoparticle-modified nanocrystalline cellulose. Trends Biotechnol. 2012;30(5): 283–90.

[61] Jiang F, Hsieh YL. Super water absorbing and shape memory nanocellulose aerogels from TEMPO-oxidized cellulose nanofibrils via cyclic freezing–thawing. J Mater Chem. 2014;2:350–9.

[62] Cao X, Ding B, Yu J, Al-Deyab SS. Cellulose nanowhiskers extracted from TEMPO-oxidized jute fibers. Carbohydr Polym. 2012;90(2):1075–80.

[63] Fujisawa S, Okita S, Fukuzumi H, Saito T, Isogai A. Preparation and characterization of TEMPO-oxidized cellulose nanofibril films with free carboxyl groups. Carbohydr Polym. 2011;84(1):579–83.

[64] Kekäläinen K, Liimatainen H, Biale F, Holzforschung JN. Nanofibrillation of TEMPO-oxidized bleached hardwood kraft cellulose at high solids content. Holzforschung. 2015;1–12.

[65] Pettersen SR. Cross-linking oxidized cellulose nanofibrils for the formation of stable hydrogel structures. Institutt for kjemisk prosessteknologi; 2013.

[66] Mishra SP, Thirree J, Manent AS, Chabot B, Daneault C. Ultrasound-catalyzed TEMPO-mediated oxidation of native cellulose for the production of nanocellulose: effect of process variables. BioResources. 2011;6(1):121–43.

[67] Li Q, Renneckar S. Molecularly thin nanoparticles from cellulose: isolation of sub-microfibrillar structures. Cellulose. 2009;16(6):1025–32.

[68] Castro-Guerrero CF, Gray DG. Chiral nematic phase formation by aqueous suspensions of cellulose nanocrystals prepared by oxidation with ammonium persulfate. Cellulose. 2014;21(4):2567–77.

[69] Xiao YT, Chin WL, Abd Hamind SB. Facile preparation of highly crystalline nanocellulose by using ionic liquid. Adv Mater Res. 2015;1087:106–10.

[70] Lazko J, Sénéchal T, Landercy N, Dangreau L, Raquez JM, Dubois P. Well defined thermostable cellulose nanocrystals via two-step ionic liquid swelling-hydrolysis extraction. Cellulose. 2014;21(6):4195–207.

[71] Hassan ML. Bagasse and rice straw nanocellulosic materials and their applications. Handbook of Polymer Nanocomposites Processing, Performance and Application. 2015. p. 47–64.

[72] Laurén P. Small-animal SPECT/CT and nanofibrillar cellulose hydrogels: a preclinical evaluation of a potential novel biomaterial application. University of Helsinki; 2013. p. 71.

[73] Wicaksono R, Syamsu K, Yuliasih I, Nasir M. Cellulose nanofibers from cassava bagasse: characterization and application on tapioca-film. Chem Mater Res. 2013;3(13): 79–87.

[74] Gatenholm P. Bone Healing Grafts Fabricated by Nanoscale Assembly of Biological Building Blocks. 2010 p. 126–32.

[75] Missoum K. Modification chimique de surface de NanoFibrilles de Cellulose (NFC) [Internet]. 2012. Available from: http://www.theses.fr/2012GRENI105/document

[76] Morais JPS, Rosa MF, Souza MSM, Alexandrino LC, Ribeiro AC. Processo de obtenção de nanocristais de celulose a partir do línter de algodão. Pesquisa e Desenvolvimento. 2012 p. 22.

[77] De Morais E, Corrêa AC, Manzoli A, Leite FL, de Oliveira CR, Mattoso LHC. Cellulose nanofibers from white and naturally colored cotton fibers. Cellulose. 2010;17(3): 595–606.

[78] Gamelas JAF, Pedrosa J, Lourenco AF, Ferreria PJ. Surface properties of distinct nanofibrillated celluloses assessed by inverse gas chromatography. Colloids Surf A Physicochem Eng Asp. 2015;469:36–41.

[79] Mariño M, Da silva LP, Durán N, Tasic L. Enhanced materials from nature: nanocellulose from citrus waste. Molecules. 2015;20:5908–23.

[80] Baheti VK, Abassi R, Militky J. Ball milling of jute fibre wastes to prepare nanocellulose. World J Eng. 2012;9(1):45–50.

[81] Baheti V, Abbasi R, Militky J A. Optimisation of ball milling parameters for refinement of waste jute fibres to nano/micro scale in dry conditions. J Text Eng. 2013;59(5): 87–92.

[82] Kumar A, Negi YS, Choudhary V, Bhardwaj NK. Characterization of cellulose nanocrystals produced by acid-hydrolysis from sugarcane bagasse as agro-waste. J Mater Physic Chem. 2014;2(1):1–8.

[83] Do Nascimiento DM, Norões AKM, Souza NF, Alexandre LC, Morais JPS, Mazzeto SE, Rosa MF. Thermal and structural characteristics of waste derived biomass for potential application in nanomaterials. 7th International Symposium on Natural Polymers and Composites.

[84] Chen Y, Wu Q, Huang B, Huang M, Ai X. Isolation and characteristics of cellulose and nanocellulose from lotus leaf stalk agro-wastes. BioResources. 2015;10(1):684–96.

[85] Kalita E, Nath BK, Agan F, More V, Deb P. Isolation and characterization of crystalline, autofluorescent, cellulose nanocrystals from saw dust wastes. Ind Crops Prod. 2014;

[86] Li Y, Zhu H, Xu M, Zhuang Z, Xu M, Dai H. High yield preparation method of thermally stable cellulose nanofibers. BioResources. 2014;9(2):1986–97.

[87] Peng Y, Gardner DJ, Han Y, Cai Z, Tshabala MA. Drying cellulose nanofibrils: in search of a suitable method. Cellulose. 2012;19(1):91–102.

[88] Fairman E. Avoiding Aggregation During Drying and Rehydration of Nanocellulose. University of Maine; 2014. p. 60.

[89] Peng Y, Gardner DJ, Han Y, Kiziltas A, Cai Z, Tshabalala MA. Influence of drying method on the material properties of nanocellulose I: thermostability and crystallinity. Cellulose. 2013;20(5):2379–92.

[90] Belbekhouche S, Bras J, Siqueira G, Chappey C, Lebrun L, Khelifi B, Marais S, Dufresne A. Water sorption behavior and gas barrier properties of cellulose whiskers and microfibrils films. Carbohydr Polym. 2011;83(4):1740–8.

[91] Majoinen J, Kontturi E, Ilkkala O, Gray DG. SEM imaging of chiral nematic films cast from cellulose nanocrystal suspensions. Cellulose. 2012;19(5):1599–605.

[92] Dankovich TA, Gray DG. Contact angle measurements on smooth nanocrystalline cellulose (I) thin films. J Adhes Sci Technol. 2011;25(6–7):699–708.

[93] Wang Y, Heim LO, Xu Y, Buntkowsky G, Zhang K. Transparent, stimuli responsive films from cellulose based organogel nanoparticles. Adv Funct Mater. 2015;25:1434–41.

[94] Chen W, Li Q, Wang Y, Yi X, Zeng J, Yu H, Liu Y, Li J. Comparative study of aerogels obtained from differently prepared nanocellulose fibers. Chem Sus Chem. 2014;7:154–61.

[95] Zhang ZL, Wang WJ, Shao ZQ, LV SY. Rheological properties of cellulose nanocrystals suspension at different pH. Nanosci Nanotechnol. 2012;

[96] Li MC, Wu Q, Song K, Lee S, Qing Y, Wu Y. Cellulose nanoparticles: structure–morphology–rheology relationship. ACS Sustain Chem Eng. 2015;3(5):821–32.

[97] Dugan JM, Gough JE, Eichhorn SJ. Bacterial cellulose scaffolds and cellulose nanowhiskers for tissue engineering. Nanomedicine. 2013;8(2):287–98.

[98] Jeong SL, Lee SE, Yang H, Jin YH, Park CS, Park YS. Toxicologic evaluation of bacterial synthesized cellulose in endothelial cells and animals. Mol Cell Toxicol. 2010;6(4): 373–80.

[99] Wang ZL, Jia YY, Shi Y, Cong DL, Chen YY, Jia SR, Zhou YL. Research on characterization and biocompatibility of nano-bacterial cellulose membrane. Chem J Chinese Univ. 2009;30(8):1553–8.

[100] Edwards JV, Prevost N, French A, Concha M, De Lucca A, Wu Q. Nanocellulose-based biosensors: design, preparation, and activity of peptide-linked cotton cellulose nanocrystals having fluorimetric and colorimetric elastase detection sensitivity. Engineering. 2013;5(9A):20–8.

[101] He X, Xiao Q, Lu C, Wang Y, Zhang X, Zhao J, Zhang W, Zhang W, Deng Y. Uniaxially aligned electrospun all-cellulose nanocomposite nanofibers reinforced with cellulose nanocrystals: scaffold for tissue engineering. Biomacromolecules. 2014;15(2): 618–27.

[102] Tasset S, Cathala B, Bizot H, Capron I. Versatile cellular foams derived from CNC-stabilized Pickering emulsions. RSC Adv. 2014;4:893–8.

[103] Dai D, Fan M, Collins P. Fabrication of nanocelluloses from hemp fibers and their application for the reinforcement of hemp fibers. Ind Crops Prod. 2013;44:192–9.

[104] Kumar S, Hofmann M, Steinmann B, Foster EJ, Weder C. Reinforcement of stereolithographic resins for rapid prototyping with cellulose nanocrystals. ACS Appl Mater Interfaces. 2012;4(10):5399–407.

[105] Ibrahim MM, El-Zawawy WK. Extraction of cellulose nanofibers from cotton linter and their composites. Handbook of Polymer Nanocomposites Processing, Performance and Application. 2015. p. 145–64.

[106] Syverud K, Pettersen SR, Draget K, Chinga-Carrasco G. Controlling the elastic modulus of cellulose nanofibril hydrogels—scaffolds with potential in tissue engineering. Cellulose. 2015;22(1):473–81.

[107] Empson YM, Ekwueme EC, Hong JK, Paynter DM, Kwansa AL. High elastic modulus nanoparticles: a novel tool for subfailure connective tissue matrix damage. Transl Res. 2014;164(3):244–57.

[108] Mahmoud KA, Mena JA, Male KB, Hrapovic S, Kamen A, Luong JHT. Effect of surface charge on the cellular uptake and cytotoxicity of fluorescent labeled cellulose nanocrystals. ACS Appl Mater Interfaces [Internet]. 2010;2(10):2924–32. Available from: http://dx.doi.org/10.1021/am1006222

[109] Zhang Y, Nypelö T, Salas C, Arboleda J, Hoeger IC, Rojas OJ. Cellulose nanofibrils: from strong materials to bioactive surfaces. J Renew Mater. 2013;1(3):195–211.

[110] Orelma H, Filpponen I, Johansson LS, Österberg M, Rojas OJ. Surface functionalized nanofibrillar cellulose (NFC) film as a platform for immunoassays and diagnostics. Biointerphases. 2012;7(61):1–12.

[111] Andrade FK, Moreira SMG, Domingues L, Gama FMP. Improving the affinity of fibroblasts for bacterial cellulose using carbohydrate-binding modules fused to RGD. J Biomed Mater Res Part A. 2010;92A(1):9–17.

[112] Fu L, Zhang J, Yang G. Present status and applications of bacterial cellulose-based materials for skin tissue repair. Carbohydr Polym. 2013;92(2):1432–42.

[113] Yang H, Gurgel P, Carbonell RG. Hexamer peptide affi nity resins that bind the Fc region of human immunoglobulin G. J Pept Rees. 2006;66:120–37.

[114] Edwards JV, Prevost N, Sethumadhavan K, Ullah A, Condon B. Peptide conjugated cellulose nanocrystals with sensitive human neutrophil elastase sensor activity. Cellulose. 2013;20(3):1223–35.

[115] Liu H, Wang D, Song Z, Shang S. Preparation of silver nanoparticles on cellulose nanocrystals and the application in electrochemical detection of DNA hybridization. Cellulose. 2011;18(1):67–74.

[116] Mangalam AP, Simonsen J, Benight, Albert S. Cellulose/DNA hybrid nanomaterials. Biomacromolecules [Internet]. 2009;10(3):497–504. Available from: http://dx.doi.org/10.1021/bm800925x

[117] Dong S, Cho HJ, Lee YW, Roman M. Synthesis and cellular uptake of folic acid-conjugated cellulose nanocrystals for cancer targeting. Biomacromolecules [Internet]. 2014;15(5):1560–7. Available from: http://dx.doi.org/10.1021/bm401593n

[118] Pereira MM, Rapsoso NRB, Brayner R, Teixeira EM, Oliveira V, Quintão CCR, Camargo LSA, Mattoso LHC, Brandão HM. Cytotoxicity and expression of genes involved in the cellular stress response and apoptosis in mammalian fibroblast exposed to cotton cellulose nanofibers. Nanotechnology. 2013;24(7):75–103.

[119] Czaja W, Krystynowicz A, Bielecki S, Jr. RMB. Microbial cellulose—the natural power to heal wounds. Biomaterials [Internet]. 2006;27(2):145–51. Available from: http://www.sciencedirect.com/science/article/pii/S0142961205007040

[120] MacNeil. Progress and opportunities for tissue-engineered skin. Nature [Internet]. 2007;445(7130):874–80. Available from: http://www.scopus.com/inward/record.url?eid=2-s2.0-33847256045&partnerID=40&md5=1fdd157f69a1cc6a2901c5930d385526

[121] Czaja WK, Young DJ, Kawecki M, Brown RM Jr. The future prospects of microbial cellulose in biomedical applications. Biomacromolecules. 2007;8(1):1–12.

[122] Fu L, Zhang Y, Li C, Zhihong W, Zhuo Q, Huang X, Qiu G, Zhou P, Yang G. Skin tissue repair materials from bacterial cellulose by a multilayer fermentation method. J Mater Chem. 2012;22:12349–57.

[123] Nwe N, Furuike T, Tamura H. Selection of a biopolymer based on attachment, morphology and proliferation of fibroblast NIH/3T3 cells for the development of a biodegradable tissue regeneration template: alginate, bacterial cellulose and gelatin. Process Biochem. 2010;45(4):457–66.

[124] Ul-Islam M, Khan T, Park JK. Water holding and release properties of bacterial cellulose obtained by in situ and ex situ modification. Carbohydr Polym. 2012;88(2):596–603.

[125] Kalia S, Dufresne A, Cherian BM, Kaith BS, Avérous L, Njuguna J, Nassiopoulos E. Cellulose-based bio- and nanocomposites: a review. Int J Polym Sci. 2011;2011:1–35.

[126] Helenius G, Backdahl H, Bodin A, Nannmark U, Gatenholm P, Risberg B. In vivo biocompatibility of bacterial cellulose. J Biomed Mater Res A. 2006;76(2):431–8.

[127] Jorfi M, Foser J. Recent advances in nanocellulose for biomedical applications. J Appl Polym Sci. 2015;132(14):1–19.

[128] Shanmuganathan K, Capadona JR, Rowan SJ, Weder C. Stimuli-responsive mechanically adaptive polymer nanocomposites. ACS Appl Mater Interfaces. 2010;2(1):165–74.

[129] Shanmuganathan K, Capadona JR, Rowan SJ, Weder C. Biomimetic mechanically adaptive nanocomposites. 2010. 35AD;1–2(212–222).

[130] Nguyen JK, Park DJ, Skousen JL, Hess-Dunning AE, Tyler DJ, Rowan SJ, Weder C, Capadona JR. Mechanically-compliant intracortical implants reduce the neuroinflammatory response. J Neural Eng. 2014;11(5):1–15.

[131] Potter KA, Jorfi M, Householder KT, Foster EJ, Weder C, Capadona JR. Curcumin-releasing mechanically adaptive intracortical implants improve the proximal neuronal density and blood–brain barrier stability. Acta Biomater. 2014;10(5):2209–22.

[132] Millon LE, Wan WK. The polyvinyl alcohol-bacterial cellulose system as a new nanocomposite for biomedical application. J Biomed Mater Res. 2006;79(2):245–53.

[133] Schumann DA, Wippermann J, Klemm DO. Artificial vascular implants from bacterial cellulose: preliminary results of small arterial substitutes. Cellulose. 2009;16(5): 877–85.

[134] Azevedo EP, Retarekar R, Raghavan ML, Kumar V. Mechanical properties of cellulose: chitosan blends for potential use as a coronary artery bypass graft. J Biomater Sci Polym Ed. 2013;24(3):239–52.

[135] Klemm D, Schumann D, Udhardt U, Marsch S. Bacterial synthesized cellulose—artificial blood vessels for microsurgery. Prog Polym Sci. 2001;26:1561–603.

[136] Berti F V, Rambo CR, Dias PF, Porto LM. Nanofiber density determines endothelial cell behavior on hydrogel matrix. Mater Sci Eng C. 2013;33:4684–91.

[137] Pertile RAN, Andrade FK, Jr. CA, Gama M. Surface modification of bacterial cellulose by nitrogen-containing plasma for improved interaction with cells. Carbohydr Polym [Internet]. 2010;82(3):692–8. Available from: http://www.sciencedirect.com/science/article/pii/S0144861710004352

[138] Bodin A, Concaro S, Brittberg M, Gatenholm P. Bacterial cellulose as a potential meniscus implant. J Tissue Eng Regen Med. 2007;1(5):406–8.

[139] Nimeskern L, Martínez Ávila H, Sundberg J, Gatenholm P, Müller R, Stok KS. Mechanical evaluation of bacterial nanocellulose as an implant material for ear cartilage replacement. J Mech Behav Biomed Mater. 2013;22:12–21.

[140] Amorim WL, Costa HO, Souza FC, Castro MG, Silva L. Experimental study of the tissue reaction caused by the presence of cellulose produced. Braz J Otorhinolaryngol. 2009;75(2):200–7.

[141] Iamaguti LS, Branda˜o CVS, Minto BW. Handbook of Polymer Nanocomposites. Processing, Performance and Application [Internet]. 2015. Available from: http://link.springer.com/10.1007/978-3-642-38649-7

[142] FalcÃ\poundso SC, EvÃ\textordfemeninencio Neto J, Coelho AR de B. Incorporation by host tissue of two biomaterials used as repair of defects produced in abdominal wall of rats. Acta Cir Bras [Internet]. scielo; 2008;23:78–83. Available from: http://www.scielo.br/scielo.php?script=sci_arttext&pid=S0102-86502008000100013&nrm=iso

[143] Nemetz AP, Loures DRR, Coelho JCU. Efeito estrutural da utilização de celulose biossintética e politetrafluoretileno expandido como substitutos do peritônio em cães. Arq Bras Cir Dig. 2001;14(2):139–42.

[144] Silva EC. Hidroxiapatita Sintética em alvéolo dentário após exodontia em Felis catus: estudo clínico, radiológico e histomorfométrico. Universidad Federal de Vicosa; 2009.

[145] Novaes AB Jr, Novaes AB. IMZ implants placed into extraction sockets in association with membrane therapy (Gengiflex) and porous hydroxyapatite: a case report. Int J Oral Maxillofac Implant. 1992;7(4):563–540.

[146] Salata LA, Craig GT, Brook IM. In vivo evaluation of a new memnbrane (Gengiflex) for guided bone regeneratien (GBR),". J Dent Res. 1995;74(3):825.

[147] Macedo NL, Matuda FS, Macedo LGS, Monteiro ASF, Valera MC, Carvalho YR. Evaluation of two membranes in guided bone tissue regeneration: histological study in rabbits. Brazilian J Oral Sci. 2004;3(8):395–400.

[148] Svensson A, Nicklasson E, Harrah T, Panilaitis B, Kaplan DL, Brittberg M, Gatenholm P. Bacterial cellulose as a potential scaffold for tissue engineering of cartilage. Biomaterials. 2005;26:419–31.

[149] Wan YZ, Hong L, Jia RS, Huang Y, Zhu Y, Wang YL, Jiang HJ. Synthesis and characterization of hydroxyapatite-bacterial cellulose nanocomposites. Compos Sci Technol. 2006;66(11–12):1825–32.

[150] Huia J, Yuanyuan J, Jiao W, Yuan H, Yuan Z, Shiru J. Potentiality of bacterial cellulose as the scaffold of tissue engineering of cornea. The 2nd Conference on Biomedical Engineering and Informatics. China; 2009.

[151] Cai Z, Hou C, Yang G. Poly(3-hydroxubutyrateco-4-hydroxubutyrate)/bacterial cellulose composite porous scaffold: preparation, characterization and biocompatibility evaluation. Carbohydr Polym. 2012;87:1073–80.

[152] Phisalaphong M, Suwanmajo T, Tammarate P. Synthesis and characterization of bacterial cellulose/alginate blend membranes. J Appl Polym Sci. 2008;107(5):3419–24.

[153] Chiaoprakobkij N, Sanchavanakit N, Subbalekha K, Pavasant P, Phisalaphong M. Characterization and biocompatibility of bacterial cellulose/alginate composite sponges with human keratinocytes and gingival fibroblasts. Carbohydr Polym. 2011;853:548–53.

[154] Cai Z, Kim J. Bacterial cellulose/poly(ethylene glycol) composite: characterization and first evaluation of biocompatibility. Cellulose. 2010;17(1):83–91.

[155] Pértile R, Moreira S, Andrade F, Domingues L, Gama M. Bacterial cellulose modified using recombinant proteins to improve neuronal and mesenchymal cell adhesion. Biotechnol Prog. 2012;28:526–32.

[156] Wang J, Wan Y, Huang Y. Immobilisation of heparin on bacterial cellulose-chitosan nano-fibres surfaces via the cross-linking technique. IET Nanobiotechnol. 2012;6:52–7.

[157] Gao C, Wan Y, Lei X, Qu J, Yan T, Dai K. Polylysine coated bacterial cellulose nanofibers as novel templates for bone-like apatite deposition. Cellulose. 2011;18:1555–61.

[158] Karaaslan MA, Tshabalala MA, Yelle DJ, Buschle-Diller G. Nanoreinforced biocompatible hydrogels from wood hemicelluloses and cellulose whiskers. Carbohydr Polym. 2011;86(1):192–201.

[159] Saska S, Teixeira LN, de Oliveira PT, Minarelli Gaspar AM, Lima Ribeiro SJ, Messaddeq Y, Marchetto R. Bacterial cellulose-collagen nanocomposite for bone tissue engineering. J Mater Chem. 2012;22:22102–12.

[160] Wan Y, Gao C, Han M, Liang H, Ren K, Wang Y, Luo H. Preparation and characterization of bacterial cellulose/heparin hydrib nanofiber for potential vascular tissue engineering scaffolds. Polym Adv Technol. 2011;22:2643–8.

[161] Wang M, Yuan J, Huang X, Cai X, Li L, Shen J. Grafting of carboxybetaine brush onto cellulose membranes via surface-initiated ARGET-ATRP for improving blood compatibility. Colloids Surfaces B Biointerfaces [Internet]. 2013;103(0):52–8. Available from: http://www.sciencedirect.com/science/article/pii/S0927776512005826

[162] Solway DR, Clark WA, Levinson DJ. A parallel open–label trial to evaluate microbial cellulose wound dressing in the treatment of diabetic foot ulcers. Int Wound J. 2011;8(1):69–73.

[163] Portal O, Clark WA, Levinson DJ. Microbial cellulose wound dressing in the treatment of nonhealing lower extremity ulcers. Wounds. 2009;21(1):1–3.

[164] Wu, L.; Zhou, H.; Sun, H.-J.; Zhao, Y.; Yang, X.; Cheng, S. Z. D.; Yang, G. Thermoresponsive bacterial cellulose whisker/poly(NIPAM-co-BMA) nanogel complexes: synthesis, characterization, and biological evaluation. Biomacromolecules. 2013;14:1078–84.

[165] Fernandes SCM, Sadocco P, Aonso-Varona A, Palomares T, Eceiza A, Silvestre AJD, Mondragon I, Freire CSR. Bioinspired antimicrobial and biocompatible bacterial cellulose membranes obtained by surface functionalization with aminoalkyl groups. ACS Appl Mater Interfaces. 2013;5:3290–7.

[166] Cassano R, Trombino S, Ferrarelli T, Nicoletta FP, Mauro MV, Giraldi C, Picci N. Hemp fiber (*Cannabis sativa* L.) derivatives with antibacterial and chelating properties. Cellulose. 2013;20:547–57.

[167] Caldeira E, Piskin E, Granadeiro L, Silva F, Gouveia ICJ. Biofunctionalization of cellulosic fibres with L-cysteine: assessment of antibacterial properties and mechanism of action against *Staphylococcus aureus* and *Klebsiella pneumonia*. Biotechnol. 2013;168:426–35.

[168] Cassano R, Trombino S, Ferrarelli T, Barone E, Arena V, Mancuso C, Picci N. Synthesis, characterization, and anti-inflammatory activity of diclofenac-bound cotton fibers. Biomacromolecules. 2010;11:1716–20.

[169] Wei B, Yang GA, Hong F S. Preparation and evaluation of a kind of bacterial cellulose dry films with antibacterial properties. Carbohydr Polym. 2011;84(1):533–8.

[170] Sun X, Zhang L, Cao Z, Deng Y, Liu L, Fong H, Sun Y. Electrospun composite nanofiber fabrics containing uniformly dispersed antimicrobial agents as an innovative

type of polymeric materials with superior antimicrobial efficacy. ACS Appl Mater Interfaces. 2010;2:952–6.

[171] Fortunati, E., Armentano, I., Zhou, Q., Iannoni, A., Saino, E., Visai, L. et al. Multifunctional bionanocomposite films of poly(lactic acid) cellulose nanocrystals and silver nanoparticles. Carbohydr Polym. 2012;87:1596–605.

[172] Dufresne A. Nanocellulose: a new ageless bionanomaterial. Mater Today. 2013;16(220–227):220–7.

[173] Song J, Tang A, Liu T, Wang J. Fast and continuous preparation of high polymerization degree cellulose nanofibrils and their three-dimensional macroporous scaffold fabrication. Nanoscale. 2013;5:2482–90.

[174] Baek C, Hanif Z, Cho SW, Kim DI, Um SH. Shape control of cellulose nanocrystals via compositional acid hydrolysis. J Biomed Nanotechnol. 2013;9(7):1293–8.

[175] Lahiji RR, Xu X, Reifenberger R, Raman A, Rudie A, Moon RJ. Atomic force microscopy characterization of cellulose nanocrystals. Langmuir. 2010;26(6):4480–8.

[176] Postek MT, Vladar A, Dagata J, Farkas N, Ming B, Sabo R, Wegner TH, Beecher J. Cellulose nanocrystals the next big nano-thing? Instrumentation, Metrology, and Standards for Nanomanufacturing II. Proceedings of SPIE—The International Society for Optical Engineering, San Diego, 2008;7042:1–11.

[177] Kolakovic R. Nanofibrillar Cellulose in Drug Delivery. University of Helsinki; 2013. p. 60.

[178] Cesca K, Cacicedo ML, Bossio VE, Castro GR, Porto LM. BaCarb™: a novel bioinorganic matrix for local drug delivery. BMC Proc. 2014;8:77.

[179] Kolakovic R, Laaksonen T, Peltonen L, Laukkanen A, Hirvonen J. Spray-dried nanofibrillar cellulose microparticles for sustained drug release. Int J Pharm. 2012;430(1–2):47–55.

[180] Trovatti E, Freire CSR, Pinto PC, Almeida IF, Costa P, Silvestre AJD, et al. Bacterial cellulose membranes applied in topical and transdermal delivery of lidocaine hydrochloride and ibuprofen: in vitro diffusion studies. Int J Pharm [Internet]. 2012;435(1): 83–7. Available from: http://www.sciencedirect.com/science/article/pii/S0378517312000063

[181] Chang CW WM. Preparation of microfibrillated cellulose composites for sustained release of H_2O_2 or O_2 for biomedical applications. ACS Sustain Chem Eng. 2013;1:1129–34.

[182] Weng L, Rostamzadeh P, Nooryshokry N, Le HC, Golzarian J. In vitro and in vivo evaluation of biodegradable embolic microspheres with tunable anticancer drug release. Acta Biomater. 2013;9(6):6823–33.

[183] Ernsting MJ, Tang WL, MacCallum N, Li SD. Synthetic modification of carboxymethylcellulose and use thereof to prepare a nanoparticle forming conjugate of docetax-

el for enhanced cytotoxicity against cancer cells. Bioconjugate Chem. 2011;22:2474–86.

[184] Wen Y, Oh JK. Dual-stimuli reduction and acidic pH-responsive bionanogels: intracellular delivery nanocarriers with enhanced release. RSC Adv. 2014;4:229–37.

[185] Justin O. Zoppe, Ville Ruottinen, Janne Ruotsalainen, Seppo Rönkkö, Leena-Sisko Johansson, Ari Hinkkanen, Kristiina Järvinen, and Jukka Seppälä. Synthesis of cellulose nanocrystals carrying tyrosine sulfate mimetic ligands and inhibition of alphavirus infection. Biomacromolecules. 2014;15(4):1534–42.

Chapter 9

Cellulose - Chitosan Nanocomposites - Evaluation of Physical, Mechanical and Biological Properties

Guillermo H. Riva, Joaquín García-Estrada, Brenda Vega, Fernando López-Dellamary, María E. Hérnandez and José A. Silva

Additional information is available at the end of the chapter

http://dx.doi.org/10.5772/61727

Abstract

This research describes the preparation of membranes with chitosan (CS) as the polymeric matrix and cellulose nanocrystals (CNC) as reinforcement. The aim was to evaluate their physical, mechanical and biological properties, and to determine their potential for biomedical use. Membranes were prepared via casting CNC suspensions in CS solution, at CNC concentrations of 0.5%, 1.0% and 2.0% (w/w) with pure chitosan as a reference. Analysis of membrane properties was performed using several techniques, such as ATR – FTIR, SEM, swelling test, maximum water absorption, dynamical mechanical analysis and *in vivo* (Winstar rats) biocompatibility and biodegradability assays for biological evaluation. Experimental results established that CNC reduced swelling rates and increased the maximum water absorption when CNC concentration was higher. Therefore, the presence of CNC in the matrix reduced Young's modulus by approximately 50% in comparison with pure chitosan. All formulations demonstrated biocompatibility and biodegradability values ranged between 4% and 21% in the 30 days after implantation. Based on these results, these membranes may be of use for biomedical applications.

Keywords: Cellulose nanocrystals, chitosan, biomedical, biocompatibility, nanocomposites

1. Introduction

1.1. Nanotechnology in actual context

The advances in nanotechnology for biomedical use are increasing and are at the forefront of scientific research. In recent years, hot spot areas such as drug transportation, tissue regeneration or nanomaterial development for cell-growth scaffolds have been constantly advancing. Biomaterials that are considered for biomedical applications must confirm to strict biological, physical and mechanical characteristics. Some biopolymers in particular, offer advantages in terms of sustainability and low environment impact compared to ceramics and metals. These attributes are the *biocompatibility* (an absence of inflammatory, cytotoxicity or invasive response in native cells, tissues or organs in vivo), *biodegradability* and *bioabsorbability* (the material and its by-products will degrade and/or be absorbed or safety eliminated from the body). More suitable properties may be the *degradation rate* (this rate must match the regeneration time of tissue in damaged zones, as well as transfer the mechanical efforts to new tissue in a timely manner), *porosity* (directly linked to mass transport and efficient tissue regeneration) and *surface morphology*. As drawbacks, the physical and mechanical properties of these materials make them less suitable than petroleum based plastics and other materials (metals, alloys and clays). As a consequence, reinforcement of the matrix is an option to counterbalance some of those drawbacks.

1.2. Biopolymers

1.2.1. Cellulose

Cellulose — the most abundant biopolymer on Earth — has an annual production of 7.5 x 10^{10} tons. This biopolymer is widely distributed in higher plants, sea animals (tunicates), and to a lesser degree in algae, fungi, bacteria, invertebrates, and is even found in protozoans such as *Dictyostelium discoideum*. In general, cellulose is a hard, fibrous, and water insoluble substance that plays an essential function in keeping the structure of cell walls in plants [1]. Cellulose can be found in its purest form in plants (i.e. cotton fibers). However, in wood, leaves and plant stalks, it is found mixed with other materials such as lignin and hemicelluloses. Cellulose nanofibers have the potential to be used in multiple ways, notably as a reinforcement material in the development of nanocomposites [2].

Thus, the preparation of biocompatible nanocomposites employing cellulose nanocrystals (CNC) as a reinforcement is a natural choice based on them being inert, biocompatible, biodegradable [3], and non-cytotoxic. They also contribute to the regeneration of damaged tissues or organs [4] and have mechanically desirable properties [5].

1.2.1.1. Types of processes used to obtain CNC

Cellulose nanocrystals (CNC) can be obtained by different techniques and processes:

a. Mechanical processes [6], i.e. using used bleached pulp of softwoods and hardwoods as a material raw to obtain nanocrystals from. The process begins with the soaking and grinding of fibers, followed by sieving and refining (for hardwoods this process is repeated several times). Finally, fibers are submitted to high pressure and homogenization processes (1000 Bar, 180 min), which are repeated until CNC is obtained. Energy consumption and Young's modulus (YM) are higher in hardwoods than softwoods. The tensile resistance in softwoods is better (75 MPa versus 63 MPa respectively).

b. Enzymatic processes [7], which have the advantage of simplifying the mechanical process of obtaining cellulose microfibers, mainly in the homogenization and disintegration stages. Acid hydrolysis was applied as a pre-treatment to the mechanical process. The sequence begins with refining and enzymatic treatment followed by a second refining and then homogenization. Endoglucanase was used as the enzyme and microfibers of 10–20 nm diameters were obtained.

c. Biological processes [8], a new method of obtaining nanocellulose using the Dutch elm disease fungus was established. The technique begins with the soaking and disintegrating of pulp in 2 liters of water, followed by 20 minutes in an autoclave with added sucrose and yeast to ensure feeding and growth. The fungus was left for 2–4 days at room temperature with mild stirring. Fibers were added to an autoclave to be washed again and the process finished with refining high speed cutting.

d. Chemical processes [1]. There are two basic methods for chemical modification: The first changes the surface energy characteristics of nanocrystals allowing an improvement in compatibility, especially when they are used with hydrophobic or non-polar matrices in nanocomposites, thereby obtaining a better dispersion in the matrix. The second method adds electric charges to the nanofibers (positive and negative) allowing a good dispersion. Acetylation of nanocrystals is an example of the first kind of modification [9], making their surfaces more hydrophobic. Kenaf fibers (*Hibiscus cannabius*) were modified using acetic anhydride before cellulose nanofibers were obtained from acetylated cellulose. This technique includes processes of disintegration, refining, cryorupture, and high pressure homogenization.

1.2.1.2. Nanocellulose for biomedical use

Nanocellulose has been called "biomaterials′ eyes" due to its potential for numerous applications in the biomedical field, including skin grafts to burn damage and wounds, growing of blood vessels, nerve reconstruction, brain membranes, and scaffolds in tissue engineering and bone reconstruction. Tissue engineering (TE) involves searching for new materials and artefacts to interact in positive ways with biological tissues. Furthermore, TI is seeking a primal artefact to cellular development *in vitro*, rearrangement and development of tissue when it will be implanted. The main attribute wanted in biopolymers with a potential biomedical use is a controllable and specific activity, to be used mainly in cellular scaffolds. Recently, many of these kinds of materials have been developed, having the required properties (physical/ chemical and mechanical) dependent mostly on the final application (tissue regeneration, drug releasing, scaffolding, etc.). The success of scaffolds depends mainly on cellular adhesion and surface growth. The chemical surface of a biopolymer can cause the cellular response to interfere with the adhesion, proliferation, migration and cellular functionalization. The interaction in cell surface it's whole important in the graft, including its rejection. For the regeneration of tissues, three fundamental aspects are important: the cells, and the bearing and growing factors. The cells synthetize the matrix to new tissues, the bearing creates a suitable environment for cell development, and growing factors promote cell regeneration. Furthermore, regeneration must be promoted and if it is necessary, the new material must be absorbed or biodegraded. Studies of the interactions of cell bearing are crucial for the feasibility of grafts. Different responses of cells can be observed from several materials, based on the ability of cells to distinguish and/or adapt to the surface of the material. This last factor is crucial, because it drives different responses such as cell proliferation, cell migration or feasibility. With issues regarding the skin, several laboratories have shown an interest in developing products that offer advantages such as the immediate mitigation of pain, close adhesion at wound surfaces and the reduction of infection rates. Nanocellulose has a large surface area that brings a better capability for water absorption and elasticity, these being the best characteristics for a recovering bandage, as microbial activity is stopped. Indeed, nanocellulose is very effective in reducing pain and promoting the granulation suitable for wound bandages. Another great advantage of nanocellulose consists in the capability to be built in any shape and size, making it ideal for covering extensive and difficult areas of the human body [10].

Hence, cellulose and CNC have been used in biomedicals. Past research has shown them to be ideal for tissue engineering, producing favorable results [3, 6]. CS membranes with nanoreinforcement must show a Young's modulus of 1,500—2,300 MPa to be suitable for biomedical use [11].

Furthermore, applications with excellent, proven results have been reported as follows [10]:

a. Pharmaceutical. Cellulose has excellent properties of compaction when it is mixed with other pharmaceutical excipients, forming dense matrices that make the administration of therapeutic drugs easy. Nanocellulose offers potential advantages as an excipient in drug release. Its large surface area and negative charge suggest that higher quantities of therapeutical drugs can be added to the surface of this material, showing the potential for a large quantity of charge and the optimal control of dosification. The proven biocompatibility of cellulose supports the use of nanocellulose for similar purposes. The hydroxide groups on the surface offer a site for surface modification to a broad range of chemical groups, using different methods. The surface modification can be used to tune the charge and drug release that are not normally linked with nanocellulose such as hydrophobic and non-ionized drugs.

b. Odontology. Nanocellulose can be used as biological barrier due to its porosity. This makes it ideal for use with infections, loss of fluids and it has an analgesic effect that allows therapeutic drugs be used easily and absorb the residual fluids during inflammatory stages, can be rejected in a controlled and painless way.

c. Ophthalmology [12]. Researchers explored the potential of nanocellulose as a scaffold and found it suitable for use in the development of tissue engineering for the cornea. They studied the growth of human stem cells in nanocellulose. The growth of corneal stem cells inside the scaffold was verified with a scanning laser microscope. The results suggested the potential of this biomaterial as a scaffold for tissue engineering of artificial corneas.

d. Vascular surgery [13]. Researchers studied artificial vascular implants of nanocellulose in two cases: The first was a microsurgery study, where nanocellulose implants were used as an artificial part of the carotid artery of rats for a year. These results showed the incorporation of nanocellulose under the formation of tissues and internal growth of active fibroblasts over a long period. In a second study, the implants were used to replace the carotid artery of pigs. After three months, the implants were retired and analyzed at both macro and microscopic levels. Seven implants (87.5%) were found in use and just one of them was blocked. This data showed that the innovative techniques of nanocellulose engineering have allowed the production of stable vascular conduits and confirmed the very notable achievement of the use of tissues for blood vessels in vivo as a part of cardiovascular programs.

e. Reconstructive / aesthetic surgery [14] Ideal for nasal reconstruction. The response of tissue in the presence of nanocellulose in nose bone was evaluated. In the study, 22 rabbits were used and in 20 of them a cellulose film was added to the nasal dorsum, with the remaining two acting as a control. After three and six months, the new bone was extracted for histopathology studies. Parameters such as blocking of blood flow, inflammation intensity and inflammation by the presence of purulent liquids were found to be stable, probably due to the surgical process itself rather than the presence of cellulose. For the other parameters, the statistical response was not significant. The nanocellulose coverage showed good compatibility and remained unchanged over time, making this material an excellent option for rebuilding new bone.

1.2.2. Chitosan

Chitosan (CS) is a biomaterial of proven use in the biomedical field due its biocompatibility, biodegradability and antibacterial activity, making it ideal for drug transportation, tissue engineering, wound healing, and antibacterial uses [15]. Furthermore, chitosan is bioactive and nontoxic. This biopolymer has a wide range of uses such as substance separation due to its barrier property and sensors, as well as food packing. Other authors describe the preparation of bionanocomposites using chitosan and different nanoreinforcements with the goal of obtaining a better mechanical performance, as well as the barrier properties and sensing detectors [16]. Chitosan is commonly amorphous and can be processed in flexible films. It is bioactive, non-toxic, and suited to biomedical applications such as pharmaceutical products like films, pearls or spheres, and gels, powders, etc. Previous works in cellulose-chitosan nanocomposites showed good results in physical, mechanical and biological tests [17, 18].

Chitosan can form flexible, clean and hard films [19] with a good oxygen barrier [20]. Furthermore, it can be used as packing material, mainly as a covering and edible film [21] extending the average life of foods [22, 23]. Chitosan can also form a semipermeable covering to modify the inner atmosphere, thereby reducing the transpiration rate of the product in the packaging [24]. Despite good results with respect to their mechanical properties, chitosan films can be brittle, making it necessary to use plasticizers to increase their flexibility [25]. Plasticizers such as glycerol can improve the processability, as well as the mechanical properties of chitosan [26]. In another study, researchers reported a concentration of 20% (w/w) of glycerol as the appropriate concentration to improve the flexibility of chitosan films [27]. To prevent this drawback of rigidity and brittleness of chitosan, the addition of reinforcement has showed to be useful in enhancing its mechanical, thermal and barrier properties. When the particles are smaller, the interaction with the matrix is better [28], with the low cost a sign of efficiency [29].

Fillers on a nanometric scale (called nanoparticles or nanoreinforcements) with good dispersion drive an interface matrix / filler, changing the molecular mobility, relaxation behavior and thermal and mechanical properties of the material [30].

A positive result in the elaboration of chitosan – CNC nanocomposites was obtained, using electrospinning for fibers of a derivative of chitosan / cellulose in an ionic liquid (IL). The chitosan / cellulose composite were electrospun in the ethanol co-solvent, using the IL to dissolve the chitosan and cellulose at the same time. Furthermore, the IL was capable of building fibers of pure chitosan / cellulose composite after the IL was removed by the ethanol. The fibers of this composite were manufactured as a three-dimensional shape, offering antibacterial activity to treat burns, bedsores and skin ulcers [31].

Nanofibers were obtained from chitosan and cellulose, with chitin used as a reinforcement material at different concentrations (from 1.25% to 5.0% w/w). This allowed the optimizing of the process conditions to obtain homogeneous and porous nanofibers. This material has a potential for use in wound bandages and skin burns [32].

Layer-by-Layer technique (LbL) is a technique to elaborate nanocomposites of chitosan and cellulose whiskers. The interactions between amine groups (chitosan) and sulfate groups (cellulose whiskers) ensure the linkages between matrix and nanoreinforcement to elaborate the films. The average thickness of each bi-layer (whiskers / chitosan) was 7 nm and each film was formed by 30 bi-layers. These materials have a wide range of uses such as packaging and biomedicals [15].

Chitosan films used for cell scaffolding in the regeneration of the tympanic membrane (type I experimental tympanoplasty) were elaborated. These films were grafted in New Zealand rabbits with successful results of tympanic tissue regeneration [33].

2. Objective

Demonstrate that CNC can improve the mechanical and physical properties of a chitosan matrix. Furthermore, determine the biocompatibility and biodegradability of CS and CNC nanocomposites via biological tests and then based on the results obtained, determine a potential use of these nanocomposites in the biomedical area.

3. Experimental

3.1. Materials

Materials used for this study were Biomedical Grade chitosan from Sigma Aldrich (Deacetylation grade 75–85%), acetic acid, alpha cellulose (Neucel Cellulose Ltd.), male laboratory rats (Wistar), cellulose acetate membranes for dialysis, vacuum oven, high resolution microscopy, dynamical-mechanical and chemical analysis. The formulations employed in this research are presented in Table 1.

http://dx.doi.org/10.5772/61727

Material	Reinforcement	Formulation (w/w)
CS	CNC	Pure CS
		CS + CNC (0.5%)
		CS + CNC (1.0%)
		CS + CNC (2.0%)

Table 1. Formulations of CS + CNC films

3.2. Methodology

- Preparation of CNC

Alpha cellulose was ground and mixed with sulfuric acid (64% concentration) for one hour under constant stirring and at a controlled temperature (approximately 50° C in a warm bath). After that, the liquor was added to deionized water, cooled at 8° C 1:10 (v/v) to stop the reaction. The liquor was centrifuged at 4000 RPM for 5 minutes, separating the liquor into two phases: solid (cellulose gel in the bottom of recipient) and liquid (with acid remainders). The liquid phase was disposed of and deionized water was added to the recipient to remove excess acid from the gel (containing CNC), prior to centrifugations (three in total). The washed gel was put in dialysis membranes in deionized water under stirring until it reached a pH of 5. After that, the CNC were submitted to ultrasound treatment for 2 minutes and finally vacuum filtered using 0.45 micron Wharton paper, and kept cooled.

- Chitosan – CNC films elaboration

Chitosan (4 g) was dissolved in acetic acid at 2% (v/v) per each formulation, under constant stirring for 2 hours. Next, CNC in different concentrations were added (0.5%, 1.0% and 2.0%) and then stirred for two more hours. The substance was cast on Petri plates affording a concentration of 0.4 mL/cm^2 to obtain the same quantity of nanocomposite on the plates. The plates were then put into a vacuum oven (28°C–70 MPa) for 96 hours. After drying, NaOH (1.0 N) was added to the Petri dishes to precipitate the films and then the CS + CNC films were washed with deionized water until they reached a pH of 7. Finally, films were dried at room temperature for 48 hours. The thickness of films was measured with a micrometer and dimensioned for physical and mechanical testing.

- Physical evaluation

When performing swelling tests, modifications in the dimensions of specimens were made to evaluate swelling changes over time. The dry weight (W_i) of each specimen was taken and then films of each formulation were put in deionized water to control the weight each minute until a constant value was achieved [34]. For maximum water retention (MWR), the dry weight (W_i) of each specimen (8 per formulation) was taken and it was then added to deionized water. Weights were controlled at 30, 60, and 120 minutes, and then 24 hours, before the final weight was obtained (W_f) [34]. Finally, the MWR was determined by the following formula:

$$\mathrm{MWR}\ (\%) = (W_f - W_i) \,/\, (W_i) \times 100$$

• Mechanical evaluation

Dynamical mechanical analysis was used in static mode to evaluate Young's modulus for all formulations, with the stress-strain test operating in the controlled force mode. Typical testing conditions were 0.1 N preload, 1 N/min ramp and a gauge length of ca. 10 mm. Strips of 0.075 mm x 5 mm x 20 mm were used. The temperature range was 37.05 ± 0.05°C and the moisture content was 98%, determined by a TA Instruments DMA 800 used in wet conditions.

• Biological evaluation

The best way to test our material was in living specimens. Biocompatibility and biodegradability tests were carried out. For these, the dry weight (W_i) of films was controlled. The biological subjects for testing were 16 Lab rats (of the Wistar breed). The animals were submitted to a surgical procedure involving grafting two portions of films for each formulation (4 rats per formulation). Each animal had two sub-cutaneous cuts (in the middle of back, on the right side for biocompatibility and on the left side for biodegradability). Every three days, the rats were controlled to prevent any infection or adverse reaction to the nanomaterial. Specimens were euthanized 30 days after surgery and the two portions of membrane were retired, lyophilized and controlled for dry weight (W_f) in order to obtain the biodegradability value:

$$\text{Biodegradability (\%)} = (W_i - W_f) / (W_i) \times 100$$

Biocompatibility was proved by SEM images at the moment the grafts were retired, the mortality rate of specimens and the non-presence of encapsulation, fibrillation or any rotten portion of membrane after 30 days.

• Characterization of nanocomposites

To characterize the morphology of films, the response and the different levels of biocompatibility and biodegradability, a JEOL JM 6300 with a double gold layer to avoid the electrical charge of the sample was used. The magnification was from around 500x up to 35,000x and the voltage employed was between 7kV and 20kV.

• IR spectroscopy

This technique was used to determine changes in functional groups of nanocomposite as a consequence of the presence of CNC in the chitosan matrix, or previous chemical treatment. Portions of thin films of all formulations were analyzed in an IP Spectrometer in transmittance mode (4,000 to 500 cm^{-1}). Other parameters were: 16 scans per spectrum, ATR mode and a resolution of 0.4 cm^{-1}. A Perkin Elmer Spectrum GX FT-IR System was used.

4. Results

4.1. CNC + chitosan films

The films obtained from the combination of chitosan and CNC showed a transparent aspect with a slight yellow color. Transparence in the films suggests a good distribution of CNC in the CS matrix. At the same time, the slight yellow color is due to the natural presence of impurities in this biopolymer [35]. The average thickness of films was 30 microns and the surface was smooth. Visible CNC agglomerates and dark points were absent from these films as can be seen in Figure 1.

These characteristics in all formulations of pure CS and CS + CNC were the same, and accorded with findings of other authors [19, 21, 35, 36]. Even with CNC concentrations over 20% (w/w), transparent films were reported, although with CNC concentrations over 60% the films showed a translucent appearance (a greyish color) and a rigid but brittle consistency [35]. Chitosan films and polyethylene oxide (PEO) employing electrospinning produced good results in transparence and homogeneity [32].

Figure 1. a. Film of CS + CNC (0.5%). b. Film of CS + CNC (1.0%) and c. Film of CS + CNC (2.0%)

4.2. Morphology of CS + CNC films

SEM images showed a good dispersion of CNC in the nanocomposites due to the affinity between the nanoreinforcement and the matrix (having similar chemical structure and hydrophilic nature), as well as a good interaction of negative charges of the sulfate groups of CNC with the amine groups of CS [35]. The presence of nanoparticles was observed in different formulations of nanocomposites, as can be seen in Figure 2. These bundles formed by several tens of CNC, with dimensions of between 150–200 nm up to 500 nm on average, and in some cases up to 1 or 2 microns, mainly in CS + CNC (2.0%). The phenomenon was common when the CNC concentration was higher (over 5% and 10%) and even when working with concentrations of 5% and 10% it was possible to see white punctuations in the matrix surface via SEM images, seen in the cross section of CNC [36]. Another possible phenomenon linked to a higher concentration of CNC in the matrix, is the formation of a polyelectrolyte macroion complex between CS and CNC. The size of these complexes could reach several microns in length. These particles are formed by CNC and surrounded by CS chains. The shape of these complexes depends on the quantity of NH_2 in CS. When the concentration of NH_2 is high, the shape of the complex tends to be spherical [37]. These particles could have potential in biomedicals for drug transportation and controlled release due to the charges on cellulose and the linkage that can be achieved with other substances and active composites.

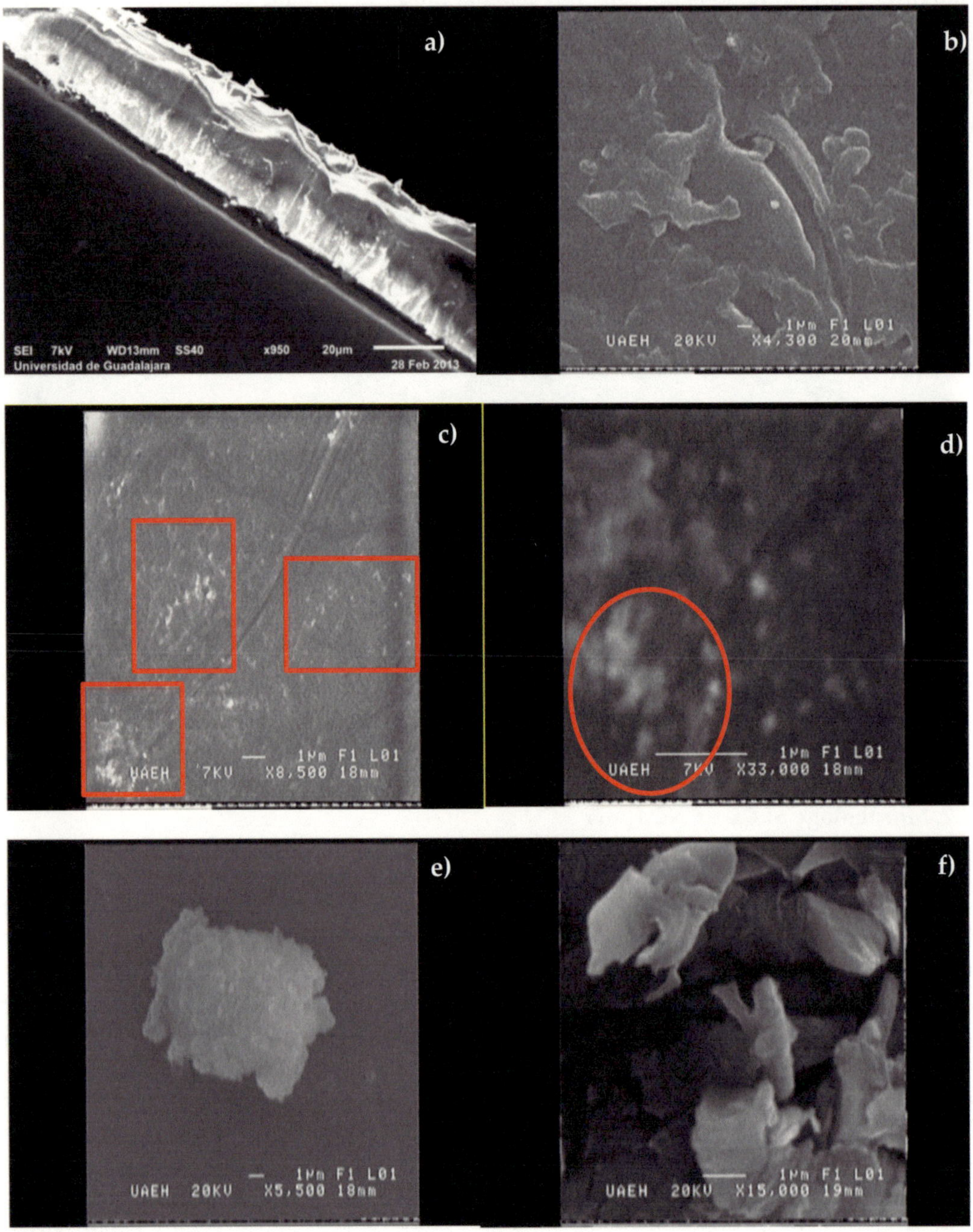

Figure 2. a SEM image of a cross section of CS + CNC (1.0%) film at 950x. b. SEM image of CS film at 4,300x. c. SEM image of clusters of CS + CNC (0.5%) film at 8,500x (circled in red), d. SEM image of clusters of CS + CNC (0.5%) at 35,000x (circled in red). e CS – CNC (2.0%) macroion at 5,500x and 2f. CS + CNC (2.0%) at 15,000x.

4.3. Physical properties

Swelling capability is presented in Figure 3. Swelling capability showed an evident change with the presence of CNC in the matrix. For water swelling, pure CS films showed saturation times of 3–4 minutes. With formulations of CS + CNC (0.5%) saturation times reduced to 1–2 minutes, while for formulations with CNC of 1% and 2%, the saturation time was 1 minute. These results show that CNC allow water entrance between CS chains as an unstructuring element, making the material more hydrophilic. This behavior was opposite to that observed in another study, which used higher concentrations of CNC (from 5% up to 60%) making the CS films less permeable, thereby reducing the swelling capability significantly [35, 36, 38]. The maximum water retention (MWR) showed a similar response with the presence of CNC, as presented in Table 2.

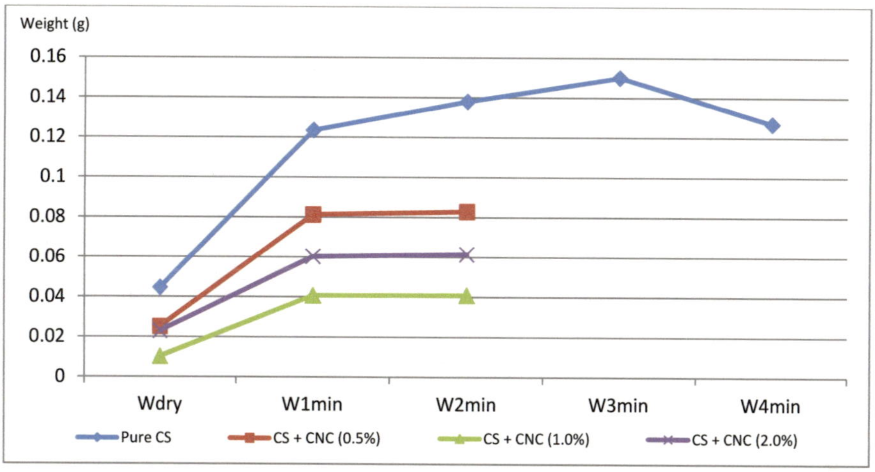

Figure 3. Swelling capability for CS + CNC nanocomposites

Formulation	Variation in time		
	W30min	**W120min**	**W24h**
Pure CS	189.72	198.86	187.49
CS + CNC (0.5%)	237.68	222.67	226.64
CS + CNC (1.0%)	346.52	277.79	258.18
CS + CNC (2.0%)	234.68	236.70	214.65

Table 2. Values of MWR for CS + CNC nanocomposites

For pure CS films, the MWR was 187% of its own weight in water, while CS + CNC (0.5%) films obtained values of 226%, CS + CNC (1.0%) values of 258% and CS + CNC (2.0%) values of 214%, as can be seen in Figure 4. In these cases, the low concentrations of CNC improve the water retention properties of films in comparison with results of other authors [15, 35] which showed a remarkable reduction in the physical properties of CS + CNC films, using concentrations from 5% up to 60%. This phenomenon can be explained by the affinity of CNC to the water, due their hydrophilic nature (and O–H linkages). Another reason is the plasticizing effect of CNC at low concentrations of CS matrix, contributing to the unstructuring effect in CS chains allowing the passage of water inside the polymer. This effect makes the film more flexible, facilitating the entry of water into the polymer chains. A similar effect was reported when adding different concentrations of glycerol (from 2% to 6%), used as a plasticizer in starch films with CS covering. Even at the highest CS concentration, the presence of glycerol increased this parameter, with values from 71.8% to 125.4% in comparison with pure CS (35.7% to 74.8%) [21].

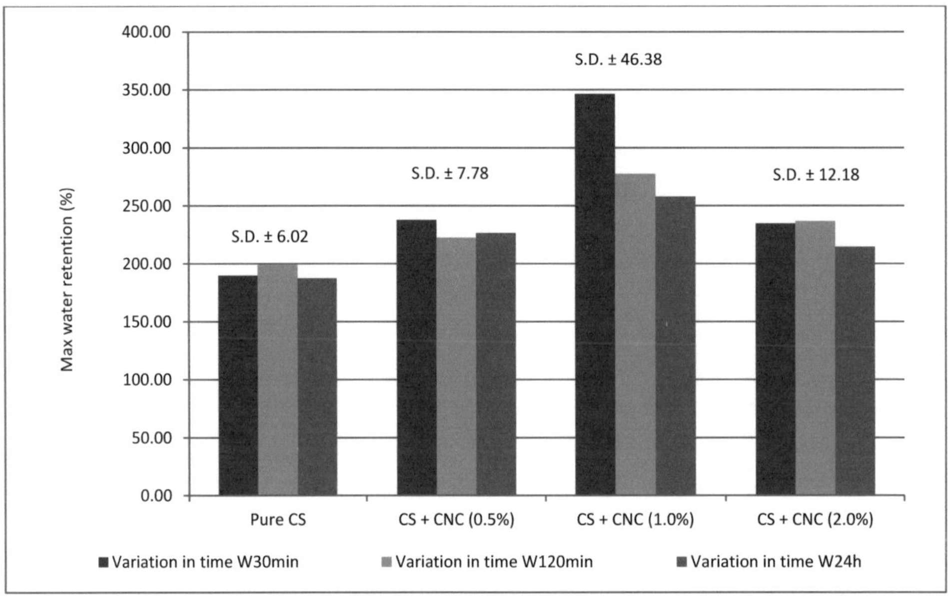

Figure 4. Values (%) of maximum water retention for pure CS and CS + CNC films

4.4. Mechanical properties

The values obtained from formulations with CNC suggest that these modified the mechanical properties in comparison with pure CS. Young's modulus (YM) reached by pure CS was 2,916 ± 933 MPa, similar to that observed in another study [35], while for formulations with CNC the values were CS + CNC (0.5%) 1,575 ± 82 MPa, CS + CNC (1.0%) 1690 ± 433 MPa and for CS + CNC (2.0%) the value was 1657 ± 4 MPa, as can be seen in Figure 5. There was a large variation in the results for YM in pure CS, with values from 1,080 MPa up to 2,852 MPa. The average for pure CS ranged between 2,300 and 2,500 MPa. This variation is because the matrix polymer has crystalline and amorphous zones, which will exhibit different resistances (high and low values). Furthermore, residual efforts stored in films as a consequence of some conditions in their elaboration process (casting solution), appears in the first stages of assay, when the machine break these efforts first before apply all the force in the film in a homogeneous way.

The graphs showed an inflexion point in effort–strain curves. This phenomenon was present in some films of CS + CNC (1.0%) and CS + CNC (2.0%).

Some explanations for the decrease of Young's Modulus in the films with CNC suggest the cause as the formation of aggregates and CS–CNC complexes due to the agglomeration of nanoparticles in one single place, generating stresses around them and becoming a potential weak point in the film. It could be deduced that CNC acted as a plasticizer of the matrix, making it less rigid. However, CS + CNC formulations showed values lower than those obtained for pure CS. However, these values were less variable than pure CS values and a further detail was observed: CNC aggregation didn't contribute to the deformation or rupture of films.

Another possible cause of the observed decrease in YM could be the low concentrations of CNC employed in this study (0.5%, 1.0% and 2.0%) in comparison with other studies that used concentrations above 5%, 10% until 60% [15, 35] that saw improvements in YM of 78% to 150% and 230% to 320% respectively. The decrease in YM must be understood as a loss of stiffness, making a more flexible material, but with low strain values (1–2%). This last can be improved if plasticizers (i.e. glycerol) are included in the formulations. The values obtained are similar with CS films with chitin concentrations of 2.7% to 37.5%, having values from 1,622 ± 377 MPa to 2,318 ±619 MPa respectively [11]. Based on these values, a potential use of these films could be in healing and wound recovery, as well as scar prevention.

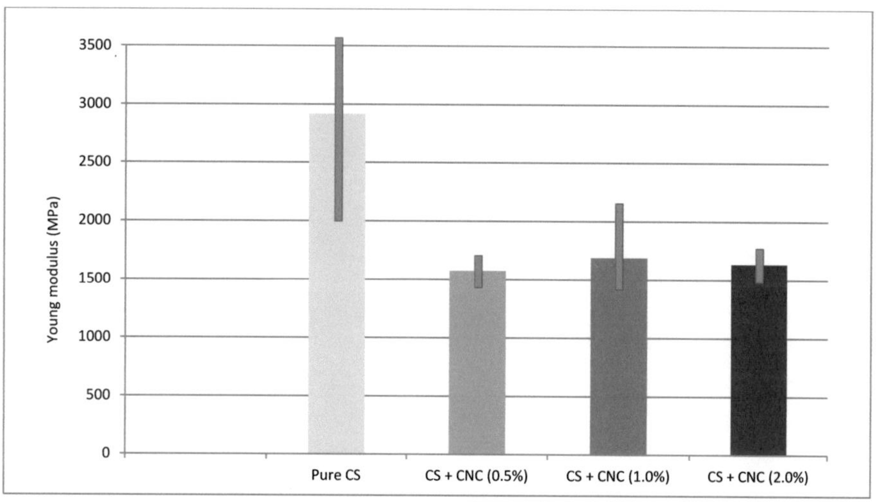

Figure 5. Young's Modulus values of pure CS and CS + CNC films

4.5. IR Spectroscopy of CS + CNC films

As can be seen in Figure 6, for pure CS the IR spectrum showed bands at 3,360 and 3,265 cm^{-1} typical of O–H and N–H bonds respectively. The band at 2,870 cm^{-1} corresponds to C–H bonds and the signal at 1,650 cm^{-1} is commonly seen with type I amides and C=O bonds linked to acetyl and amine groups, and 1,560 cm^{-1} corresponds to amide groups (N–H). Bands of 1,420 and 1,375 cm^{-1} correspond to residual CH_3N–acetyl glucosamine and –CH_2 groups respectively. Finally, bands of 1,060 and 1,030 cm^{-1} confirm the presence of C–O bonds.

Regarding nanocomposites, formulations of CS + CNC of 0.5%, 1.0% and 2.0% didn't show variations in their spectrums when compared with a pure CS spectrum, as shown in Figure 7. This is due to the similarity between cellulose and chitosan in terms of their chemical structure. For that reason, the bands for both polysaccharides are the same, except for the presence of an amine group with chitosan [15]. Another cause is the low concentration of CNC, which could be undetected by FTIR. A similar phenomenon has been observed in chitin, chitosan and glycerol nanocomposites [38].

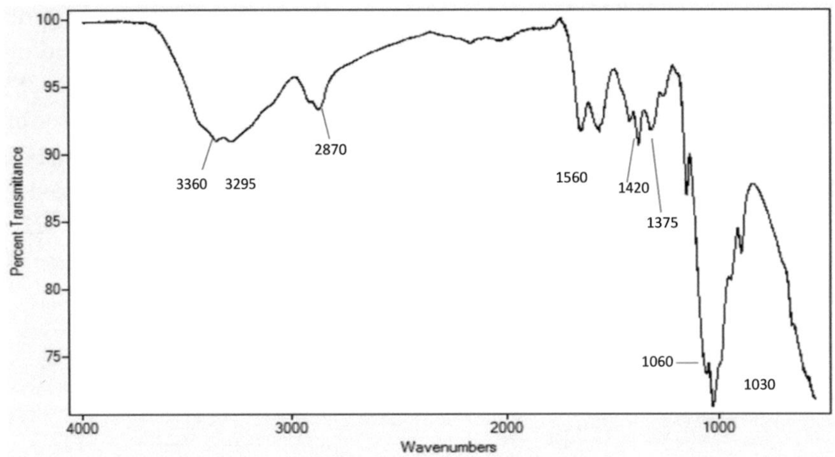

Figure 6. IR spectrum of pure CS film

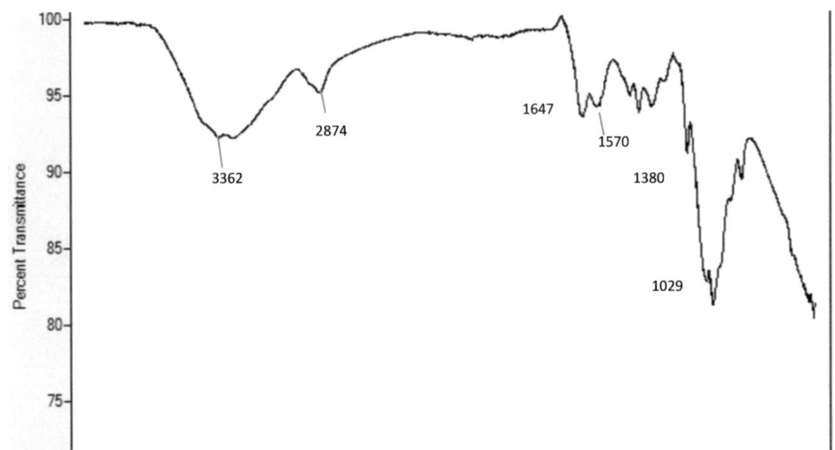

Figure 7. IR spectrum of CS + CNC (2.0%) film

4.6. Biological testing

Biological results showed 100% biocompatibility for all formulations in 16 rats. After 30 days, the specimens had not shown an adverse reaction, septicemia or death in the presence of CNC. At the point the grafts were retired, no negative influence was observed over the surface of muscles or sub-cutaneous tissues (no evidence of coagulation, fibrosis, and/or rotten or dead tissues), as can be seen in Figure 8. A further typical sign of CS in living tissues is the slight yellow color over the muscle or tissue. In a few cases rolled grafts were found, due mainly to movement of the specimen during the 30-day period. A further possible cause could be that the graft was not fixed with sutures or staples in the biological test, in order to prevent a different reaction or influence over the grafts.

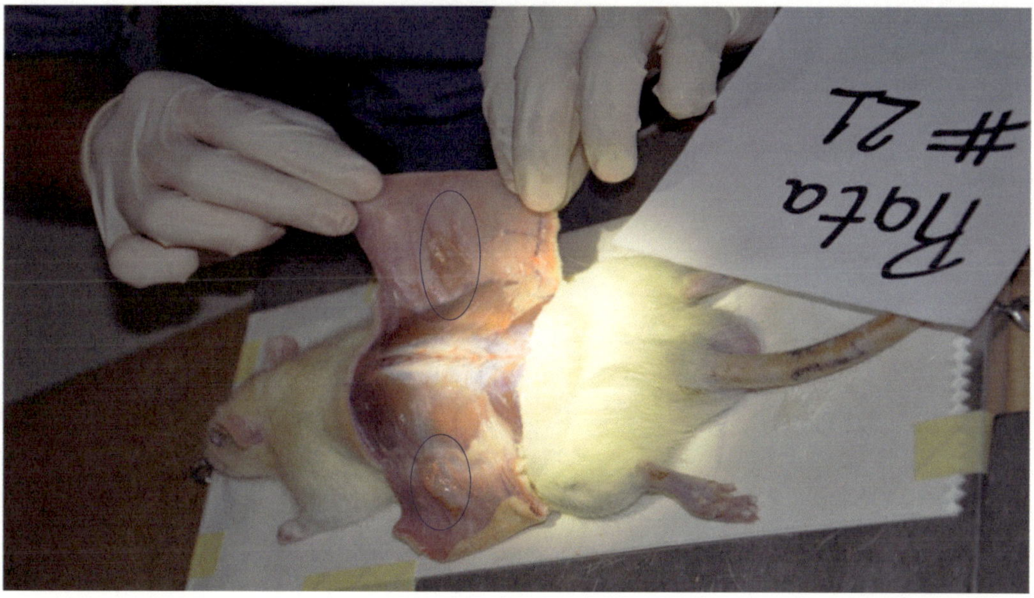

Figure 8. Surgical stage of retiring the portions of CS + CNC (0.5 %) (circled in blue), to evaluate biocompatibility and biodegradability in the specimen.

The biodegradability values were partial but positive for all formulations, as presented in Table 3. In pure CS, the values reached between 6.21% and 8.55% in parallel. One sample had 100% biodegradability and another sample had more weight than the initial weight, suggesting the presence of new tissue as fascia and/or loose connective tissue over the graft (as presented in Figure 9), or cells or blood inside the membrane (biodegradability value: -12.89%). The CS + CNC (0.5%) had the most weight gain in their films in comparison with other formulations (-13.63%, -12.64% and -55.83% respectively). Just one sample of this group had a low biodegradability value (7.82%). For CS + CNC (1.0%) the values reached 0.71% to 21.02%. The highest values of biodegradability were shown by CS + CNC (2.0%) being 19.59% to 60.37% respectively) with just one specimen that gained weight.

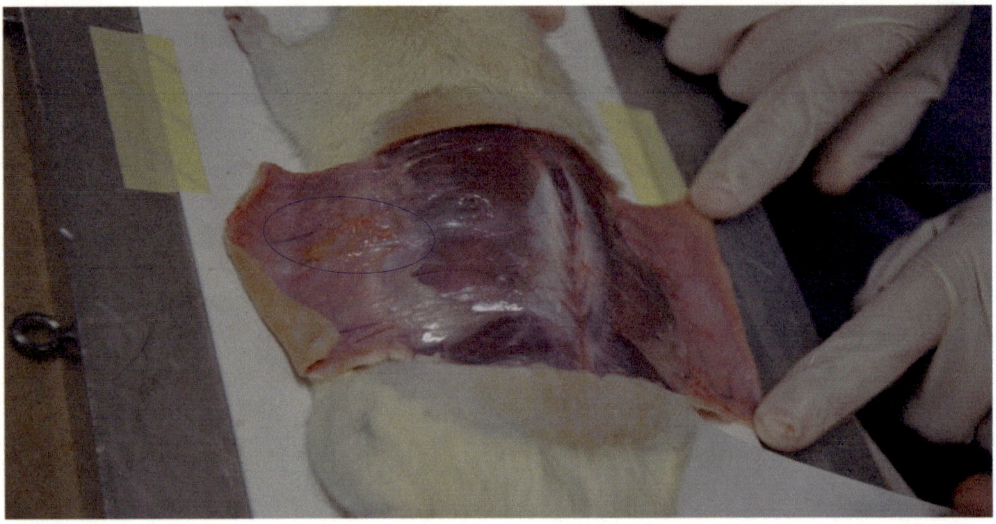

Figure 9. Presence of loose connective tissue and blood vessels (circled in blue) formed over the portion of CS + CNC (1.0%)

	Biodegradability values (%)			
	Pure CS	**CS+CNC (0.5%)**	**CS+CNC (1.0%)**	**CS+CNC (2.0%)**
Specimen 1	**-12.89**	**-13.48**	8.33	19.59
Specimen 2	8.55	**-12.64**	0.71	60.37
Specimen 3	100.00	7.82	21.02	**-4.92**
Specimen 4	6.21	**-55.83**	5.52	40.52

Table 3. Biodegradability values for CS + CNC nanocomposites

Indeed, a relationship between the concentration of CNC and higher values of biodegradability can be deduced. The unstructuring effect of CNC in CS chains allows the inflow of water, cells and living fluids in the membrane. This relationship is important because it informs predictions and design of future research into different presentations of nanocomposites with higher concentrations of CNC, in order to study control of the rate of biodegradation for specific biomedical uses (depending on the part or organ in the body to be used).

The presence of loose connective tissue and fascia were evident for all formulations, as well as the formation of small blood vessels surrounding the graft. Based on the principle of time of biodegradability of new tissue, these films presented different values after 30 days. For example, CS + CNC (0.5%) showed the best results for new scaffold tissue in three formulations. In the other cases, we can estimate the total time for biodegradation to be 3–6 months (less time with higher concentrations of CNC based on these results). This could be useful in the future development of possible biomedical uses of these films.

Another aspect to consider with respect to biodegradation values is the type of films. In this study, non porous films were elaborated. The porosity rate and the size of pores directly affect the biodegradability of films, because they allow the easy passage of water, blood, cells and other fluids. From their condition, these films could be useful for tissue scaffolding as can be seen in Figures 10 and 11 (SEM images). A typical pattern of degradation of chitosan shown as hexagonal borders was observed in all formulations, at different stages. The formation of fascia in different stages was evident.

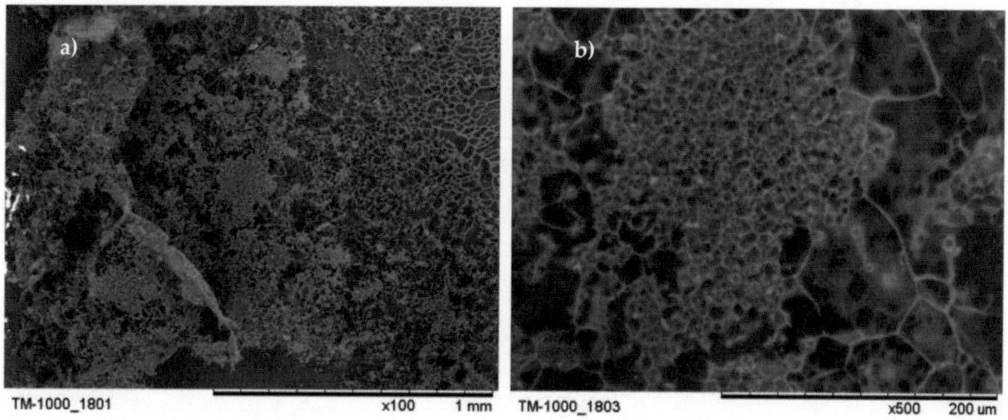

Figure 10. a. Presence of loose connective tissue in CS + CNC (0.5%) films at 100x and Fig. b. Hexagonal pattern of chitosan degradation in CS + CNC (0.5%) at 500x.

A pattern of biodegradation of CS films is supported by the fact that the biodegradation rate of CS has a linear relationship with the Degree of Deacetylation (DD). When the DD value (from 0 to 100) is closer to 100, the material shows a slow rate of biodegradation [39]. Other aspects to consider are that CS is degraded by proteases (mainly lysozymes) that attack N-acetyl glucosamine linkages making the process faster, but these linkages have less presence in CS chains when DD values are higher. Also, in this state, lysozyme activity is low and the biodegradation rate is slow. Furthermore, CS is a semi crystalline polymer. In amorphous zones, lysozyme activity is intense and when the positive charges increase, the interactions between cells with CS are better, thereby improving biocompatibility [39].

CS scaffolds with a low Degree of Acetylation (DA) and low molecular weight have a higher rate of biodegradation. Indeed, CS scaffolds with a low DA present lower biodegradation times, smaller pores, better mechanical properties, moderate water absorption and more intense cell activity than CS scaffolds with a higher DA [40]. A pore size of between 60 and 90 microns is suitable to allow lysozymes inside the structure and to act in polymer chains.

The CS type used in this research has a DD between 75–85% and a medium molecular weight related to low biodegradation values. In CS + CNC films, the CNC had an unstructuring effect, creating spaces between CS chains, thereby allowing water (and lysozymes) access into the scaffold structure, and showing larger values of biodegradation when CNC concentration was higher.

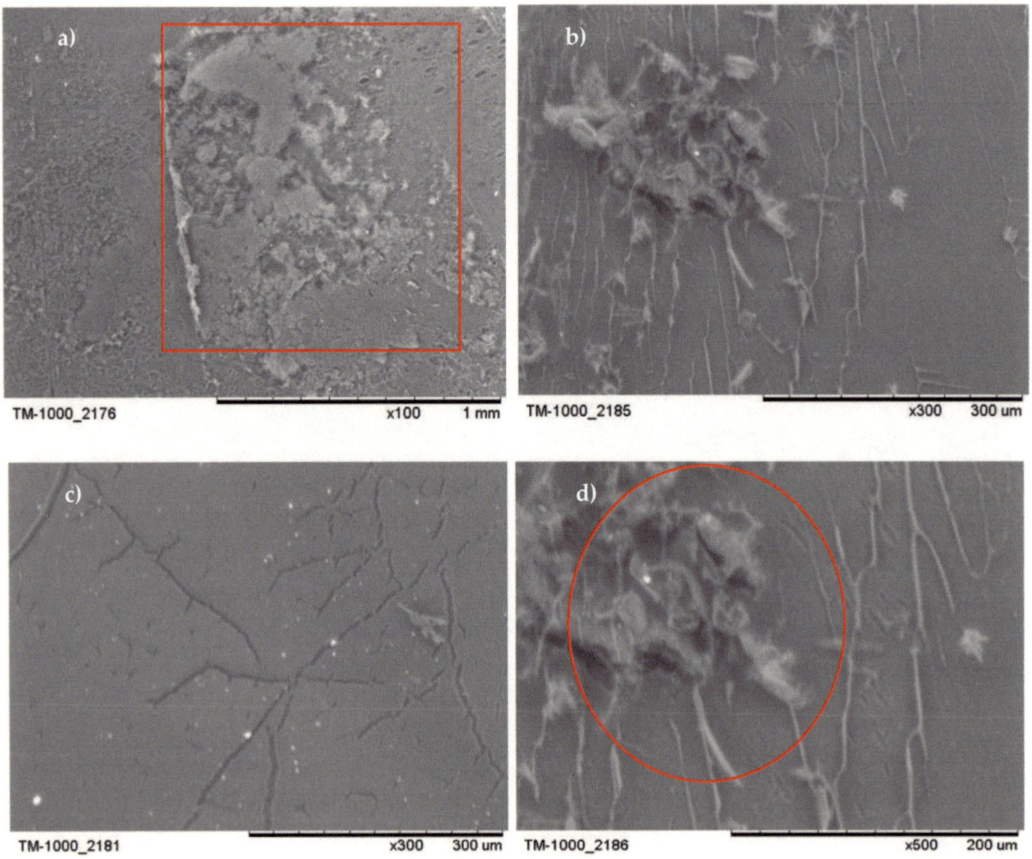

Figure 11. a. Presence of loose connective tissue (circled in red) in CS + CNC (1.0%) films at 100x and b. 300x. Fig. 11c First steps of degradation of chitosan d. Presence of fascia (circled in red) in CS + CNC (1.0%) films at 500x

The attachment, morphology and proliferation of cells on CS scaffolds is highly dependent on the type of cell lines, the source and characteristics of CS, the methods of CS scaffold preparation, and the characteristics of the chitosan scaffolds. In addition, chitosan with a low molecular weight and low DA has excellent potential as a scaffolding material for a variety of tissue regeneration systems [40].

5. Conclusions

- The addition of CNC improves physical properties, increasing their capability of swelling and water capitation. Mechanically, the main factor was the unstructuring effects over the chitosan chains. In biological tests, CNC show positive results for biocompatibility and biodegradability.

- CS + CNC nanocomposites showed physical, mechanical and biological properties suitable for biomedical use in recovering / wound healing, tissue scaffolding and scar prevention.
- There is a direct relationship between higher concentrations of CNC and higher values of biodegradability.
- The formation of the CS-CNC macroion complex in this study could have potential for controlled drug release taking into account other presentations of product (microspheres, nanopowders or nanoparticles).

Acknowledgements

Guillermo H. Riva, wishes to express his appreciation to the following institutions: German Exchange Students Agency (DAAD) for the scholarship and kind support during his M.Sc. Studies in Mexico; to the University of Guadalajara (UdG) for his formation as a M.Sc., for sponsorship and the financial support of this study, and Neucel Specialty Cellulose Limited (BC, Canada) for donating the alpha cellulose used in this study.

Author details

Guillermo H. Riva[1], Joaquín García-Estrada[2], Brenda Vega[2], Fernando López-Dellamary[1], María E. Hérnandez[3] and José A. Silva[1*]

*Address all correspondence to: jasilva@dmcyp.cucei.udg.mx

1 University of Guadalajara, Wood, Cellulose and Paper Research Department "Karl AugustinGrellmann" Carretera Guadalajara, Jalisco, C.P., México

2 Biomedical Research Center of Occidente (CIBO), Guadalajara, Jalisco, C.P., México

3 University of Guadalajara, Exact Sciences and Engineering Faculty, Colonia Olímpica, Guadalajara, Jalisco, C.P., México

References

[1] Habibi, Y., L. A. Lucia, Rojas, O. J. Cellulose Nanocrystals: Chemistry, Self-Assembly, and Applications. Chemical Reviews. 2010; 110(6): 3479-3500 DOI 10.1021/cr900339w

[2] Alemdar, A. and M. Sain. Biocomposites from wheat straw nanofibers: Morphology, thermal and mechanical properties. Composites Science and Technology. 2008; 68(2): 557-565. DOI:10.1016/j.compscitech.2007.05.044

[3] Sannino, A., Demitri, C., Madaghiele, M. Biodegradable Cellulose-based Hydrogels: Design and Applications. Materials. 2009; 2: 353-373. DOI:10.3390/ma2020353

[4] Shanmuganathan, K., J. R. Capadona, et al. Biomimetic mechanically adaptive nanocomposites. Progress in Polymer Science. 2010; 35(1–2): 212-222. DOI:10.1016/j.progpolymsci.2009.10.005

[5] Jonoobi, M., J. Harun, et al. Mechanical properties of cellulose nanofiber (CNF) reinforced polylactic acid (PLA) prepared by twin screw extrusion. Composites Science and Technology. 2010; 70(12): 1742-1747. DOI: 10.1016/j.compscitech.2010.07.005

[6] Oksman, K., A. P. Mathew, Bondeson, D., Kvien, I. Manufacturing process of cellulose whiskers/polylactic acid nanocomposites. Composites Science and Technology. 2006; 66(15): 2776-2784. DOI:10.1016/j.compscitech.2006.03.002

[7] Siddiqui, N. Characterization of Mechanically and Enzymatically Produced Cellulose Nanofibers from Wood Pulp. [thesis]. The Graduate School: Maine, University of Maine. 2008.

[8] Janardhnan, S., Sain, M. Targeted disruption of hydroxyl chemistry and crystallinity in natural fibers for the isolation of cellulose nano-fibers via enzymatic treatment. BioResources. 2011; 6(2): 1242-1250. DOI: 10.15376/biores.6.2.1242-1250

[9] Jonoobi, M., Harun, J., Mathew, A.P., Hussein, M.Z., Oksman, K. Preparation of cellulose nanofibers with hydrophobic surface characteristics. Cellulose. 2010; 17: 299-307. DOI 10.1007/s10570-009-9387-9

[10] Kalia, S., A. Dufresne, Cherian, B., Keith, B. S., Avérous, L., Njuguna, J. Cellulose-Based Bio- and Nanocomposites: A Review. International Journal of Polymer Science. 2011, (Article ID 837875): 35. DOI:10.1155/2011/837875

[11] Shelma R., Willi Paul and Sharma C.P. Chitin Nanofibre reinforced Chitosan Films for Wound Healing Application. Trends Biomater. Artif. Organs. 2008; 22(2): 111-115

[12] Hui, J., Yuanyuan, J., Jiao, W., Yuan, H., Yuan, Z., Shiru, J. Potentiality of Bacterial Cellulose as the Scaffold of Tissue Engineering of Cornea. 2nd International Conference on Biomedical Engineering and Informatics (BMEI 2009). DOI 10.1109/BMEI.2009.5305657

[13] Schumann, D., J. Wippermann, *et al.* Artificial vascular implants from bacterial cellulose: preliminary results of small arterial substitutes. Cellulose. 2009; 16(5): 877-885. DOI 10.1007/s10570-008-9264-y

[14] Amorim, W. L., Costa, H. O., Coehlo de S., F., Germanos de C., M., da Silva, L. Experimental study of the tissue reaction caused by the presence of cellulose produced by Acetobacter xylinum. Brazilian Journal of Otorhinolaringology. 2009; 75(2): 200-207. DOI 10.1590/S0034-72992009000200008

[15] de Mesquita, J. P., Donnici, C. L., Pereira, F. V. Biobased Nanocomposites from Layer-by-Layer Assembly of Cellulose Nanowhiskers with Chitosan. Biomacromolecules. 2010; 11(2): 473-480. DOI 10.1021/bm9011985

[16] Darder, M., López-Blanco, M., Aranda, P., Aznar, A.J., Bravo, J., Ruiz-Hitzky, E. Microfibrous chitosan-sepiolite nanocomposites. Chemical Materials. 2006; 18: 1602-1610. DOI 10.1021/cm0523642

[17] Gupta, K. C. and M. N. V. R. Kumar. Structural changes and release characteristics of crosslinked chitosan blends in response to solution pH. Journal of Macromolecular Science, Part A. 1999; 36(5-6):827-841.DOI:10.1016/S1381-5148(00)00038-9

[18] [18} Madihally, S. V. and H. W. T. Matthew. Porous chitosan scaffolds for tissue engineering. Biomaterials. 1999; 20(12): 1133-1142. DOI:10.1016/S0142-9612(99)00011-3

[19] Azeredo, H. M., L. H. Mattoso, Avena-Bustillos, R. Ceotto, G., Munford, M., Wood, D., Mc Hugh, T. Nanocellulose reinforced chitosan composite films as affected by nanofiller loading and plasticizer content. Journal of Food Science. 2010; 75(1): N1-7. DOI: 10.1111/j.1750-3841.2009.01386.x

[20] Jeon, Y. J., Kamil, J.Y.V.A., Shahidi, F. Chitosan as an edible invisible film for quality preservation of herring and Atlantic cod. Journal of Agriculture and Food Chemistry. 2002; 50: 5167–5178. DOI: 10.1021/jf011693l

[21] Bangyekan, C., D. Aht-Ong, Aht-Ong, D., Srikulkit, K. Preparation and properties evaluation of chitosan-coated cassava starch films. Carbohydrate Polymers. 2006; 63(1): 61-71. DOI:10.1016/j.carbpol.2005.07.032

[22] Durango, A. M., Soares, N.F.F., Andrade, N.J. Microbiological evaluation of an edible antimicrobial coating on minimally processed carrots. Food Control. 2006; 17: 336-341. DOI:10.1016/j.foodcont.2004.10.024

[23] Campaniello, D., Bevilacqua, A., Sinigaglia, M., Corbo, M. R. Chitosan: antimicrobial activity and potential applications for preserving minimally processed strawberries. Food Microbiology. 2008; 25(8): 992-1000. DOI:10.1016/j.fm.2008.06.009

[24] Nisperos-Carriedo. Edible coatings and films based on polysaccharides. Edible coatings and films to improve food quality. 1994. J. M. Krochta, Baldwin, Nisperos-Carriedo. Lancaster, U. K., Technomic. 305-335.

[25] Suyatma, N. E., Copinet, A., Coma, V., Tighzert, L. Mechanical and barrier properties of biodegradable films made from chitosan and poly(lactic acid) blends. Journal of Polymer Environment. 2004; 12: 1-6. DOI: 10.1023/B:JOOE.0000003121.12800.4e

[26] Suyatma, N. E., Tighzert, L., Copinet, A. Effects of hydrophilic plasticizers on mechanical, thermal, and surface properties of chitosan films. Journal of Agriculture and Food Chemistry. 2005; 53(10): 3950-3957. DOI 10.1021/jf048790+

[27] Di Gioia, L., Guilbert, S. Corn Protein-based Thermoplastic Resins: Effect of some Polar and Amphiphilic Plasticizers. Journal of Agriculture and Food Chemistry. 1999; 47: 1254-1261. DOI: 10.1021/jf980976j

[28] Ludueña, L. N., Alvarez, V.A., Vasquez, A. Prossesing and microstructure of PCL/ clay nanocomposites. Materials Science and Engineering A. 2007; 460-461: 121-129. DOI:10.1016/j.msea.2007.01.104

[29] Sorrentino, A., Gorrasi, G., Vittoria, V. Potential perspectives of bionanocomposites for food packaging applications. Trends in Food Science & Technology. 2007; 18(2): 84-95. DOI:10.1016/j.tifs.2006.09.004

[30] Vaia, R. A., Wagner, H.D. Framework for nanocomposites. Materials Today. 2004; 7: 32-37. DOI:10.1016/S1369-7021(04)00506-1

[31] Park, T.-J., Y. Jung, Choi, S. W., Park, H., Kim, H., Kim, E., Lee, S. H., Kim, J. H. Native chitosan/cellulose composite fibers from an ionic liquid via electrospinning. Macromolecular Research. 2011; 19(3): 213-215. DOI:10.1007/s13233-011-0315-0

[32] Jacobs, V., Patanaik, A and Anandjiwala, R.D. Electrospun chitosan nanofibre membranes for antimicrobial application: role of electrospinning processing parameters. European Cells and Materials. 2010; Vol. 19. Suppl. 1, pp 4. ISSN 1473-2262.

[33] Macías, H. Quitosana como andamiaje celular en regeneración de membrana timpánica en conejo Nueva Zelanda (timpanoplastía experimental tipo I) [thesis]. Departamento de Neurociencias. Guadalajara, Universidad de Guadalajara 2010.

[34] ASTM D 587 - 98 Standard Test Method for Swelling and Water Retention (1998)

[35] Fernandes, S., Freire, C., Silvestre, A., Neto, C. P., Gandini, A., Berglund, L., Salmén, L. Transparent chitosan films reinforced with a high content of nanofibrillated cellulose. Carbohydrate Polymers. 2010; 81: 394-401. DOI:10.1016/j.carbpol.2010.02.037

[36] Khan, A., Khan, R. A., Salmieri, S., Le Tien, C., Riedl, B., Bouchard, J., Chauve, G., Tan, V., Kamal, M. R., Lacroix, M. Mechanical and barrier properties of nanocrystalline cellulose reinforced chitosan based nanocomposite films. Carbohydrate Polymers. 2012; 90: 1601– 1608. DOI:10.1016/j.carbpol.2012.07.037

[37] Wang, H., Roman, M. Formation and properties of chitosan-cellulose nanocrystal polyelectrolite-macroion complexes for drug delivery applications. Biomacromolecules. 2011; 12: 1585-1593. DOI: 10.1021/bm101584c

[38] Mathew, A. P., Laborie, M.P., Oksman, K. Cross-linked chitosan/chitin crystal nanocomposites with improved permeation selectivity and pH stability. Biomacromolecules. 2009; 10(6): 1627-1632. DOI: 10.1021/bm9002199

[39] Croisier, F., Jerome, C. Chitosan based biomaterials for Tissue Engineering. European Polymer Journal. 2013; (49) 780 – 792. DOI: 10.1016/j.eurpolymj.2012.12.009.

[40] Han, T., Nwe, N., Furuike, T., Tokura, S., Tamura, H. Methods of N – acetylated chitosan scaffolds and it's *in vitro* biodegradation by lysozyme. J. Biomedical Science and Engineering, 2012; (5): 15-23. DOI: 10.4236/jbise.2012.51003.

Chapter 10

Nanocellulosic Materials in Tissue Engineering Applications

Tamilselvan Mohan, Silvo Hribernik, Rupert Kargl and Karin Stana-Kleinschek

Additional information is available at the end of the chapter

http://dx.doi.org/10.5772/61344

Abstract

This chapter deals with an overview of design and fabrication of three-dimensional (3D) scaffolds for tissue engineering (TE) applications using the electrospinning technique. A general introduction to cellulose, a short overview of sources and methodology for the production of cellulose nanocrystals (CNCs), and principles of tissue engineering and the electrospinning technique will be given. Applications for CNCs are manifold and range from super water absorbent, drug delivery, packaging, personal care to pharmaceuticals. However, in this chapter the application in tissue engineering will be discussed in detail.

Keywords: Cellulose nanocrystals, tissue engineering, electrospinning, three-dimensional scaffold

1. Introduction

Cellulose, the most ubiquitous natural biopolymer, is considered as a virtually inexhaustible source of raw material for the increasing demand of environmentally friendly, sustainable and biocompatible products [1, 2]. Despite its ubiquity in essentially every traditional and long-established scientific and industrial field, cellulose permanently occupies prominent positions in the most advanced and emerging technologies, as well [3]. It has been used either in native or regenerated form in several applications including paper, board, fibers, films, textiles, packaging materials, hygienic products, and serves as a support material in chromatography and many other life science applications[1]. The most common attributes when describing the importance of cellulose are hydrophilicity, high reactivity, low fouling properties, renewability, nontoxicity, and the compatibility with many other organic and inorganic materials. This facilitates the creation of hybrid materials and nanocomposites with unique properties such as conductive fibers, papers with printed circuits, flame retardancy, and wound dressing materials [1].

Cellulose is the major component in the rigid cell walls of wood, other bio-plants and certain bacteria, algae, fungi, and some marine animals (e.g., tunicates). Cellulose is a linear syndiotactic homopolymer consisting of D-anhydroglucopyranose units (AGU), which are covalently connected through β-(1 → 4)-glycosidic bonds between the carbon atoms C(1) and C′(4) of adjacent glucose units, resulting in a cellobiose unit (see Figure 1), which is the main building block of the cellulose polymer [4, 5]. The elemental composition of cellulose (carbon: 44-45%, hydrogen: 6.0-6.5%, and the rest being oxygen) was first revealed by the French chemist, Anselme Payen in 1838 [6]. Each of the AGU units consists of three free hydroxyl groups (at C-2, C-3 and C-6), which usually determine the reactivity of the cellulose polymer. The terminal hydroxyl groups at both ends of the cellulose chains (C-1 and C-4) are chemically different in nature. The C-1 at the right end of the chain is an aliphatic aldehyde group with reducing property, whereas C-4 at the left end is an alcoholic hydroxyl group with nonreducing property (see Figure 1). Cellulose exists in six different polymorphs, namely cellulose I, II, III, and IV. Among these, cellulose I occurs in two allomorphs, that is, I_α and I_β. Cellulose II or regenerated cellulose is formed by treating native cellulose I with either strong alkali solution or by precipitating cellulose I with Schweitzer's reagent (tetraamminediaquacopper dihydroxide $[Cu(NH_3)_4(H_2O)_2](OH)_2$). Cellulose III ($III_I$ and III_{II}) and cellulose IV (IV_I and IV_{II}) are obtained from either cellulose I or cellulose II with suitable modification [7-9]. Hearle et al. developed a two-phase model or fringed fibril model assuming that cellulose I contains low-ordered so-called amorphous and highly ordered, that is, crystalline regions [10]. This means that cellulose I is not perfectly crystalline. The intramolecular hydrogen bonds are formed between O-3-H and O-5′and also between O-2-H and O-6′ of one AGU unit and the adjacent AGU unit of the same polymer chain. The intermolecular hydrogen bonds are formed between O-6-H and O-3 ′of the AGU units of one polymer chain and the neighboring chain. Eventually, the different hydrogen bonding modes are responsible for the formation of stiff and rigid linear chains, which can further assemble into microfibrils with different chain lengths and orientation. When subjected to proper combinations of mechanical, chemical, and/or enzymatic treatments, these highly ordered crystalline regions present within the microfibrils can be isolated [11]. The isolated crystalline part is referred to as cellulose nanocrystals (CNCs). Owing to its nanoscale dimension and intrinsic physicochemical properties such as excellent biodegradability, a low eco-toxicity, biocompatibility, high specific strength and modulus, high surface area, and unique optical properties, CNCs are promising renewable biomaterials that can be used as a reinforcing component in the preparation of high-performance nanocomposite scaffold materials. Many new highly porous 3D scaffold materials with attractive properties can be prepared by the physical incorporation of CNC into a natural or synthetic polymer matrix, which will be discussed in section 5.

http://dx.doi.org/10.5772/61344

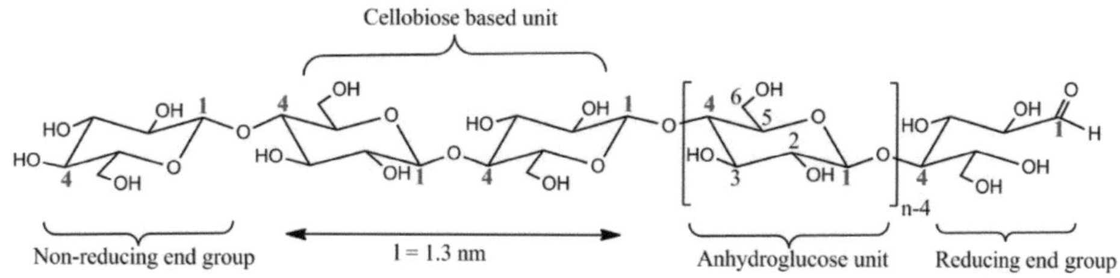

Figure 1. Molecular structure of the cellulose macromolecule.

2. Cellulose nanocrystals – Sources and methodology of preparation

Cellulose is at the forefront in scientific development in the areas of medicine, drug delivery, and sensoric applications, to name just a few, with a pronounced shift of employing cellulose substrates in nanometric dimensions, especially cellulose nanocrystals (CNCs). An array of attractive properties in combination with above-mentioned attributes are the main reason for an immense body of work, conducted with CNCs can be obtained following either of the two approaches: bottom-up by biosynthesis or top-down by disintegration of plant materials [11, 12]. Acid hydrolysis of cellulose substrates is the most widely used procedure for obtaining CNCs. The driving force of CNC formation in acidic media is the difference in solubility of structural regions in bulk cellulose, namely amorphous and crystalline entities. While amorphous regions of the cellulose substrates dissolve and hydrolyze in acidic solutions, crystalline domains readily reassemble to form nanocrystals, nanorods, or nanowhiskers (see Figure 2) [13]. Sulfuric acid (H_2SO_4) is a common hydrolyzing agent, which also imparts the resultant nanocrystals with surface sulfate groups, making their colloidal dispersions very stable due to present charges (accordingly, hydrochloric acid, for instance, yields CNC particles with a weaker charge with poor colloidal stability).

A wide variety of suitable cellulose raw materials can be employed for preparation of CNCs; consequently, different physical properties can be expected. An interesting and, so far to a certain extent, underutilized source for production of cellulose nanocrystals are fibers derived from oil palm trunks. Processed using acid hydrolysis with H_2SO_4, prepared CNCs possess an increase in crystallinity and this feature, coupled with removal of lignin and hemicelluloses contributed to enhanced thermal stability [18]. Despite its abundance, corn husks are also representatives of a class of natural materials which have not been extensively appropriated for the production of cellulose nanocrystals. Isolation of CNCs with acid hydrolysis of pretreated corn husk verified the suitability of the material as a precursor for cellulose nanocrystals, with emphasis on treatment time, which affects the degree of crystallinity, particle size, and thermal stability of the product [19]. Another environmentally sustainable approach for the production of cellulose nanocrystals is the use of wastepaper with its abundance and inexpensiveness. Acid hydrolysis with H_2SO_4 of alkali-treated and bleached cellulose particles yielded CNCs with 100-300 nm in length and 3-10 nm in diameter and

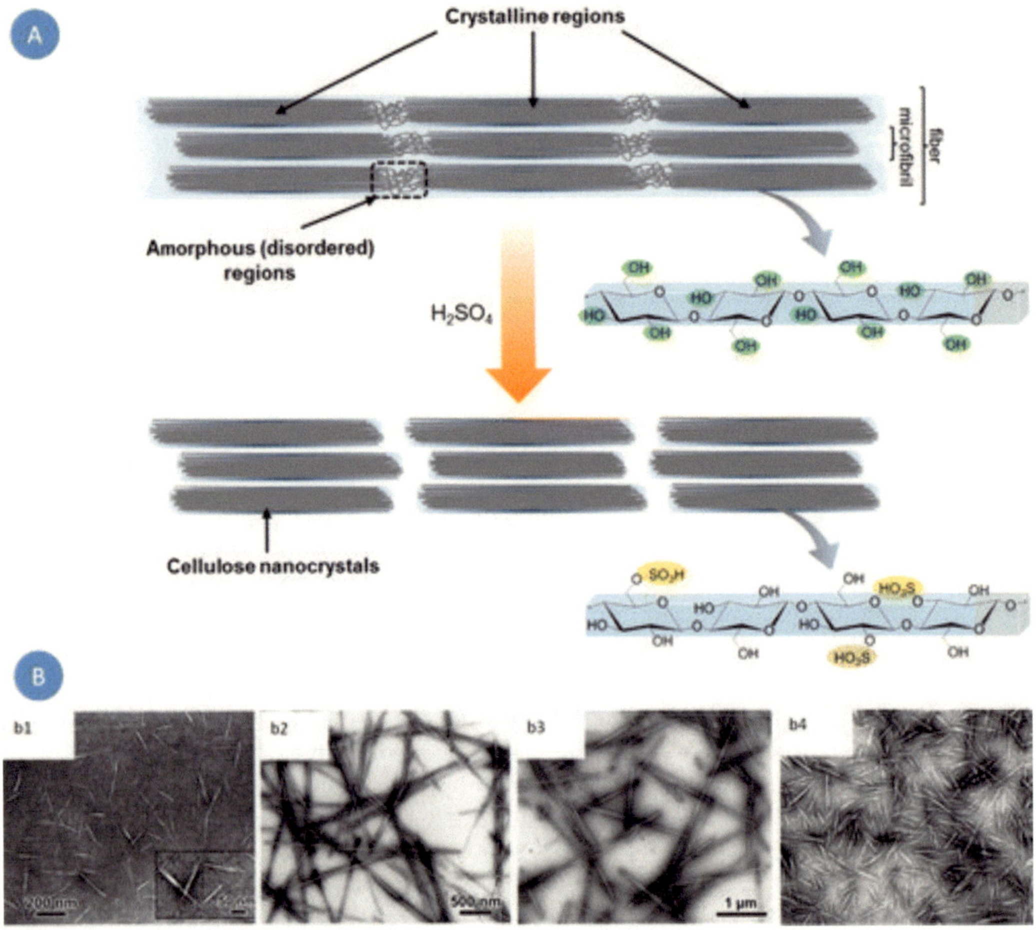

Figure 2. A) Schematics of cellulose fibers depicting crystalline and amorphous regions, and cellulose nanocrystals (with hydroxyl and sulfate groups) after sulfuric acid hydrolysis of the disordered amorphous regions. Reprinted with permission from ref [14]. Copyright 2011 Royal Society of Chemistry. (B) TEM micrographs of dispersion of cellulose nanocrystals derived from different sources: (b1) microcrystalline cellulose (Avicel), (b2) tunicate[15], (b3) green algae (*Cladophora sp.*) [16], and (b4) ramie [17]. (A) Reprinted with permission from ref [17]. Copyright 2008 Royal Society of Chemistry. (B) Reprinted with permission from refs [15] and [16]. Copyright 2012 and 2008 American Chemical Society.

crystallinity degree of approx. 76% [20]. An alternative source for production of cellulose nanocrystals is also bacterial cellulose, which is produced by acetic acid bacteria in either synthetic or nonsynthetic media via oxidative fermentation [21]. Preparation of bacterial cellulose-derived nanocrystals with different sulfate contents and, consequently, widely different colloidal properties was reported by varying the type of acid used as hydrolytic agent [22]. Whether hydrochloric acid (HCl), its mixture with H_2SO_4 or H_2SO_4 alone were used, prepared crystals exhibited zeta potential values of -5, -40, and -46 mV; in return, surface character of particles influenced their colloidal stability and interactions with other materials (e.g., xyloglucan). Kenaf bast fibers were also employed for the preparation of CNC colloidal suspensions, with cellulose extraction (a combination of alkaline treatment and bleaching) followed by acid hydrolysis [23]. Sugarcane bagasse served as a source for CNCs, which were prepared with a sulfuric acid treatment [24]. Softwood and hardwood were used in addition to nonwood sources (cotton linters and cattail, i.e., Typha) and red algae as a representative of marine pulp for the production of CNCs by H_2SO_4 hydrolysis. Dimensions of the obtained CNCs can be grouped according to the material class, with wood-source-derived particles having length from approx. 170 nm to approx. 180 nm, cattail and cotton-based particles from approx. 250 nm to approx. 278 nm, while red-algae-based CNCs exhibit the longest particles, i.e., approx. 432 nm. Nonwood CNCs exhibit superior thermal stability to wood-derived ones. Behavior, performance, and interactions with host material of CNC-embedded products rely on particles' geometry, i.e., 3D size, their aspect ratio, crystallinity, amount and type of surface groups, their distribution in a host matrix, and rheological properties [25]. Besides its above-mentioned intrinsic properties, CNCs also have superior mechanical properties, and a surface covered with numerous hydroxyl groups that enables different chemical modifications like oxidation esterification, silyation, and polymer grafting. These amazing physicochemical properties and widespread application prospect of CNCs (in personal care, foods, pharmaceuticals, as drug carriers, anticoagulant materials, and in TE) have attracted significant interest from both research scientists and industrialists.

3. Tissue engineering – Introduction to basic principles

Tissue engineering (TE) is an interdisciplinary field that applies the principles of engineering and life science toward the development of smart biological substitutes (e.g., 3D scaffolds), which potentially restore, maintain, and improve tissue functions that are malfunctioned or have been lost by different pathological conditions or as a whole organ of the human body [26]. The field of TE still remains a fast-growing area with high-potential treatments for many kinds of disease states. In TE, the cells are usually cultured/seeded onto biodegradable scaffolds that mimic the extracellular matrix (ECM), which exhibits a hierarchical architecture with structural features ranging from nano- to macrometer scale [27]. In general, cell microenvironment is very complex, and it constitutes ECM proteins, bioactive growth factors, receptors, and neighboring cells (see Figure 3). Behavior, function, and fate of cells are determined by their interaction with biochemical and biophysical cues within their surrounding microenvironment [28]. In particular, the interaction between biochemical cues, for example, receptor coupling to ECM proteins or cytokines, and biophysical cues such as modulus and fibrillar structure play a decisive role in cell fate decision [28, 29]. These cell–ECM interactions are very dynamic, since cells interact with and respond to ECM signals and accordingly can change their microenvironments. Therefore, the understanding and interlinking of bidirectional cross talk between the microenvironment and resident cells are highly important for developing efficient strategies to regenerate tissues [28]. Generally, ECM differs in composition and spatial assembly of their protein components such as collagen, elastin, proteoglycans, and adhesion molecules depending on the types of body tissues. It further inherently assembles to keep a specific tissue morphology and to supply specific instructive cues for the cells of the different organs [30].

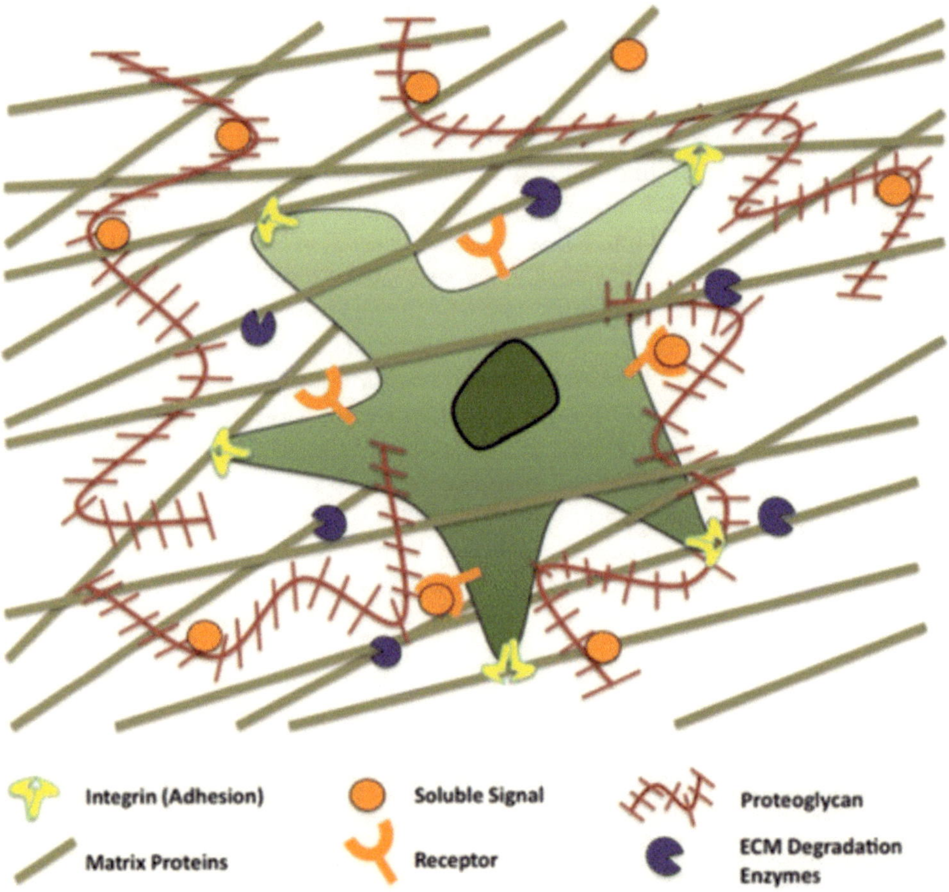

Figure 3. Schematic of the extracellular matrix. Fibrous matrix proteins (e.g., collagen, fibrin, elastin) provide structural and mechanical cues to direct cell behavior; soluble signals are sequestered by proteoglycans (proteins with polysaccharide moieties) and interact with cell surface receptors to direct cell migration, proliferation, and differentiation; integrins (transmembrane receptors) bind to matrix proteins for cell adhesion; ECM degradation enzymes (e.g., matrix metalloproteinases, serine proteinases, plasmin) cleave matrix components during cell motility and matrix remodeling. Reprinted with permission from ref [31]. Copyright 2012 Elsevier.

A unique but the simplest approach in TE and its related field regenerative medicine comprises the implantation of a scaffold that possesses inherent properties to actively promote the body's inborn capacity of healing and self-repair/growth of cells and tissue regeneration [27]. Scaffolds in general exhibit several attractive features that can influence their performance, and therefore the scaffolds should be carefully designed with all essential physicochemical properties for each targeted tissue. The scaffold materials, in general, should have the capacity to mimic the form and function of ECM in order to promote the cell–biomaterial interactions, cellular invasion, attachment, infiltration or migration, differentiation, and proliferation, thus rendering the lost tissue regeneration [27]. They should also permit sufficient transport of gases and nutrients, and allow regulatory factors to achieve cell survival [27]. Physiochemical properties of the scaffold materials, for instance, the roughness, topography, functional groups, porosity, pore size, pore interconnectivity, and surface area to volume ratio are the key aspects that should be considered in the design of scaffolds, because they play a pivotal role in tissue growth, vascularization, and nutrient supply [28, 32-38]. Furthermore, the scaffolds should be mechanically strong and easily biodegradable at a controlled rate under the condition of interest, and provide a low degree of inflammation and cytotoxicity [39].

Biodegradability of a scaffold under a given condition is an important issue in TE, and thus materials for scaffolds design should be carefully chosen. The material should be absorbable and subjectable to degradation. The rate of degradation should match the time of tissue formation, that is, until the scaffold material is mechanically stable and possesses all its intrinsic physicochemical properties and the injured tissue is completely replaced by healthy tissue and its functions are restored [35, 39]. Taking these points into consideration, the selection of the scaffold materials and the preparation methodology is highly dependent on the requirements of the desired applications and given properties of the scaffold. Further, the most difficult challenge in TE is to create a scaffold with specific physical, chemical, mechanical, and biological properties. As stated above, the mechanical (dimensional) stability of the scaffold is imperative in TE, and it should yield similar performance to the tissues to be repaired or regenerated. In this way, the scaffold would give a better mechanical integration of the construct in the biological tissue. In addition, the mechanical signals (i.e., the performance of the scaffold under a given condition) are known to create impact on the "cell fate" [40] and therefore a consideration is usually given to the role of mechanotransduction during cell expansion and differentiation [41, 42]. Overall, a precise modulation of these mentioned properties is highly challenging as biological tissues present a vast spectrum of mechanical properties, which are governed by their anatomical functions. For example, many tissues, in general, are viscoelastic with nonlinear, anisotropic, and heterogeneous mechanical properties [43].

The materials (synthetic and/or from natural origin) and fabrication methods that enable the manufacturing of scaffolds with defined micro- and nanostructure length scales for different TE applications are manifold. As natural and derivatized biopolymers starch, carrageenan, alginate, chitin, chitosan, gellan gum, hyaluronic acid, chondroitin sulfate, lingocellulose, cellulose acetate [44], carboxymethyl cellulose [45], cellulose acetate propionate [46], hydroxyproply cellulose [47] are used for creating scaffolds and cell culturing owing to their inherent bioactivity, low toxicity, high biocompatibility and in most cases biodegradability [48]. Proteins such as collagen, gelatin, elastin, silk fibroin, and fibrin are also employed for the same purpose [48-50]. In the case of the synthetically based category, aliphatic polyesters such as poly(L-lactic acid), polycaprolactone, poly(glycolide), and poly-[D,L-(lactide-co-glycolide)], polyethylene oxide, polyvinyl alcohol, poly(N-isopropylacrylamide), polyacrylic acid, and poly(2-hydroxyethylmethacrylate) are widely employed for preparing TE scaffolds since they exhibit excellent biocompatibility, biodegradability, and considerable mechanical stability [34].

While significant advances were made over the years, it is still impossible to design and engineer TE scaffolds having all desired properties from a conventional single polymeric material. In this regard, in order to better mimic the ECM hierarchical structure and for efficient growth of cells, electrospun nanocomposite fiber mats generated from multicomponent systems have gained a significant interest for different TE applications in the recent years, due to their good mechanical properties, nanotopography, and well-defined porosity [28, 32-34]. In order to overcome some limitations such as weak mechanical properties, lack of electrical conductivity, the absence of adhesive and microenvironment-defining moieties, or the incompetence to enable cells to self-assemble to 3D tissues [51], incorporation of nanofillers as reinforcing agents have become a popular approach in the past years. Several inorganic and organic nanofillers have been used together with polymeric materials for the preparation of TE scaffolds targeting at regeneration of different tissues of interest. As nanofillers, hydroxyapatite, carbon nanotubes, metal nanoparticles (e.g., silver, gold, and iron oxide), etc. have been used for the controlled immobilization of several biologically active molecules as well as to construct complex 3D tissues of different cell types [51-53]. Despite the number of advantages of metal or carbon-based nanofillers, there are some challenges and issues that remain to be addressed, including processability, toxicity, biocompatibility, and biodegradability. In this context, CNCs in particular, are highly attractive and important. They have emerged as potential nanofillers for preparation of biobased nanostructured composite materials, due to their aforementioned intrinsic properties [54]. Moreover, as specified before, CNCs exhibit low elongation at break, high aspect ratio, high surface area, and high crystallinity, which make them excellent candidates as load-bearing components for obtaining high-strength scaffold materials [55, 56]. For fabricating 3D scaffolds with nanofibrillar structures and with nanotopography to better mimic ECM dimensions, the electrospinning technique has been used extensively during the last few years, which will be discussed in the next section.

4. Electrospinning technique – Introduction to basic principles

Electrospinning is a straightforward, cost-effective, and robust technique, which is nowadays frequently applied in engineering well-defined nanocomposites fiber mats/scaffolds that hold promise in serving as a synthetic ECM. It utilizes electrical forces to create fine polymer fibers with diameters ranging from a few micrometers down to a few nanometers using polymer solutions of both natural and synthetic polymers [57]. This technique has been known for over 60 years in the textile industry for manufacturing nonwoven fiber fabrics, but has received a great importance only in the past decade. This is not only because of its versatility in spinning a wide range of polymeric fibers with controlled pore structure, but also due to its ability to consistently produce fibers in the submicron range that is otherwise difficult to achieve by using standard mechanical fiber spinning techniques [58-62]. Among several other advantages provided by this technique, it also gives the possibility to manipulate the spun fiber composition and produce highly reproducible fiber mats with micro- or nanotopography while enabling to tailor the fiber diameter, porosity, orientation, density, and high surface-area-to-volume ratio. Owing to these advantages, electrospun nanofibers have been widely examined for their use in numerous fields including nanocatalysis, protective clothing, filtration, biomedical, pharmaceutical, optical electronics, healthcare, as well as in tissue engineering [63-68]. Techniques such as electrostatic precipitators and pesticide sprayers work similarly to the electrospinning process; however, the latter process is mainly based on the principle that the applied strong mutual electrical repulsive forces overcome weaker forces of surface tension

http://dx.doi.org/10.5772/61344

in the charged polymer liquid. There are, currently, two standard electrospinning setups – vertical and horizontal. In general, the electrospinning process is conducted at ambient temperature and pressure. The typical electrospinning apparatus setup is depicted in Figure 4.

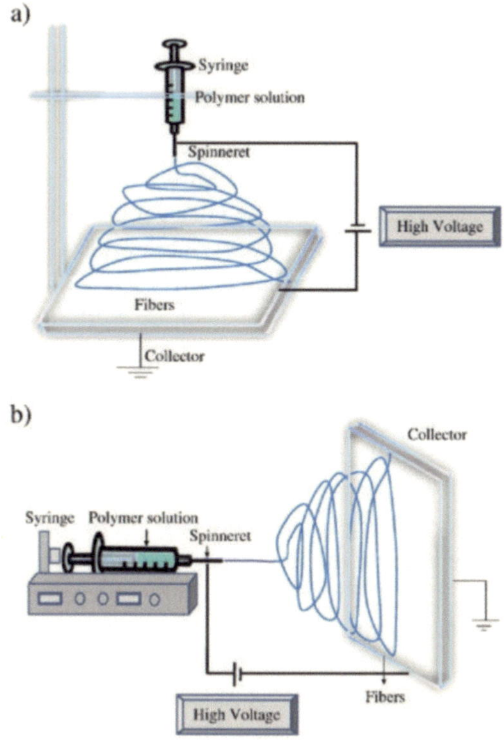

Figure 4. Schematic representation illustrating the set up of an electrospinning apparatus. A) typical vertical set up, b) horizontal setup. Reprinted with permission from ref [59]. Copyright 2010 Elsevier.

A basic electrospinning setup consists of three major components such as a high-voltage power supply; a spinneret, i.e., a pipette tip or a plastic syringe; and a grounded collecting plate that is usually a metal screen, plate, or rotating mandrel, and the technique uses electrical forces for stretching the solubilized polymer as it falls and solvent evaporates out. In this technique, prior to the spinning the dissolved polymer solution or melt of interest is introduced into the capillary tube and then a high-voltage electrical field is applied between a metallic nozzle of a syringe and a metallic collector. As a result, the polymer solution or melt, held by its surface tension at the end of the capillary tube is subjected to a high electrical field, and this high-voltage source induces charge of a certain polarity on the liquid surface of the polymer solution or melt. At end of the needle tip, the polymer solution deforms from a pendant droplet to a conical shape (Taylor cone). When the strength of the electrical force is stronger than the surface tension of the polymer solution, a jet with polymer solution is formed and ejected subsequently from the cone surface and travels to the opposite polarity, i.e., metallic collector. As soon as the polymer jet solution travels, the solvent evaporates in the air, together with stretching and acceleration of the polymer jet, leading to the deposition of an extremely thin polymer (nano-)fiber on the collector. The polymer jet is only stable at the end of the capillary tube and after that the instability begins, and thus electrospinning process offers a simplified method for the manufacturing of ultrathin nanofibers [59-62].

5. Cellulose-based scaffold – Materials' characterization and role in cell growth

Despite several advantages mentioned above in different fields, the application of electrospinning and spun nanofibers in the field of TE is as yet not realized up to its full potential, and there are still some challenges that need proper attention. The diameters of electrospun nanofibers are usually orders of magnitude smaller when compared to the sizes of cells. As a result, the cells are able to organize over the fibers of the scaffold or spread and attach to the initially adsorbed proteins or cells at multiple focal points [62]. Regardless of its numerous attractive properties, scaffolds pose some other limitations such as nonuniform cellular distribution and lack of cellular migration/infiltration with increasing depth under normal passive seeding conditions. This is because the electrospun scaffolds or mats are often characterized by entangled fibers and a densely packed membrane structure that give obvious limitation to cellular migration/infiltration and populate in the interior of the scaffolds. These drawbacks could potentially hamper the development and potential applicability of electrospun fibers, in particular for 3D tissue or organs. In this context, the issue of cellular infiltration into the fiber architecture is gaining more and more attention owing to its potential in stagnating further applications of electrospun scaffolds materials in various TE applications. To overcome these limitations, an extensive set of fabrication techniques, such as wet electrospinning, laser or UV photolithography, salt leaching, polymer blending, solvent casting or a combination of nanofibers and microfibers, have been used to obtain scaffolds having enlarged pore size, and interconnectivity, and subsequently promote cellular migration/infiltration [69-71]. Up to now, the electrospinning technique has been increasingly exploited in the production of nanofibrous scaffolds or mats from the widest range of polymeric materials (natural and synthetic) for tissue engineering applications (see section 3). Besides the limitation to cellular migration/infiltration, another important major concern with spun 3D scaffolds, which are made from either mono or bicomponent system, is mechanical stability, since they are often not strong enough for many tissue engineering applications, as described above. One way to improve the mechanical properties of the electrospun scaffold is to integrate some reinforcing agents or fillers, and for such purpose a range of materials is used, which can be found elsewhere.

Among other materials, incorporation of stiff and rodlike CNCs with high aspect ratios as reinforcing agent into the electrospun fiber mats have been investigated as an effective and alternative method in numerous studies in fabricating a high-strength and highly porous composite scaffold with multiscaled features, which would better mimic the natural hierarchical organization of ECM [72-74]. 3D porous scaffolds consisting of naturally occurring CNCs as reinforcing agents are important and attractive for many TE applications due to their vast number of above-mentioned advantages [73,74]. Electrospinning technique allows CNCs to orient along the fiber axis, thus yielding unidirectional fiber-reinforced composites for applications that require high-fiber stiffness, uniform morphology, and reduced fiber diameter [72-74]. Until now, numerous electrospun composite materials incorporated with CNCs have been produced from natural and synthetic polymers, and employed in the production of 3D tissues for implantation. Jia et al. investigated the potential applicability of CNCs in combination with microcrystalline cellulose (MC) as fillers in electrospun cellulose acetate (CA) nanofiber mats for vascular tissue scaffolds [75]. It has been found that the electrospun fibers are in the range of submicron and the incorporation of CNC and MC increased the porosity of the scaffolds, for example, a mean diameter of 1016 ± 572 nm and porosity of 80±9% are obtained at a ratio of 1:1 (MC:CNC, 10 wt.%). To understand the impact of CNC and MC on the viability of rat aortic vascular smooth muscle cells (VSMC) within the generated scaffolds were tested in vitro. It turned out that the cell viability, adhesion, migration, and proliferation with the scaffold integrated with CNC and MC are significantly improved compared with those of CA alone, confirming the synergistic enhancement by the micro- and nanoscaled features within the 3D fibrous mesh structure. In the recent work, He et al. fabricated the electrospun all-cellulose nanocomposite materials reinforced with 20% CNC, which are uniaxially aligned using a rotating drum as collector [76]. The obtained scaffold material showed well-dispersed and substantially orientated CNCs along the fiber axis, and, further, the inclusion of CNCs induced a positive effect in creating scaffolds with uniform morphology and ordered distribution of nanofibers with diameter ranging from 212 to 221 nm. The resulting final material with ordered microstructure demonstrated a remarkable improvement in the tensile properties; for instance, the tensile strength and the elastic modulus are increased by 102% and 172% in the fiber alignment direction, due to enhanced interfacial bonding between CNCs and the regenerated cellulose matrix employed in the production of the electrospun scaffold. The applicability of the material is tested for the growth of human dental follicle cells (cDFCs), and the results illustrated that the fiber mats are nontoxic and the cells are able to attach, infiltrate, and subsequently proliferate in the entire scaffold, thereby inducing ordered organization of cDFCs along the fiber alignment direction (Figure 5). The authors have further proposed that these microstructured electrospun scaffolds can be used in several other TE strategies, for example, in the development of blood vessel, tendon, or nerves, where the mechanical performance and cell orientation are critical issues [77].

Huang et al. manufactured mechanically highly stable electrospun scaffold composed of nanofibers from silk protein such as fibroin and CNCs (diameter: 20–40 nm, length: 400–500 nm) as nanofillers that are extracted from *Morus alba L.* branch bark [73]. Results showed that in the resulting material, the nanoparticles CNCs are well dispersed and considerably oriented along the fiber axis in the silk fibroin matrix, as reported by the author. The electrospun fiber mats reinforced with CNCs (2%) showed almost twofold increase in the mechanical properties such as the tensile strength and the Young's modulus compared to virgin (unreinforced) silk fibroin nanofiber. Interestingly, the electrospun nanofibers that were obtained using bacterial cellulose nanocrystals as nanofiller and silk fibroin showed similar increase in the mechanical properties [73, 78]. Unfortunately, the importance of electrospun scaffolds derived from fibroin/CNCs is not recognized as cell growth material in TE to date.

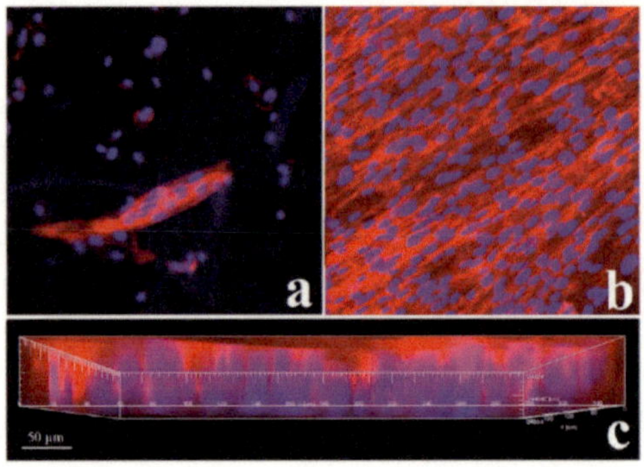

Figure 5. Confocal laser scanning microscopy images of hDFCs loaded in electrospun cellulose/CNCs nanocomposite nanofibers: (a) cultured for 3 days, (b) cultured for 7 days, and (c) the 3D view of electrospun nanofibers with cells. Scale bar: 50 μm. Reprinted with permission from ref [76]. Copyright 2014 American Chemical Society.

While the scaffolds manufactured from fully renewable cellulose are highly captivating and important for TE applications, the electrospinning of these natural origins is always demanding and needs special care in choosing the right solvent and viscosity of the polymer solution. It can be resolved, however, by blending CNCs with synthetic biopolymers to meet the right solution viscosity and consistency in electrospinning. In this regard, in the recent years, synthetic biopolymers, in particular, PLA in combination with CNCs as nanofiller, have received a vast interest in the production of electrospun nanocomposite fibers [79]. For example, in the work of Xian et al., PLA electrospun nanofiber mats incorporated with CNCs prepared from avicel (microcrystalline) cellulose are fabricated [79]. It has been found that the strength of the nanofibers is improved by 30% with loading of 1 wt.% CNCs. The latter acted as nucleating agent of PLA crystallization leading to increased crystallinity of PLA in the resulting nanocomposite fibers. In another work by Shi et al., PLA/CNCs electrospun fiber mats from a solvent mixture consisting of N,N'-dimethylformamide (DMA) and chloroform were prepared and characterized in terms of in vitro degradation properties [80]. The obtained nanocomposite mats showed 5-fold and 22-fold increase in tensile stress and Young's modulus upon the addition of 5 wt.% CNCs, and they undergo faster degradation in phosphate-buffered saline (PBS) solution than that in neat PLA mats. These types of materials exhibiting superior mechanical and fast degradation properties are particularly interesting for making TE scaffolds targeted at short-term application CNCs.

Owing to the fact that CNCs are highly polar and show high tendency for the formation of hydrogen bonding promoting agglomeration, dispersing aqueous or nonpolar-based CNCs in a PLA matrix is highly tedious and time consuming. To overcome this hurdle, Li et al. proposed an alternative approach based on a water-in-oil (W/O) emulsion system consisting of a dispersed phase of CNCs aqueous suspension and an immiscible continuous phase of PLA solution [81]. The electrospun fiber mats prepared from these system showed either a core-shell or hollow structure depending on the emulsion droplet size, and the CNCs are aligned along the core or on the wall of the hollow cylinder (Figure 6). Upon the addition of 5 wt.% CNCs, a drastic increase in stiffness (Young's modulus), i.e., up to 549% and maximum increase of tensile strength of 90% are obtained for the electrospun composite fiber mats. As described by the authors, besides the reinforcement offered by the CNCs, these hollow structures, as multifunctional system, can be potentially employed in several applications such as for the attachment of biologically active molecules, controlled drug delivery, and as scaffolds.

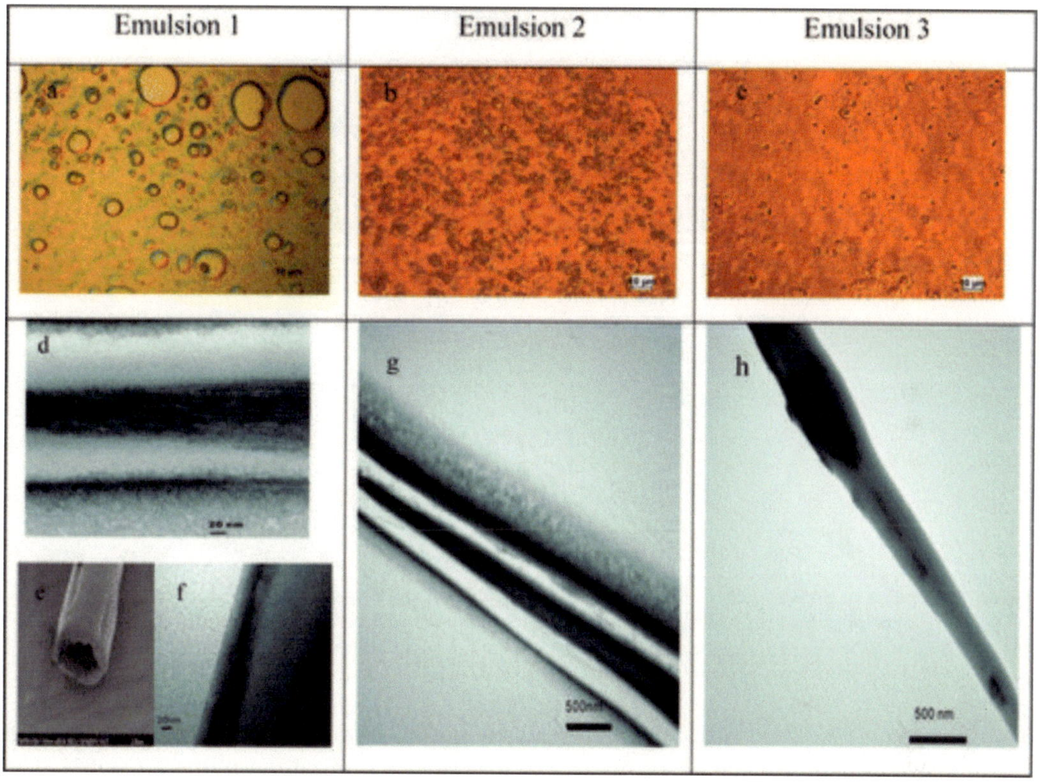

Figure 6. Correlations between emulsion droplet size (a ~6 μm, b ~3 μm, and c <1 μm) and electrospun fiber structure (d and g represent core-shell; e and f, hollow cylinder; and h, CNC aggregation). Reprinted with permission from ref [81]. Copyright 2013 American Chemical Society.

Zhou et al. used PLA grafted with maleic anhydride (MPLA) in order to advance the interfacial tension between the hydrophobic PLA matrix and the hydrophilic CNCs [82]. The fiber mats electrospun from MPLA and CNCs (5 wt.%) showed reduced diameter and polydispersity with increased CNCs content, and improved thermal and mechanical properties (e.g., tensile strength increased up to more than 10 MPa). The suitability of scaffolds for in vitro degradation and cytocompatability are tested using human adult adipose (derived from mesenchymal stem cells, hASCs; see Figure 7). Furthermore, it has been demonstrated that the cells could attach and proliferate in the entire scaffold (not only on the surface but also deep inside the fiber mats) and induce ordered cellular organization in fiber alignment direction. Results showed that the scaffolds are highly biocompatible, biodegradable, and cytocompatible, demonstrating that they have high potential to be used in bone tissue engineering [82].

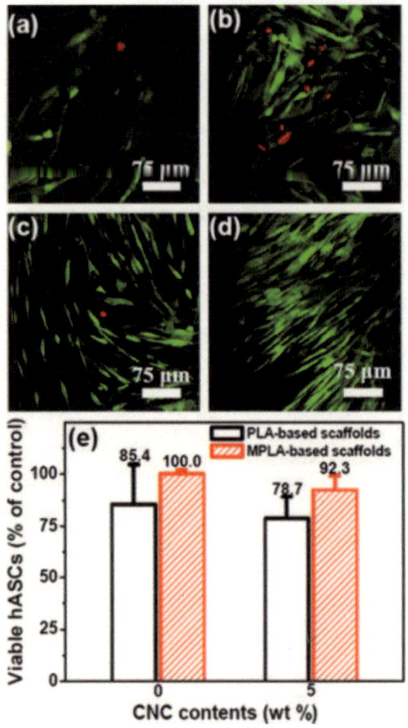

Figure 7. Response of hASCss to PLA/CNC and MPLA/CNC nanofibrous scaffolds after 7 days of culture. Fluorescence micrographs of stained cells consisting of live (green) and dead (red) cells for PLA (a); PLA/CNC-5 (b); MPLA (c); and MPLA/CNC-5 (d) scaffolds, scale bar represents 75 μm; proliferation viability of cells (e). Reproduced with permission from ref [82]. Copyright 2013 American Chemical Society.

Nanocomposite fiber mats based on biodegradable poly(ε-caprolactone), PCL, widely used in TE, is also prepared by means of electrospinning by incorporating with CNCs [83]. The inclusion of CNCs with increasing concentrations into the PCL matrix induced considerably positive effects on the mechanical, thermal, and surface properties. The tensile strength and modulus are increased to 36% (with 10 wt.% CNCs) and 167% (with 15 wt.% CNCs). In another study, electrospun PCL nanofibers are reinforced with CNCs, which are obtained from ramie cellulose fibers. In this case, in order to improve the interfacial tension with the PCL matrix, the surface of CNCs is chemically modified with low molecular weight PCL diol [84]. Although this approach yielded unexpected results on the morphology of the nonwoven mats in which the individual nanofibers became annealed during the electrospinning process, the mechanical properties of the nanofibers are significantly improved after reinforcing with 2.5 wt.% CNCs. For example, the Young's modulus and the tensile strength are increased up to 1.5-fold and many folds compared to those of pure PCL mats.

As CNCs are dispersible in water, a direct mixing of CNCs with several water soluble polymers is feasible. As a consequence, several research groups have attempted to develop multifunctional electrospun scaffolds based on water-soluble polymer matrices applicable for TE applications, such as polyethylene oxide (PEO) [85, 86] and polyvinyl alcohol (PVA) [74, 84].

In the case of nanofiber mats electrospun from PVA and CNCs aqueous dispersion, the incorporation of CNCs displayed a considerable reinforcement effect, in particular, on the aligned electrospun PVA fiber mats compared with the isotropic ones. For instance, the aligned mats demonstrated a 35% and 45% increase in Young's modulus and tensile strength compared to the isotropic mats. Moreover, aligned PVA/CNCs fibers yielded a lower diameter compared to the fibers that are produced from PVA alone. With addition of 15 wt.% CNCs the tensile strength and Young's modulus are further increased to 95% and 118% for the aligned PVA/CNCs, clearly demonstrating that the CNCs with high aspect ratio can align inside the polymer matrix during electrospinning because of the higher shear force of the polymer jet, thereby causing tremendous increase in the physical properties of the material [74]. In a similar work carried out by Peresin et al., the reinforcement effect of CNCs on electrospun PVA with different concentrations of acetyl groups and CNCs is studied, and the results showed a significant improvement in the mechanical properties [84]. It is suggested that the higher hydrolysis degree of PVA caused a strong interaction between PVA and CNCs, led to effective reinforcement of the fiber mats, which can be related to the reinforcing effect of the dispersed phase of CNCs, via the percolation network held by hydrogen bonds. The PEO/CNCs electrospun nanocomposite fiber mats are also prepared by using different concentration of wood-based CNCs (up to 20 wt.%). In this case, upon increasing the CNCs concentration in electrospinning solution, more uniform and finer nanofibers are formed [85], due to an enhanced electrical conductivity of the spinning solution. Furthermore, the tensile strength and Young's modulus are substantially improved 152% and 180% upon the addition of nanofiller CNCs. This enhanced mechanical property is due to the efficient stress transfer from PEO to CNCs stemming from their strong hydrogen bonding interactions and uniform dispersion, as well as high alignment of CNCs in the electrospun nanocomposite fibers.

Generally, the incorporation of CNCs can decrease the fiber diameter of the electrospun nanocomposite fiber mats, and this can impose some limitations of these system in TE applications. One such example is cellular infiltration which can be hindered by decreasing the scaffolds pore size/fiber diameter. Zhou et al. reported that 24% decrease in fiber diameter for the electrospun nancomposite PEO/CNCs when incorporated with 20 wt.% [85]. Whereas only a 19% reduction in fiber diameter is observed for the nanocomposite fiber mats electrospun from MPLA/CNCs upon incorporation of 5 wt.% CNCs. This limitation can be alleviated by adopting the electrospinning technique to increase the pore size of the scaffolds [82]. It is known that the mechanical stability of the electrospun scaffolds is weakened by the increased porosity and pore size. In this case, incorporation of nanofiller CNCs to reinforce or impart higher mechanical strength of the nanofibrous scaffolds is highly useful strategy, while offering effective and appropriate 3D scaffolds for the accommodation and sufficient migration or infiltration of cells.

6. Summary and outlook

During the past few years, the need to utilize biomaterials such as cellulose nanocrystals for tissue engineering application has increased dramatically. CNCs as a potential nanofiller and reinforcement material have created a significant interest especially in the design and fabrication of three-dimensional porous scaffold materials for the growth of 3D tissue. Scaffolds prepared from the CNC-based nanocomposites using electrospinning technique have demonstrated to be highly suitable to better mimic the extracellular matrix. Regardless of the considerable numbers of different polymeric (natural and synthetic) materials used for scaffold preparation, incorporation of CNCs has significantly improved the mechanical properties, porosity, and thus the adhesion, migration, and proliferation of the cells. Even though only a very few and no in vivo biological tests have been carried out in order to access the performance of scaffolds for practical applications, several studies are in progress to develop novel multifunctional biomaterials with tailored properties for tissue engineering applications.

Author details

Tamilselvan Mohan[1*], Silvo Hribernik[2], Rupert Kargl[2] and Karin Stana-Kleinschek[2]

*Address all correspondence to: tamilselvan.mohan@uni-graz.at; tamilselvan.mohan@gmail.com

1 Institute of Chemistry, University of Graz, Graz, Austria

2 Institute of Materials and Design, University of Maribor, Maribor, Slovenia

References

[1] Klemm D, Heublein B, Fink H-PBohn A: Cellulose: Fascinating Biopolymer and Sustainable Raw Material. *Angewandte Chemie Int Ed.* 2005;44:3358-3393. DOI:10.1002/anie.200460587.

[2] D. Klemm, B. Philipp, T. Heinze, U. HeinzeWagenknecht W: In Comprehensive Cellulose Chemsitry: Fundamental and Analytical Methods. Vol. 1. 1998: Wiley-VCH.

[3] Lin NDufresne A: Nanocellulose in biomedicine: Current status and future prospect. *Eur Polym J.* 2014;59:302-325. DOI:http://dx.doi.org/10.1016/j.eurpolymj.2014.07.025.

[4] Parthasarathi R, Bellesia G, Chundawat S P S, Dale B E, Langan P, Gnanakaran S: Insights into Hydrogen Bonding and Stacking Interactions in Cellulose. *J Physic Chem A.* 2011;115:14191-14202. DOI:10.1021/jp203620x.

[5] Krassig H A: *Cellulose: Structure, Accessibility and Reactivity.* Amsterdam [u.a.]: Gordon and Breach. 1996.

[6] Anselme P: Troisieme memire sur le development vegetaux. *Comptes rendus.* 1838;7:1052-1125.

[7] Stipanovic A J, Sarko A: Packing Analysis of Carbohydrates and Polysaccharides. 6. Molecular and Crystal Structure of Regenerated Cellulose II. *Macromolecules*. 1976;9:851-857. DOI:10.1021/ma60053a027.

[8] Hess K, Kissig H: Zur kenntnis der hochtemperatur-Modifikation der Cellulose (Cellulose IV). *Zietschrift Physikalische Chemie* B. 1941;49:235-239.

[9] Marrinan H J, Mann J: Infrared Spectra of the Crystalline Modifications of Cellulose. *J Polym Sci*. 1956;21:301-311. DOI:10.1002/pol.1956.120219812.

[10] Hearle J W S: A Fringed Fibril Theory of Structure in Crystalline Polymers. *J Polym Sci*. 1958;28:432-435. DOI:10.1002/pol.1958.1202811722.

[11] Habibi Y, Lucia L A, Rojas O J: Cellulose Nanocrystals: Chemistry, Self-Assembly, and Applications. *Chem Rev*. 2010;110:3479-3500. DOI:10.1021/cr900339w.

[12] Lee K-Y, Aitomäki Y, Berglund L A, Oksman K, Bismarck A: On the Use of Nanocellulose as Reinforcement in Polymer Matrix Composites. *Composites Sci Technol*. 2014;105:15-27. DOI:http://dx.doi.org/10.1016/j.compscitech.2014.08.032.

[13] Lu P, Hsieh Y-L: Preparation and Properties of Cellulose Nanocrystals: Rods, Spheres, and Network. *Carbohydrate Polym*. 2010;82:329-336. DOI:http://dx.doi.org/10.1016/j.carbpol.2010.04.073.

[14] Moon R J, Martini A, Nairn J, Simonsen J, Youngblood J: Cellulose Nanomaterials Review: Structure, Properties and Nanocomposites. *Chem Soc Rev*. 2011;40:3941-3994. DOI:10.1039/C0CS00108B.

[15] Elazzouzi-Hafraoui S, Nishiyama Y, Putaux J-L, Heux L, Dubreuil F, Rochas C: The Shape and Size Distribution of Crystalline Nanoparticles Prepared by Acid Hydrolysis of Native Cellulose. *Biomacromolecules*. 2008;9:57-65. DOI:10.1021/bm700769p.

[16] Osorio-Madrazo A, Eder M, Rueggeberg M, Pandey J K, Harrington M J, Nishiyama Y, Putaux J-L, Rochas C, Burgert I: Reorientation of Cellulose Nanowhiskers in Agarose Hydrogels under Tensile Loading. *Biomacromolecules*. 2012;13:850-856. DOI: 10.1021/bm201764y.

[17] Habibi Y, Goffin A-L, Schiltz N, Duquesne E, Dubois P, Dufresne A: Bionanocomposites Based on Poly([varepsilon]-caprolactone)-Grafted Cellulose Nanocrystals by Ring-Opening Polymerization. *J Mater Chem*. 2008;18:5002-5010. DOI:10.1039/B809212E.

[18] Lamaming J, Hashim R, Sulaiman O, Leh C P, Sugimoto T, Nordin N A: Cellulose Nanocrystals Isolated from Oil Palm Trunk. *Carbohydrate Polym*. 2015;127:202-208. DOI:http://dx.doi.org/10.1016/j.carbpol.2015.03.043.

[19] Mendes C A d C, Ferreira N M S, Furtado C R Gde Sousa A M F: Isolation and Characterization of Nanocrystalline Cellulose from Corn Husk. *Mater Lett*. 2015;148:26-29. DOI:http://dx.doi.org/10.1016/j.matlet.2015.02.047.

[20] Danial W H, Abdul Majid Z, Mohd Muhid M N, Triwahyono S, Bakar M BRamli Z: The Reuse of Wastepaper for the Extraction of Cellulose Nanocrystals. *Carbohydrate Polym*. 2015;118:165-169. DOI:http://dx.doi.org/10.1016/j.carbpol.2014.10.072.

[21] Esa F, Tasirin S MRahman N A: Overview of Bacterial Cellulose Production and Application. *Agriculture Agricul Sci Proced*. 2014;2:113-119. DOI:http://dx.doi.org/10.1016/j.aaspro.2014.11.017.

[22] Pirich C, de Freitas R, Woehl M, Picheth G, Petri D S, Sierakowski M: Bacterial Cellulose Nanocrystals: Impact of the Sulfate Content on the Interaction with Xyloglucan. *Cellulose*. 2015;22:1773-1787. DOI:10.1007/s10570-015-0626-y.

[23] Kargarzadeh H, M. Sheltami R, Ahmad I, Abdullah I, Dufresne A: Cellulose Nanocrystal: A Promising Toughening Agent for Unsaturated Polyester Nanocomposite. *Polymer*. 2015;56:346-357. DOI:http://dx.doi.org/10.1016/j.polymer.2014.11.054.

[24] Van Hai L, Son H, Seo Y: Physical and Bio-composite Properties of Nanocrystalline Cellulose from Wood, Cotton Linters, Cattail, and Red Algae. *Cellulose*. 2015;22:1789-1798. DOI:10.1007/s10570-015-0633-z.

[25] Li M-C, Wu Q, Song K, Lee S, Qing Y, Wu Y: Cellulose Nanoparticles: Structure–Morphology–Rheology Relationships. *ACS Sustain Chem Engin*. 2015;3:821-832. DOI: 10.1021/acssuschemeng.5b00144.

[26] Brahatheeswaran Dhandayuthapani, Yasuhiko Yoshida, Toru MaekawaKumar D S: Polymeric Scaffolds in Tissue Engineering Application: A Review. *Int J Polym Sci*. 2011;2011. DOI:10.1155/2011/290602.

[27] Place E S, Evans N D, Stevens M M: Complexity in Biomaterials for Tissue Engineering. *Nat Mater*. 2009;8:457-470. DOI:http://www.nature.com/nmat/journal/v8/n6/suppinfo/nmat2441_S1.html.

[28] Keung A J, Kumar S, Schaffer D V: Presentation Counts: Microenvironmental Regulation of Stem Cells by Biophysical and Material Cues. *Ann Rev Cell Develop Biol*. 2010;26:533-556. DOI:doi:10.1146/annurev-cellbio-100109-104042.

[29] Lutolf M P, Hubbell J A: Synthetic Biomaterials as Instructive Extracellular Microenvironments for Morphogenesis in Tissue Engineering. *Nat Biotech*. 2005;23:47-55.

[30] Kim H N, Jiao A, Hwang N S, Kim M S, Kang D H, Kim D-H, Suh K-Y: Nanotopography-Guided Tissue Engineering and Regenerative Medicine. *Adv Drug Deliv Rev*. 2013;65:536-558. DOI:http://dx.doi.org/10.1016/j.addr.2012.07.014.

[31] Wade R J, Burdick J A: Engineering ECM Signals into Biomaterials. *Mater Today*. 2012;15:454-459. DOI:http://dx.doi.org/10.1016/S1369-7021(12)70197-9.

[32] Bauer A L, Jackson T L, Jiang Y: Topography of Extracellular Matrix Mediates Vascular Morphogenesis and Migration Speeds in Angiogenesis. *PLoS Comput Biol*. 2009;5:e1000445. DOI:10.1371/journal.pcbi.1000445.

[33] Edalat F, Bae H, Manoucheri S, Cha J, Khademhosseini A: Engineering Approaches Toward Deconstructing and Controlling the Stem Cell Environment. *Annals Biomed Engin*. 2012;40:1301-1315. DOI:10.1007/s10439-011-0452-9.

[34] Place E S, George J H, Williams C K, Stevens M M: Synthetic Polymer Scaffolds for Tissue Engineering. *Chem Soc Rev*. 2009;38:1139-1151. DOI:10.1039/B811392K.

[35] Hutmacher D W: Scaffolds in Tissue Engineering Bone and Cartilage. *Biomaterials*. 2000;21:2529-2543. DOI:http://dx.doi.org/10.1016/S0142-9612(00)00121-6.

[36] Hutmacher D W, Schantz J T, Lam C X F, Tan K CLim T C: State of the Art and Future Directions of Scaffold-Based Bone Engineering from a Biomaterials Perspective. *J Tissue Engin Regen Med*. 2007;1:245-260. DOI:10.1002/term.24.

[37] Karageorgiou V, Kaplan D: Porosity of 3D Biomaterial Scaffolds and Osteogenesis. *Biomaterials*. 2005;26:5474-5491. DOI:http://dx.doi.org/10.1016/j.biomaterials.2005.02.002.

[38] Rezwan K, Chen Q Z, Blaker J J, Boccaccini A R: Biodegradable and Bioactive Porous Polymer/Inorganic Composite Scaffolds for Bone Tissue Engineering. *Biomaterials*. 2006;27:3413-3431. DOI:http://dx.doi.org/10.1016/j.biomaterials.2006.01.039.

[39] Holzapfel B M, Reichert J C, Schantz J-T, Gbureck U, Rackwitz L, Nöth U, Jakob F, Rudert M, Groll J, Hutmacher D W: How Smart Do Biomaterials Need to Be? A Translational Science and Clinical Point of View. *Adv Drug Deliv Rev*. 2013;65:581-603. DOI:http://dx.doi.org/10.1016/j.addr.2012.07.009.

[40] Discher D E, Janmey P, Wang Y L: Tissue Cells Feel and Respond to the Stiffness of their Substrate. *Science*. 2005;18:1139-1143. DOI:10.1126/science.1116995.

[41] Chen C S: Mechanotransduction – A Field Pulling Together? *J Cell Sci*. 2008;121:3285-3292. DOI:10.1242/jcs.023507.

[42] DeForest C AAnseth K S: Advances in Bioactive Hydrogels to Probe and Direct Cell Fate. *Ann Rev Chem Biomol Engin*. 2012;3:421-444. DOI:10.1146/annurev-chembioeng-062011-080945.

[43] Hollister S J: Scaffold Design and Manufacturing: From Concept to Clinic. *Adv Mater*. 2009;21:3330-3342. DOI:10.1002/adma.200802977.

[44] Entcheva E, Bien H, Yin L, Chung C-Y, Farrell M, Kostov Y: Functional Cardiac Cell Constructs on Cellulose-Based Scaffolding. *Biomaterials*. 2004;25:5753-5762. DOI:http://dx.doi.org/10.1016/j.biomaterials.2004.01.024.

[45] Barbieri D, Yuan H, de Groot F, Walsh W Rde Bruijn J D: Influence of Different Polymeric Gels on the Ectopic Bone Forming Ability of an Osteoinductive Biphasic Calcium Phosphate Ceramic. *Acta Biomaterialia*. 2011;7:2007-2014. DOI:http://dx.doi.org/10.1016/j.actbio.2011.01.017.

[46] Gomes G d S, Almeida A T d, Kosaka P M, Rogero S O, Cruz Á S, Ikeda T I, Petri D F S: Cellulose Acetate Propionate Coated Titanium: Characterization and Biotechnological Application. *Mater Res*. 2007;10:469-474.

[47] Kitamura M, Ohtsuki C, Iwasaki H, Ogata S-i, Tanihara M, Miyazaki T: The Controlled Resorption of Porous α-Tricalcium Phosphate Using a Hydroxypropylcellulose Coating. *J Mater Sci: Mater Med*. 2004;15:1153-1158. DOI:10.1023/B:JMSM.0000046399.40310.47.

[48] Mano J F, Silva G A, Azevedo H S, Malafaya P B, Sousa R A, Silva S S, Boesel L F, Oliveira J M, Santos T C, Marques A P, Neves N M, Reis R L: Natural Origin Biodegradable Systems in Tissue Engineering and Regenerative Medicine: Present Status and Some Moving Trends. *J. R. Soc. Interface*. 2007;4:999-1030.

[49] Fernandes E M, Pires R A, Mano J F, Reis R L: Bionanocomposites from Lignocellulosic Resources: Properties, Applications and Future Trends for their Use in the Biomedical Field. *Progr Polym Sci*. 2013;38:1415-1441. DOI:http://dx.doi.org/10.1016/j.progpolymsci.2013.05.013.

[50] Gomes S, Leonor I B, Mano J F, Reis R L, Kaplan D L: Natural and Genetically Engineered Proteins for Tissue Engineering. *Progr Polym Sci*. 2012;37:1-17. DOI:http://dx.doi.org/10.1016/j.progpolymsci.2011.07.003.

[51] Dvir T, Timko B P, Kohane D S, Langer R: Nanotechnological Strategies for Engineering Complex Tissues. *Nat Nano*. 2011;6:13-22.

[52] Armentano I, Dottori M, Fortunati E, Mattioli S, Kenny J M: Biodegradable Polymer Matrix Nanocomposites for Tissue Engineering: A Review. *Polym Degrad Stability*. 2010;95:2126-2146. DOI:http://dx.doi.org/10.1016/j.polymdegradstab.2010.06.007.

[53] Okamoto M, John B: Synthetic Biopolymer Nanocomposites for Tissue Engineering Scaffolds. *Progr Polym Sci*. 2013;38:1487-1503. DOI:http://dx.doi.org/10.1016/j.progpolymsci.2013.06.001.

[54] Dugan J M, Gough J E, Eichhorn S J: Bacterial Cellulose Scaffolds and Cellulose Nanowhiskers for Tissue Engineering. *Nanomedicine*. 2013;8:287-298. DOI:10.2217/nnm.12.211.

[55] Dreher K L: Health and Environmental Impact of Nanotechnology: Toxicological Assessment of Manufactured Nanoparticles. *Toxicol Sci*. 2004;77:3-5. DOI:10.1093/toxsci/kfh041.

[56] El-Ansary A, Al-Daihan S, Bacha A, Kotb M:, *Toxicity of Novel Nanosized Formulations Used in Medicine*, in *Oxidative Stress and Nanotechnology*, D. Armstrong and D. J. Bharali, (Eds.). 2013, Humana Press. p. 47-74.

[57] Agarwal S, Wendorff J H, Greiner A: Use of Electrospinning Technique for Biomedical Applications. *Polymer*. 2008;49:5603-5621. DOI:http://dx.doi.org/10.1016/j.polymer.2008.09.014.

[58] Kai D, Liow S S, Loh X J: Biodegradable Polymers for Electrospinning: Towards Biomedical Applications. *Mater Sci Engin: C*. 2014;45:659-670. DOI:http://dx.doi.org/10.1016/j.msec.2014.04.051.

[59] Bhardwaj N, Kundu S C: Electrospinning: A Fascinating Fiber Fabrication Technique. *Biotechnol Adv*. 2010;28:325-347. DOI:http://dx.doi.org/10.1016/j.biotechadv.2010.01.004.

[60] Reneker D H, Yarin A L, Fong H, Koombhongse S: Bending Instability of Electrically Charged Liquid Jets of Polymer Solutions in Electrospinning. *J Appl Phys*. 2000;87:4531-4547. DOI:doi:http://dx.doi.org/10.1063/1.373532.

[61] Schreuder-Gibson H, Gibson P, Senecal K, Sennett M, Walker J, Yeomans W, Ziegler D, Tsai P P: Protective Textile Materials Based on Electrospun Nanofibers. *J Adv Mater*. 2002;34:44-55.

[62] Pham Q P, Sharma U, Mikos A G: Electrospinning of Polymeric Nanofibers for Tissue Engineering Applications: A Review. *Tissue Engin*. 2006;12:1197-1211. DOI:10.1089/ten.2006.12.1197.

[63] Luu Y K, Kim K, Hsiao B S, Chu B, Hadjiargyrou M: Development of a Nanostructured DNA Delivery Scaffold via Electrospinning of PLGA and PLA–PEG Block Copolymers. *J Controlled Release*. 2003;89:341-353. DOI:http://dx.doi.org/10.1016/S0168-3659(03)00097-X.

[64] Subbiah T, Bhat G S, Tock R W, Parameswaran S, Ramkumar S S: Electrospinning of nanofibers. *J Appl Polym Sci*. 2005;96:557-569. DOI:10.1002/app.21481.

[65] Ramakrishna S, Fujihara K, Teo W-E, Yong T, Ma Z, Ramaseshan R: Electrospun nanofibers: solving global issues. *Mater Today*. 2006;9:40-50. DOI:http://dx.doi.org/10.1016/S1369-7021(06)71389-X.

[66] Cui W, Zhou S, Li X, Weng J: Drug-loaded biodegradable polymeric nanofibers prepared by electrospinning. *Tissue Engin*. 2006;12:1070-1070.

[67] Barnes C P, Sell S A, Knapp D, C, Walpoth B H, Brand D, Bowlin G L: Preliminary Investigation of Electrospun Collagen and Polydioxanone for Vascular Tissue Engineering Applications. *Int J Electrospun Nanofibers Applic*. 2007;1:73-87.

[68] Welle A, Kröger M, Döring M, Niederer K, Pindel E, Chronakis I S: Electrospun Aliphatic Polycarbonates as Tailored Tissue Scaffold Materials. *Biomaterials*. 2007;28:2211-2219. DOI:http://dx.doi.org/10.1016/j.biomaterials.2007.01.024.

[69] Gomes M, Malafaya P, Reis R: Methodologies for Processing Biodegradable and Natural Origin Scaffolds for Bone and Cartilage Tissue-Engineering Applications. In *Biopolymer Methods in Tissue Engineering*, A. Hollander and P. Hatton, (eds.). 2004, Humana Press. p. 65-75.

[70] Ma P X: Biomimetic Materials for Tissue Engineering. *Adv Drug Delivery Rev.* 2008;60:184-198. DOI:http://dx.doi.org/10.1016/j.addr.2007.08.041.

[71] Ma P X: Scaffolds for Tissue Fabrication. *Mater Today*. 2004;7:30-40. DOI:http://dx.doi.org/10.1016/S1369-7021(04)00233-0.

[72] Changsarn S, Mendez J D, Shanmuganathan K, Foster E J, Weder C, Supaphol P: Biologically Inspired Hierarchical Design of Nanocomposites Based on Poly(ethylene oxide) and Cellulose Nanofibers. *Macromol Rapid Commun*. 2011;32:1367-1372. DOI: 10.1002/marc.201100183.

[73] Huang J, Liu L, Yao J: Electrospinning of Bombyx mori Silk Fibroin Nanofiber Mats Reinforced by Cellulose Nanowhiskers. *Fibers Polym*. 2011;12:1002-1006. DOI:10.1007/s12221-011-1002-7.

[74] Lee J, Deng Y: Increased Mechanical Properties of Aligned and Isotropic Electrospun PVA Nanofiber Webs by Cellulose Nanowhisker Reinforcement. *Macromol Res.* 2012;20:76-83. DOI:10.1007/s13233-012-0008-3.

[75] Jia B, Li Y, Yang B, Xiao D, Zhang S, Rajulu A V, Kondo T, Zhang L, Zhou J: Effect of Microcrystal Cellulose and Cellulose Whisker on Biocompatibility of Cellulose-Based Electrospun Scaffolds. *Cellulose*. 2013;20:1911-1923. DOI:10.1007/s10570-013-9952-0.

[76] He X, Xiao Q, Lu C, Wang Y, Zhang X, Zhao J, Zhang W, Zhang X, Deng Y: Uniaxially Aligned Electrospun All-Cellulose Nanocomposite Nanofibers Reinforced with Cellulose Nanocrystals: Scaffold for Tissue Engineering. *Biomacromolecules*. 2014;15:618-627. DOI:10.1021/bm401656a.

[77] Dugan J M, Collins R F, Gough J EEichhorn S J: Oriented Surfaces of Adsorbed Cellulose Nanowhiskers Promote Skeletal Muscle Myogenesis. *Acta Biomater*. 2013;9:4707-4715. DOI:http://dx.doi.org/10.1016/j.actbio.2012.08.050.

[78] Park D J, Choi Y, Heo S, Cho S Y, Jin H-J: Bacterial Cellulose Nanocrystals-Embedded Silk Nanofibers. *J Nanosci Nanotechnol*. 2012;12:6139-6144. DOI:10.1166/jnn.2012.6371.

[79] Xiang C, Joo Y L, Frey M W: Nanocomposite Fibers Electrospun from Poly(lactic acid)/Cellulose Nanocrystals. *J Biobased Mater Bioenergy*. 2009;3:147-155. DOI:10.1166/jbmb.2009.1016.

[80] Shi Q, Zhou C, Yue Y, Guo W, Wu Y, Wu Q: Mechanical Properties and In Vitro Degradation of Electrospun Bio-nanocomposite Mats from PLA and Cellulose Nanocrystals. *Carbohyd Polym*. 2012;90:301-308. DOI:http://dx.doi.org/10.1016/j.carbpol.2012.05.042.

[81] Li Y, Ko F KHamad W Y: Effects of Emulsion Droplet Size on the Structure of Electrospun Ultrafine Biocomposite Fibers with Cellulose Nanocrystals. *Biomacromolecules*. 2013;14:3801-3807. DOI:10.1021/bm400540v.

[82] Zhou C, Shi Q, Guo W, Terrell L, Qureshi A T, Hayes D J, Wu Q: Electrospun Bio-Nanocomposite Scaffolds for Bone Tissue Engineering by Cellulose Nanocrystals Reinforcing Maleic Anhydride Grafted PLA. *ACS Appl Mater Interf*. 2013;5:3847-3854. DOI:10.1021/am4005072.

[83] Sheng L, Jiang R, Zhu Y, Ji Y: Electrospun Cellulose Nanocrystals/Polycaprolactone Nanocomposite Fiber Mats. *J Macromol Sci*, Part B. 2013;53:820-828. DOI: 10.1080/00222348.2013.861311.

[84] Zoppe J O, Peresin M S, Habibi Y, Venditti R A, Rojas O J: Reinforcing Poly(ε-caprolactone) Nanofibers with Cellulose Nanocrystals. *ACS Appl MaterInterf*. 2009;1:1996-2004. DOI:10.1021/am9003705.

[85] Zhou C, Chu R, Wu R, Wu Q: Electrospun Polyethylene Oxide/Cellulose Nanocrystal Composite Nanofibrous Mats with Homogeneous and Heterogeneous Microstructures. Biomacromolecules. 2011;12:2617-2625. DOI:10.1021/bm200401p.

[86] Park W-I, Kang M, Kim H-SJin H-J: Electrospinning of Poly(ethylene oxide) with Bacterial Cellulose Whiskers. Macromol Symp. 2007;249-250:289-294. DOI:10.1002/masy.200750347.